SOCIETY FOR EXPERIMENTAL BIOLOGY
seminar series: 28

BIOCHEMISTRY OF PLANT CELL WALLS

T0291472

Biochemistry of plant cell walls

Edited by
C. T. BRETT
Botany Department, The University, Glasgow
and
J. R. HILLMAN
Botany Department, The University, Glasgow

CAMBRIDGE UNIVERSITY PRESS
Cambridge
London   New York   New Rochelle
Melbourne   Sydney

CAMBRIDGE UNIVERSITY PRESS
Cambridge, New York, Melbourne, Madrid, Cape Town, Singapore, São Paulo, Delhi

Cambridge University Press
The Edinburgh Building, Cambridge CB2 8RU, UK

Published in the United States of America by Cambridge University Press, New York

www.cambridge.org
Information on this title: www.cambridge.org/9780521103633

First published 1985
This digitally printed version 2009

A catalogue record for this publication is available from the British Library

ISBN 978-0-521-30487-0 hardback
ISBN 978-0-521-10363-3 paperback

CONTENTS

Contents

LIST OF CONTRIBUTORS

M. ANDREAE
Abteilung Cytologie, Pflanzenphysiologisches Institut,
Universität Göttingen, Untere Karspule 2, D-3400 Göttingen,
F.R.G.

M. BENZIMAN
Department of Biological Chemistry, Institute of Life
Sciences, The Hebrew University of Jerusalem, 91904 Jerusalem,
Israel

C.T. BRETT
Department of Botany, University of Glasgow, Glasgow G12 8QQ,
U.K.

A.J. BUCHALA
Institut de Biologie végétale et de Phytochimie, Université,
CH 1700, Fribourg, Switzerland

T. CALLAGHAN
Department of Biological Chemistry, Institute of Life
Sciences, The Hebrew University of Jerusalem, 91904 Jerusalem,
Israel

E. GRIEF
John Innes Institute, Colney Lane, Norwich NR4 7UH, England

S. LEVY
Zoology Department, University of Bristol, Woodland Road,
Bristol BS8 1UG, England

G. MACLACHLAN
Department of Biology, McGill University, 1205 Dr. Penfield
Avenue, Montreal, Quebec, Canada H3A 1B1

Y. MASUDA
Department of Biology, Faculty of Science, Osaka City
University, Sumiyoshi-ku, Osaka 558, Japan

H. MEIER
Institut de Biologie végétale et de Phytochimie, Université,
CH 1700 Fribourg, Switzerland

A.C. NEVILLE
Zoology Department, University of Bristol, Woodland Road,
Bristol BS8 1UG, England

D.H. NORTHCOTE
Department of Biochemistry, University of Cambridge, Tennis
Court Road, Cambridge CB2 1QW, England

M.A. O'NEILL
AFRC Food Research Institute, Norwich NR4 7UA, England

J. PHILLIPS
John Innes Institute, Colney Lane, Norwich NR4 7UH, England

J.S.G. REID
Department of Biological Science, University of Stirling,
Stirling FK9 4LA, Scotland

K. ROBERTS
John Innes Institute, Colney Lane, Norwich NR4 7UH, England

D.G. ROBINSON
Abteilung Cytologie, Pflanzenphysiologisches Institut,
Universität Göttingen, Untere Karspule 2, D-3400 Göttingen,
F.R.G.

A. SAUER
Abteilung Cytologie, Pflanzenphysiologisches Institut,
Universität Göttingen, Untere Karspule 2, D-3400 Göttingen,
F.R.G.

R.R. SELVENDRAN
AFRC Food Research Institute, Norwich NR4 7UA, England

P. SHAW
John Innes Institute, Colney Lane, Norwich NR4 7UH, England

E. SMITH
John Innes Institute, Colney Lane, Norwich NR4 7UH, England

B.J.H. STEVENS
AFRC Food Research Institute, Norwich NR4 7UA, England

K.W. WALDRON
Department of Botany, University of Glasgow, Glasgow G12 8QQ,
U.K.

K.C.B. WILKIE
Chemistry Department, University of Aberdeen, Old Aberdeen
AB9 2UE, Scotland

R. YAMAMOTO
Department of Biology, Faculty of Science, Osaka City
University, Sumiyoshi-ku, Osaka 558, Japan

UNITS, SYMBOLS AND FORMULAE, ABBREVIATIONS

Standard chemical formulae and certain S.I. units, prefixes and symbols have been used in this volume. The following, widely accepted, units have also been employed but are not, *sensu stricto*, S.I. units:-

Quantity	Unit	Symbol	S.I. Equivalent
Concentration of a substance	molarity	M	mol, where 1 M = 1 mol dm^{-3}
Volume	litre	l	m^3, where 1 l = 1 dm^3 = $10^{-3} m^3$
Time	day hour minute (second	d h min sec)	second(s)
Radioactivity	Curie	Ci	Bq, where 1Ci = 37GBq; 1μCi = 37kBq; 1Bq = 1 nuclear transformation (disintegration) per second

Abbreviations:

AA	amino acid
AG	arabinogalactan
AGP	arabinogalactan protein
aq.	aqueous
Ara	arabinose
Asn	asparagine
ATP	adenosine triphosphate
BSA	bovine serum albumin
ca.	circa, approximately
CB	cellobiose
CCO	cytochrome c oxidase
CCR	cytochrome c reductase
CDTA	cyclohexanediamine tetraacetic acid
cf.	compare
CIMS	chemical ionisation mass spectrometry
CTAB	cetyl trimethyl ammonium bromide
CWM	cell wall material

2-D	two-dimensional
2,4-D	2,4-dichlorophenoxyacetic acid
DBU	1,5-diazabicyclo-[5,4,0]-undec-5-ene
DCB	2,6-dichlorobenzonitrile
DEAE	diethylaminoethyl
DMSO	dimethylsulphoxide
DNP	dinitrophenol
dol	dolichol
dol-pp	dolichol pyrophosphate
d.p.	degree of polymerisation
DP	plastic compliance
DPA	days post-anthesis
DTT	dithiothreitol
E	outer face of membrane
EDTA	ethylenediaminetetraacetic acid
EGTA	ethyleneglycol-bis-(β-aminoethyl ether)-tetraacetic acid
EI-MS	electron-impact mass spectrometry
EM	electron microscope
EPPS	N-(2-hydroxyethyl)-piperazine-N'-3-propane sulphonic acid
ER	endoplasmic reticulum
f	furanose
FAB-MS	fast-atom-bombardment mass spectrometry
FITC	fluorescein isothiocyanate
FT-NMR	fourier-transform nuclear magnetic resonance
Fuc	fucose
g	gravity
G	elastic modulus
GA	Golgi apparatus
Gal	galactose
GalU	galacturonic acid
gc	gas chromatography
gc-ms	combined gas chromatography and mass spectrometry
GDP	guanosine diphosphate
glc,GLC	gas-liquid chromatography
Glc	glucose
GlcNAc	N-acetylglucosamine
GlcpA	glucuronic acid (pyranose form)
GlcU	glucuronic acid
GlcUA	glucuronic acid
G-1-P	glucose-1-phosphate
GTC	guanidinium thiocyanate
GTP	guanosine triphosphate
HC-1	hemicellulose 1
HC-2	hemicellulose 2
HC-A	hemicellulose A
HC-B	hemicellulose B
HOAc	acetic acid
HPLC	high-performance liquid chromatography
HRGP	hydroxyproline-rich glycoprotein
hyp	hydroxyproline
IAA	indole-3-acetic acid

IgG	immunoglobulin G
ITP	inosine triphosphate
k	Boltzmann constant
K_a	Concentration giving half-maximum activation
K_m	Michaelis constant
LM	light microscope
Man	mannose
MBC	methyl-benzimidazole-2-yl-carbamate
McAb	monoclonal antibody
Me	methyl
m.p.	melting point
m.s.	mass spectrometry
MW	molecular weight
m/z	mass-to-charge ratio
NAA	naphthalene acetic acid
NAD	nicotinamide adenine dinucleotide
NMR	nuclear magnetic resonance
p	pyranose form
p	para
P	inner face of membrane
PAGE	polyacrilamide gel electrophoresis
PAL	phenylalanine-ammonia lyase
PAW	phenol-acetic acid-water
PBS	phosphate-buffered saline
p.c.	paper chromatography
PEG	polyethylene glycol
PM	plasma membrane
pro	proline
RG	rhamnogalacturonan
Rha	rhamnose
S	stress
SDC	sodium dodecyl sulphate
ser	serine
SF	supernatant fraction
SLS	sodium lauryl sulphate
SN, S/N	supernatant
So	initial stress
t	time
T	absolute temperature
TBS	tris-buffered saline
TCA	trichloroacetic acid
TFA	trifluoroacetic acid
TLC	thin-layer chromatography
T_m	maximum stress-relaxation time
T_o	minimum stress-relaxation time
Tris	tris(hydroxymethyl)aminomethane
UDP	uridine diphosphate

UDPG   uridine diphosphate glucose
UTP   uridine triphosphate

V_{max} maximum velocity of reaction

w/v   weight to volume

Xyl   xylose

γ strain

η viscosity

λ flow distance

r   relaxation time

Ψ_w water potential

PREFACE

      The last two decades of research into the plant cell wall
have revealed the complexity of its molecular structure, and at the same
time it has become clear that the cell wall performs unexpectedly varied
and subtle roles in the life of the cell and the whole organism.   This
volume, like the S.E.B. Symposium from which it arose, seeks to relate
the complex molecular structure of the wall to some of its functions,
and to explore the mechanism by which the cell synthesises and controls
the molecular organisation of the wall.   The chapters have as their
theme the molecular approach to the cell wall, and they highlight not
only the technical expertise which this field now requires but also the
great opportunities which are now open to those who are willing to master
those techniques in order to probe the structure, function and
biosynthesis of cell-wall macromolecules.

      The S.E.B. Seminar Series Symposium on the Biochemistry of
Plant Cell Walls, held in the University of Glasgow on 10 and 11 July
1984, provided the foundation for this book.   Besides the contributions
which gave rise to the chapters that follow, interesting and valuable
papers were presented by P. Albersheim, S.C. Fry, B.G. Bowes,
L.E. Waterkeyn, A.F.D. Kennedy, J. Kemp, J.C. Good, M.C. Jarvis,
D.A. Brummell and E.E. Farmer.   We were very pleased to welcome
R.D. Preston to the conference to chair the opening session and to
participate in the discussion.   The conference was greatly helped by
grants from the Plant Developmental Physiology Group of the S.E.B.,
from I.C.I. and from Unilever, and we wish to express our gratitude to
these bodies.

      The editors are also most grateful to the many individuals
who have helped with the conference and with the production of the book.
The officers of the S.E.B. gave much assistance, notably
Dr. J. Noble-Nesbitt, Dr. D. Dorsett and Dr. A.J. Trewavas.

Dr. D.M. Neil organised the whole S.E.B. summer meeting with impeccable
efficiency.   The contributing authors were commendably prompt in
submitting their chapters, and Dr. R. Pellew of the C.U.P. expedited
publication of the volume.   Mrs. M.M. Gourlay is thanked for her great
skill and accuracy in typing the camera-ready text, and Mr. N. Tait and
Mrs. A. Sutcliffe have given much assistance with the figures.

C.T. Brett

J.R. Hillman

1 NEW PERSPECTIVES ON NON-CELLULOSIC CELL-WALL POLYSACCHARIDES
(HEMICELLULOSES AND PECTIC SUBSTANCES) OF LAND PLANTS

K.C.B. Wilkie
Chemistry Department, University of Aberdeen, Old Aberdeen,
AB9 2UE, Scotland

Key words: hemicellulose, pectic substances, xylan,
arabinan, galacturonan, Hakomori methylation, Haworth
methylation, β-elimination,4-0-methyl-D-glucuronic acid,
uronic acid, D-galacturonic acid, D-glucuronic acid,
Italian ryegrass.

There are several reviews of hemicelluloses and of
polysaccharides obtained from them (Timell 1964, 1965; Stephen 1983;
Wilkie 1979; Whistler & Richards 1970; Bailey 1973).   It is hoped that
this chapter will complement these and offer a new, and useful,
perspective.   The objectives are to identify specific areas both of
current (and earlier) practice and of thought, which would benefit from
attention when planning studies, or when considering old and new
interpretations of observations.   Most studies on hemicelluloses have
been on anatomically complex parts (e.g. leaves, wood, stem, etc.) of
plants of economic interest such as those from trees (Timell 1964, 1965),
grasses, cereals (Wilkie 1979) and legumes.   There is a considerable
knowledge of the hemicelluloses from limited parts of the plant kingdom
and very little knowledge of those from other very substantial parts of
that kingdom.   Over the past sixty years there has been an expansion of
knowledge of the hemicelluloses but generally a failure to integrate
studies, and to subtract from, as well as reliably to add to, the
information upon which interpretations of the nature of the hemicelluloses
is based.   New evidence and interpretations are, of course, not
necessarily either better or more significant than old unless more
accurate, inclusive, and informative.   It is particularly difficult to
establish the structures of glycans having uronic acid units such as the
hemicellulosic xylans and the pectic substances.   Problems relating to
such studies will be discussed in some detail.
        The term hemicellulose is defined explicitly, or more

commonly implicitly, in many ways.   It does not delimit an exact
polysaccharide material from any one plant source.   There is no single
definition which has gained universal acceptance since the original
coinage of the term by Schulze (1891).   Many polysaccharide
chemists accept a definition based in essence on their, and others',
practice at the bench.   Definitions differ.   A composite one is:

> Hemicelluloses are non-cellulosic polysaccharides other than
> starches and fructosans found in the aerial, and normally
> lignified, parts of the organs of higher land plants from
> which they can be extracted by dilute aqueous alkali after
> the removal of lignin.   They can then be precipitated from
> neutral or slightly acidified solution by the addition of
> ethanol or acetone.

Chemically similar materials from the endosperm which are,
however, soluble in water, and materials from sub-aerial parts (e.g.
roots) are also included within the term hemicellulose.

It is reasonable to extend the definition to include -

a) non-endospermic material of similar chemical type to that
   delimited even if it can be extracted by water

b) material not precipitated and remaining in solution

c) material remaining in the plantstuff because it is
   physically or chemically trapped

d) material which has suffered chemical degradation and not
   been recovered because of its changed structure

e) material deriving from parts of the plants prior to their
   lignification.

(There is confusing, and inappropriate, application of the term
hemicellulose to polysaccharides not in higher land plants e.g. in the
algae).

Some of the non-cellulosic polysaccharide materials
accompanying the hemicelluloses are termed pectic substances
(Aspinall 1970; Worth 1967).   They appear to consist of a triad of
glycans (polysaccharides) which may be interlinked in the natural state
but which have partly or completely separated during extractive
treatments.   They are particularly abundant in non-lignified tissues
such as those of fruits where the cell-walls are thin.   The claim that
pectic substances are ubiquitous in plants may be true but is not proven.
Pectic substances, on the other hand, could easily be lost, chemically

destroyed, or overlooked in studies of mature parts of plants where
hemicelluloses from thick cell walls could dilute them beyond easy
detection.

   The pectic substances are still inadequately understood but
appear to consist of D-galacturonan, or L-rhamno-D-galacturonan, basal
(main) chains carrying a variety of oligosaccharide, or of single sugar,
side units.   The (1→4) linked D-galactopyranuronic acid units are
commonly partially methyl esterified in their natural state and may be
linked (1→2) to a lower proportion of L-rhamnopyranose units in the
main chain.   These glycans are either accompanied by, or in the plant
bonded to, other glycans including L-arabinans and D-galactans.

   Methods used to isolate non-cellulosic glycans may lead to
some of the plant's population of glycan molecules being classified
both as hemicelluloses and as pectic substances.   Other material may
satisfy neither classification and, not being sought by the procedures
developed to isolate either, may evade isolation and study.   The
definition of pectic substances is partly based on procedures used to
isolate them and partly on the chemical nature of the material expected
to be obtained.   There has been an evolution and expansion of use of
the terms hemicellulose and pectic substance dictated by practice and
qualified by expectations about the types of glycan expected to be
present in each (Wilkie 1983).

   Confusion and uncertainty is caused when terms are used that
are ill-defined or, as in the case of hemicellulose, when terms have
considerable, and unrecognised, variability in their definition.   Within
a research group there is commonly a working, or bench, definition along
the lines of "If you use the procedure in the way described you will
succeed in isolating a material categorised as a hemicellulose or pectic
substance".   In another group, or at another time, or in a different
discipline (such say as botany, agriculture or chemistry) or with
different procedures considered more suitable for work on a particular
plantstuff, or for many other reasons, the materials defined at the
bench as hemicellulose, or pectic substance, will differ without the
differences being appreciated as being due to the methods employed and
the attitude governing the choice of methods.

   The concern over the definition of the term hemicellulose is
not semantic.   The object is to draw attention to the fact that the
molecular population of hemicelluloses isolated and studied greatly

depends on the methods used to isolate them and on preconceptions, and
misconceptions, about the glycans present in a plant.   Modification of
methods, or inattention to details now known to alter the outcome of the
use of methods, leads to selective, or accidental, modification of the
population isolated.

In the following, pectic substances will be included as
hemicellulose as the various definitions delimit no clearcut boundary
between them.   The expansion of the term provides a useful working
definition presupposing very little about the chemical nature, the
biological role, or the properties of the materials present in, and to
be isolated from, plantstuffs.   A major objective of a broad definition
is to ensure that all polysaccharide material is sought and, as far as
possible, recovered for study.   The nature of material present in a
plantstuff should not be anticipated too precisely or material not
corresponding to expectation may be inadequately sought and studied.

There has been little comment on the elusive, and easily lost,
parts of the molecular population included in the above extended
definition of hemicelluloses.   Why, indeed should there be?   One is
not concerned about the design of 'the Emperor's new clothes' if he is
not observed to be displaying any; by analogy there is little concern
over hemicellulosic material so elusive as to evade detection,
isolation and study by the methods routinely employed.   Does this
matter?   Yes.   If the architecture and development of the cell and cell-
wall, and the tissues and organs, are to be better understood in terms of
their molecular components, then it is necessary to ensure that all
structural features are identified.   Proportionately minor features may
have major significance in enabling us to understand the conformation
of the hemicellulosic molecules and their interrelationships with other
molecules of the same, or different, type.   If, for example, in large
glycan molecules there were only low proportions of branch-points
serving to link them covalently either to one another, or to, or through,
other molecules then these minor features would be of major importance.

Explicit, or implicit, comparisons were, and often still are,
made between hemicelluloses from plants which are taxonomically remote
from one another.   In these cases it is not surprising that the most
abundant hemicellulose in each often clearly differs.   Comparisons are
also made between the hemicelluloses from plants which are closely
related, and between those from different parts of the same plant.

Similar comparisons are made between glycans obtained from these
hemicelluloses by fractionation.   In these last two cases it often
appears that there are intriguing variations in details of fine
structure of glycans that are otherwise similar.   Are different
observations on the hemicelluloses and glycans from the same plantstuff,
for example, to be regarded as complementary or contradictory?   Quite
often it is still impossible to answer this question.   A view is
expressed later on this point.

Quantitative values for hemicelluloses from different plants,
or parts of the same plant, are often unsuitable for direct comparison.
If essentially different procedures are used to isolate, or to study,
hemicelluloses (or fractionated hemicelluloses or 'pure' glycans) from
plantstuffs then variations will be introduced by the researcher.   Unless
these variations are identified, conclusions are devalued and may be
incorrect.   Comparisons between aspects of fine structure of glycans,
or of hemicelluloses of plants of different species, are not necessarily
as meaningful as they appear.   (Fine structure refers to structural
details such as the presence and nature of side-units or chains on a
main chain, and the proportions of these units).   The most reliable
comparisons may be made in those cases where studies of hemicelluloses
from different plantstuffs were on taxonomically related, and
anatomically similar, plants provided the studies were carried out with
competence in precisely the same way.   This is most likely when studies
are by the same person or by those in a close-knit group.

There are many chemomechanical and biological factors which
affect the composition of the molecular population comprising the
hemicelluloses, or glycans (such as xylans) derived from them by
fractionation.   The concept of purity of, say, a xylan or other glycan,
is a mental amalgam relating to the material actually in the plantstuff
and to the material sought applying the research scientist's criteria of
purity which are much influenced by his expectations on the basis of his
or others' earlier studies.   Uncertainties of interpretation are built
into the words pure, hemicellulose, pectic substances and into other
practice-dependent, and attitude-dependent, terms.   They are, as
indicated, highly subjective but this important aspect of studies is not
normally remarked on (Wilkie 1983).

Apparently fundamental statements about the hemicelluloses,
and about glycans derived from them, must be viewed in a perspective

differentiating between well-established facts and doubtful 'facts',
and between reasonably conclusive speculation, and speculation which,
however interesting and imaginative, is none the less not proven.
Speculation is both an outcome of, and a necessary stimulus to, research
and is essential if the correct structural models are to be elucidated.
Speculative ideas, however, when published should not through longevity,
elegance, or repetition (or lack of repetition), be allowed to acquire
any enhancement in factual status.

      The following points are phrased retrospectively and are
designed to bring certain aspects of earlier studies to attention.   It
is suggested that the points should be considered (even if only to be
dismissed) when planning new studies and before harvesting the plantstuff.

a. Was a single plant or a collection of plants studied?

b. Did the plantstuff have a more or less unknown history
    (harvest, stage of growth, physical or chemical treatment,
    and conditions of storage) before arrival in the laboratory?

c. What was the stage of maturity and development of the plant?

d. Which parts were chosen for study and how subtly, or otherwise,
    were they dissected or excised?   Was it clear, for example,
    whether leaves and stems etc. were separated or were they
    acceptably, for the requirements of the study, considered
    together?

e. Had the plantstuff been chemically processed shortly after
    harvest?   (Plants may have suffered deterioration,
    affecting their hemicellulosic components or the ease of
    their extraction, if they were harvested in, say, a
    tropical climate and there were considerable delays in
    processing them).

f. Were enzymes inactivated, or was the material quickly
    stored at a low temperature?

g. Had the plantstuff possibly suffered microbiological, or
    other attack?   (Microbiological attack is deduced to
    have affected one plantstuff from which a xylan was
    isolated.   It almost certainly affected many other
    plantstuffs and it is inferred that the composition of
    the hemicellulosic populations of molecules may also have
    been affected.   Fungal attack, or something akin to
    ensilaging, could have degraded glycans in numerous plantstuffs

between collection and processing).

h. Had the plantstuff been dried and any of the hemicellulose
    been rendered less soluble than before such treatment?
    (Polysaccharides, like proteins, can be denatured and
    suffer an alteration of their physical properties including
    their solubility and extractability).

i. Did the researcher have preconceptions about the nature of
    the hemicellulose, or 'pure' glycan, that would be isolated?
    If so, in what way did this influence the procedures used in
    seeking to isolate the materials?

j. Was material known to have been rejected, and if so, was it
    believed to be similar to, or distinct from, that isolated?

k. Were pretreatments given before the use of alkali to extract
    the hemicelluloses?   (Were treatments used, for example, to
    remove pectic substances?)

l. Was hemicellulosic material bound to, and lost with, lignin
    and degraded lignin during the delignification procedure?
    (Some water-soluble materials dissolve into delignification
    liquors and are not normally recovered or investigated)
    (Buchala et al. 1971; Harwood 1952).

m. How complete was the removal of hemicelluloses (Buchala et al.
    1972) by the alkali and how complete the recovery of
    hemicellulosic material by precipitation?   (The extent of
    precipitation depends on many factors including time and
    temperature) (Blake & Richards 1971).

n. What did the investigator seek to obtain by fractionation?
    Did he deliberately, or more commonly incidentally, or
    even accidentally, reject material without study which
    could have influenced his conclusions about the nature of
    the hemicellulose, or glycan, in that particular plant?

o. What criteria were used to delimit a pure glycan or a
    fraction worthy of study?

p. Were the periods of chemical treatments, or their vigour, such
    that losses due to degradation may have taken place?

q. Was the presence of fluctuating soluble reserve polysaccharides
    (e.g. fructan) satisfactorily allowed for?

r. Were aspects of molecular structure, or of the overall
    composition of the molecular population, assumed to be, or

proved to be, invariant at different stages of plant growth?
If not, were growth-related variations investigated and taken
into account when considering conclusions?   If growth-related
variations were not taken into account what degree of
significance can be attached to the quantitative determinations
and to the conclusions about structure?
In many publications there is inadequate information on many of the
above points because they were not considered to be of importance at
the time.

        Stages in a routine and basic structural analysis of a
hemicellulose are outlined below.   Other procedures may be used to
amplify, or to confirm, observations and to aid in their interpretations.
Several references are to articles on specific methods which refer to
primary and other sources.   The methods described should not be
regarded as ideal but as routine methods requiring modification to
improve their suitability for the task to be undertaken;   they may be
unsuitable even if modified.

1   inactivation of enzymes

2   drying of plantstuff

3   pretreatments including removal of pectic substances and
      delignification

4   extraction of hemicelluloses

5   precipitation of hemicelluloses

6   drying of hemicelluloses

7   fractionation of hemicelluloses and isolation of pure
      glycans

8   methylation of hemicellulose or glycan

9   hydrolysis of methylated hemicellulose or methylated glycan

10   separation of the methylated components in the hydrolysate
      or methanolysate

11   preparation and identification of derivatives of methylated
      sugars

        What happens during these stages is less simple than the
wording indicates.   In what follows, in addition to giving a guide to
practice in the laboratory, attention is drawn to aspects easily
overlooked which may be considered worthy of consideration when
adopting or modifying procedures and when interpreting current or
earlier results and their significance.   The major sections are those

concerned with fractionation of hemicelluloses and with methylation
analysis of derived hemicellulosic glycans.

1.   It may be considered desirable to inactivate the enzymes
(Montgomery & Smith 1955) and to minimize the danger of microbiological
degradation except when the plantstuff can be processed almost
immediately after collection or harvest.   Inactivation is particularly
important where, as in tropical climates, there is an enhanced
possibility of enzymic (or microbiological) degradation.   Inactivation
of enzymes can be effected by boiling plantstuffs with aqueous ethanol
or ethanol.   Aqueous ethanol dissolves some of the carbohydrate
(acceptably removing naturally occurring oligosaccharides) but
dissolving some hemicellulose thereafter not normally recovered for
study - arabinose-rich glycans may partially dissolve in aqueous
ethanol (Aspinall et al. 1953).   If there are proteoglycan
hemicelluloses they may be denatured and their dissolvability, which
must be distinguished from their solubility, so altered that they will
not dissolve as completely as they would have had inactivation been
omitted.

2.   Drying of a plantstuff is in general not to be recommended if it
can be processed speedily.   (An aliquot can be dried for quantitation).
Drying at room temperature extends the time during which enzymes, if
still active, could cause changes.   Drying at a high temperature may
decrease the dissolvability of some of the hemicellulose either due to
adsorption of the hemicellulose on cellulose microfibrils, or due to
denaturation of the glycan itself or of any covalently bound protein.

3.   A preliminary treatment is often given to extract the materials
classified as pectic including galacturonans and rhamnogalacturonans
(Zitko & Bishop 1965; Roudier & Eberhard 1965).   Extraction may be
carried out using aqueous ammonium oxalate, or sodium diethylenedinitrilo-
tetraacetate, or sodium hexametaphosphate.   Use of the last appears to
cause least extensive degradation of the pectic polyuronides most likely
to be destroyed during aqueous alkaline extraction.

          Delignification (Wise et al. 1946; Whistler & BeMiller 1963)
is carried out to oxidise and to dissolve the then more soluble degraded
lignin.   The plant material is commonly milled before being delignified.
(Milling is desirable not only to ease the penetration of the chemical
liquor but also to facilitate the escape of the oxidised lignin which
might otherwise be trapped by membranes much as it would be in a

dialysis bag).   The residue after delignification is called a
holocellulose.   Delignification involves use of hot aqueous acetic acid
and sodium chlorite which generate chlorine and chlorine dioxide.   The
named procedure may effect chemical changes in addition to that stated
(namely the removal of lignin).   Treatments with acid-chlorite under
mild conditions (short periods and moderate temperatures) will lead to
the dissolution and loss of some hemicellulose particularly from non-
woody plantstuffs, such as grasses.   Woody tissues are normally
delignified under more vigorous conditions.   A low proportion of
(modified) lignin remains in the holocellulose (Buchala et al. 1971).
Attempts to remove it completely are accompanied by increased
dissolution and loss of part of the hemicellulosic material - some loss
of $(1 \rightarrow 3)$ and $(1 \rightarrow 4)$ heterolinked D-glucans takes place when grasses are
delignified under these conditions (Harwood 1952; Buchala et al. 1972).
Routine delignification procedures for annual plants and for woody
tissues appear to provide adequate, although not optimized, conditions.
During delignification in warm or hot aqueous acetic acid it is possible
that some acid-labile sugar units will be hydrolysed.   Those most
susceptible to hydrolysis and loss are the sugar units having furanose
rings such as L-arabinofuranosyl units - present in most, and probably
in all, xylans, in any genuine arabinans, in arabinogalactans, and in
the side-chains and units of galacturonans.
4.   Extraction of hemicelluloses is usually carried out by treatment of
the holocellulose with aqueous alkali under nitrogen (Whistler & Feather
1965) (to reduce the danger of aerial alkaline oxidation (Aspinall et al.
1961; Whistler & BeMiller 1958)).   Minor variations in extraction
procedures may lead to considerable variations in the composition of the
population of hemicellulosic molecules isolated.   There may be
differences between the material extracted and that left unextracted.
Possibly the non-extracted hemicellulose is covalently bound to residual
lignin (Whistler & BeMiller 1958) or adsorbed on the cellulose fibrils
in the holocellulose residue (Morrison 1973, 1974; Neilson & Richards
1978).   The residual cellulose and associated materials are together
known as α-cellulose.   It is common to treat the holocellulose several
times with alkali of one strength, say 4% KOH, and to recover the
material extracted.   Such a procedure would not lead to the removal of
all of the hemicellulosic material.   Under these conditions, the
hemicelluloses isolated should not be regarded at the time, or on later

reconsideration, as fully representative of the molecular population of the hemicellulose removed from, or present in, the plant.   'Pure' glycans also should not be regarded as necessarily fully representative of all molecules of that glycan either in the isolated hemicellulose or as it occurs in the plant.

If the holocellulose were treated immediately with strong, rather than with weak, alkali this would lead to more complete and speedier extraction.   It would, however, be at the expense of unnecessarily exposing all of the hemicellulosic material, including that which could have been extracted under milder conditions, to alkali-induced degradative changes.   Hemicelluloses extracted by weak alkali may have material retaining features which would not have survived strong alkali.

Alkali causes the so-called 'peeling' reaction (Morrison 1973), that is modification and progressive stripping, and loss, of units of $(1 \to 3)$ and $(1 \to 4)$ linked glycan chains starting at the latent aldehydic group of the so-called reducing end unit of the main chain. Alkaline peeling can be prevented, or minimized, by adding sodium borohydride to the extractant to reduce the reducing end unit of the glycan to a glycitol unit (Morrison 1974).   If borohydride is not added, then under the strongly alkaline conditions of a subsequent Haworth methylation (See 8.) further peeling would take place in the early stages.

It is desirable to treat the holocelluloses successively with alkali of increasing strength until no further hemicellulosic material is found to be extracted.   In between extractions the holocellulose residue should not be left exposed to the dangers of aerial oxidation but should be immediately immersed in the next volume of alkali under nitrogen.

Aqueous alkaline treatments will saponify any natural esters present (e.g. acetyl or diferulyl groups) and they will be lost (Neilson & Richards 1978).   The apparent absence of certain ester groups should be considered against the background of whether they would have survived the procedures used in studies at that time.   Other types of esters are those involving esterification of the carboxylic acid group of uronic acid units.   Such esters certainly include methyl esters of D-galacturonic acid units (in galacturonans).   There may be esters of D-glucuronic acid and of 4-O-methyl-D-glucuronic acid which are

substituent units on non-endospermic xylans.   The nature of any such
esters is unknown but it is reasonable to speculate that they might
include methyl esters.   It is also speculated that uronic acid esters
may have formed due to interaction between the carboxylic groups of
uronic acid units and any of the proximate hydroxyl groups of sugar
units in the same, or in other, molecules.   [Even if saponified, such
esters (which may be regarded as lactones) could reform in non-aqueous,
or neutral aqueous, solvents].   The chemical history of a glycan or
hemicellulose may greatly alter its present nature.

When uronic acid units are esterified either naturally or
during any stage in studies where base is used, then there are liable to
be profound changes affecting the material under examination and often
leading to its extensive, indeed even to its complete, loss.   If a
uronic acid is esterified and also has a substituent on its 4-0-position,
then prior to saponification the unit may undergo so-called β-elimination
(See 8.).   Substituents on non-endospermic xylans include a 4-0-methyl
group (on D-glucuronic acid units), and in rhamnogalacturonans there are
either D-galacturonic acid or L-rhamnose units on the 4-0-positions of
other D-galacturonic acid units.   The uronic acid units on β-elimination
remain glycosidically attached to adjacent units, provided they were not
also uronic acid units which had undergone β-elimination.   The uronic
esters are transformed into hexenuronate (4-deoxy-D-threo-hexopyranuronic
acid) units (Shimizu 1981) which formerly were never, and recently are
rarely, sought.   (Their former presence is indicated either by
4-deoxyhexitols, or their derivatives, obtained on treatment of sugars
in neutralised hydrolysates with sodium borohydride).

During alkaline treatments any supermolecular complexes
remaining in the holocellulose may break apart to give the isolable
species we regard as hemicelluloses and glycans.   Separation could be
dependent on β-elimination.   When this takes place the substituents
detached from the C-4 position of uronate units may include glycans,
sugar-chains, or monosaccharides.   Those of low degree of polymerisation
(d.p.) in particular, are liable to be lost completely at one or other
stages of most routine analyses (e.g. during dialysis).   Extensive
degradation of (1→4) linked polyuronides, such as galacturonans and of
rhamnogalacturonans having similar linkages, would take place.   (An
indication of the presence of rhamnogalacturonan main chains before
alkaline treatment might be the presence of residual rhamnose units at

the 'reducing end' of the, by then, liberated glycans, or oligosaccharides).   Hemicelluloses may be extracted from plantstuffs by
dimethyl sulphoxide without saponification (Bouveng & Lindberg 1965).
5.   Precipitation of hemicelluloses normally involves taking alkaline
extracts to neutrality or, commonly, making them slightly acidic with
acetic acid before adding an excess of ethanol or acetone to obtain the
precipitable hemicellulosic material (Whistler & Sannella 1965).   The
preliminary slight acidification is unlikely to cause any significant
further acid hydrolysis (See 3.).   At high concentrations of either
acetone or ethanol, inorganic acetates may precipitate and, being
flocculant, could be visually mistaken by the unwary as polysaccharide.
Silicates are present in some plants (e.g. in horsetail and rice and
many grasses) and they have precipitated under the conditions used for
the recovery of hemicelluloses (Fialkiewicz & Wilkie, unpublished).

Even after apparently successful precipitation procedures
have been used, some hemicellulose remains in solution and is rarely
sought or recovered for examination.   It is important to note the
observations of Richards (Blake & Richards 1971) about the conditions
for reproducible precipitation from a solution of a polysaccharide.
Precipitation may not be complete for several hours after addition of
the precipitants.   Unprecipitated material can be partly recovered by
desalting the solution and then removing water by freeze-drying.   The
solutions have very high concentrations of salts but the salts can be
removed by ultrafiltration, preferably using hollow fibre ultrafilters.
Molecules of low d.p. will be lost according to the ultrafilter used.
Alternatively salts can be removed by dialysis.   Dialysis is time-
consuming and cumbrous for large volumes and some loss of molecules of
low d.p. again takes place.   During dialysis or ultrafiltration losses
will routinely take place of hexenuronates, and of substituent
oligosaccharides of low d.p. and sugar units released on hexenuronate
formation.   Aqueous solutions are freeze-dried to avoid problems of
removal of water from increasingly viscous solutions if distillation
were used, and also to avoid any dangers of heat denaturation and
consequential decreased redissolvability of the hemicellulose extracted.
An alternative procedure for the removal of potassium acetate (in those
cases where potassium hydroxide had been used as extractant) is to
dissolve it in methanol which allows much of the (then insoluble)
carbohydrate, which would otherwise have been lost, to be recovered.

6.   Hemicelluloses should not be allowed to dry in air or they are
likely to become 'horny' (denatured) and not be able to redissolve
completely in the liquids in which they were previously soluble.   Where
possible it is best to keep hemicelluloses wet and, if to be kept wet
with water for a moderate time, to add a bactericide.   Hemicelluloses
can be kept under ethanol apparently without alteration of solubility
properties.   If, as is often the case, hemicelluloses must be dried,
then it is recommended that they be dissolved or suspended in water and
the water then removed by freeze-drying.   If this is not practical then
they may be steeped in successive volumes of ethanol with very vigorous
stirring, recovered by intermediate centrifugation and steeped in three
or four volumes of ether, to convert what is often a slimy gel to a
powder;   this should be dried on a porous tile in air, avoiding
granulation, leading to changed dissolvability, by continuous
pulverisation while drying.

7.   Fractionation of hemicelluloses (Whistler & Sannella 1965; Jones &
Stoodley 1965; Meier 1965; Gaillard 1961; Blake et al. 1971) and
isolation of pure glycans are carried out in similar ways but the
criteria governing the conditions used varies with the objective.
Fractionation of hemicelluloses is carried out with three main
objectives.   Firstly, to isolate any distinctively different glycans.
Secondly, to establish whether or not particular features are always
associated with, and therefore possibly indicated to be part of, a
particular glycan.   Thirdly, to determine whether in what is at first
regarded as a glycan there are molecular sub-populations that are so
distinctive that they also should be regarded as different, and
possibly separable, glycans.   Differences between such glycans might
be in degree of substitution of the main chains, in aspects of fine
structure (the presence or absence of particular structural features),
and in the range and distribution of molecules having different d.p.'s.
Fractionation is often designed to pursue a particular material as it is
enriched in units of one or more sugars at the expense of simultaneous
impoverishment with respect to other types of sugar unit.   This is often
assessed by examining sugar hydrolysates either directly by paper
chromatography (p.c.) or, after conversion to volatile derivatives (e.g.
glycitol acetates), by g.c.   Fractionation, or purification of glycans,
leads to losses of materials and often to incorrectly based and biased
conclusions about the hemicelluloses from particular plants and parts of

plants.   It is desirable to carry out complementary studies which avoid,
at least in part, losses occurring during the main fractionation.   Study
of unfractionated material gives an opportunity to detect all aspects of
structure present somewhere in the plant hemicellulosic population.
Studies of fractions obtained by different procedures from a total
hemicellulose can be used to seek information on probable structural
features present in the same glycans.   Studies of total hemicelluloses
increase the significance of quantitative comparisons between
hemicelluloses from plantstuffs of different species and of plants at
different stages of growth.

8.   Methylation of polysaccharides has been carried out by a limited
number of frequently used methods including those named after Haworth
(use of aqueous sodium, or potassium, hydroxide and dimethyl sulphate)
(Haworth 1915; Hirst & Percival 1965), Purdie (use of methyl iodide and
silver oxide) (Hirst & Percival 1965; Purdie & Irvine 1903), and Hakomori
(use of dimsyl sodium, i.e. sodium methylsulphinyl-methanide, in dimethyl
sulphoxide followed by use of methyl iodide) (Sandford & Conrad 1966;
Conrad 1972; Hakomori 1964).   The last named method was introduced in
1964 and since then has been the one mainly used - in those countries
where it is possible to ship in sodium hydride!

                   Methylation methods have as their objective the dissolution
of the polysaccharide and the methylation of every free hydroxyl group.
Methylation can be achieved if the polysaccharide can pass into solution
and there be subjected to treatments with strong base to ionise all free
hydroxyl groups which can then be converted to methyl ethers.   After
depolymerisation (see 9.) of the fully methylated glycan the positions
of former glycosidic links can be deduced.   This can be done without
ambiguity except when there is uncertainty whether the linkage is to a
4-$\underline{O}$-position and that ring is pyranose or the linkage to a 5-$\underline{O}$-position
and that ring is furanose.   In nature most sugars occur in the pyranose
ring form but L-arabinose is generally furanose and D-galactose
occasionally furanose.   Conclusions can only be regarded as reliable
if Haworth methylations were driven to completion.   If, after
methylation, any free hydroxyl groups have not been converted to
methoxyl groups then, unless such undermethylation is acknowledged,
erroneous interpretation will lead to conclusions about imagined points
of glycosidic attachment of side-chains, or of sugar units, or of
branches structurally similar to the main chain.   Some studies involving

use of Haworth methylation procedures yield materials which are not
adequately demonstrated to have been fully methylated.   In such cases
conclusions about branch-points must be suspect.   The presence of
branches on some homoglycans may now be considered to have been
inadequately 'proved' by methylation analysis studies.   It is essential
to consider in some detail the two main methods of methylation, namely
those of Haworth and of Hakomori.   The main emphasis will be placed on
what can happen during studies of hemicelluloses or glycans having
uronic acid units.

        In the Haworth method, the polysaccharide is dissolved, as
far as possible, in strong aqueous alkali (KOH or NaOH).   Methylation
involves a repetitive series of additions of aqueous sodium or potassium
hydroxide followed by slow intermediate additions of dimethyl sulphate.
Towards the end of this procedure, commonly extending over weeks or even
months, the then fairly fully methylated polysaccharide becomes much less
soluble, and the reaction unacceptably slow.   The Purdie procedure may
then be used with methyl iodide acting both as solvent and as methylating
agent in the presence of silver oxide.

        The Hakomori method in contrast to that of Haworth is
normally extremely rapid and, except, it is alleged, with glycans having
uronic acid units, leads to complete methylation in a single two-step
treatment.   If the methylation of polysaccharides having uronic acid
units is not completed after one such treatment then it is normally
considered desirable to reject the material and to repeat the process.
(This advice, it is suggested, should now be questioned as the nature of
the degradative changes which take place during subsequent Hakomori
methylations is largely understood, or capable of being understood, and
the information that can be obtained by repeated Hakomori methylation
sheds light on structural features in acidic glycans).   In the Hakomori
procedure the polysaccharide is dissolved, or swollen, in dimethyl
sulphoxide (one of the few liquids, apart from water, aqueous solutions,
tetrahydrofuran, Cadoxen (Jayme & Lang 1963) and, exotically, liquid
ammonia, in which polysaccharides will often dissolve).   The polysaccharide is then treated with dimsyl anion in dimethyl sulphoxide to
ionise the hydroxyls fully so that they will be converted to methyl ethers
when methyl iodide is added.

        The differences in speed of the Haworth and Hakomori
reactions is normally regarded as being due to the relative strengths of

the bases effecting ionisation.   Dimsyl anion in dimethyl sulphoxide
(an aprotic dipolar solvent) is much more efficient in removing protons
than is aqueous alkali.   It is now suggested that the differences in
basicity, although clearly of great importance, may not adequately
account for what happens during Haworth methylations of acidic glycans -
such as of acidic xylans and galacturonans.

As stated earlier, if uronic acid units are present in the
hemicelluloses in the plant they may, or may not, be naturally esterified.
(D-Galacturonic acid units in galacturonans are often methyl esterified.)
Uronic acid units either have sugar units (as in galacturonans) or some
methyl ether groups (as in xylans) on their 4-position.   If such units
are also naturally esterified then they are liable to undergo β-
elimination on addition of base.   If the base is aqueous alkali there
will be competitive reactions between saponification and β-elimination.
In an ester the hydrogen on C-5 of the uronate is labile and will be
removed by strong base if there is a substituent other than hydrogen on
O-4.   The action of a non-aqueous base, such as dimsyl anion, would be
to cause β-elimination as shown in Fig. 1 leading to the loss of whatever
had been on the C-4-position - whether a methoxyl group, a sugar unit, or
a glycan chain.   If, however, the acid is not esterified then the
carboxylate group will not promote loss of the C-5 hydrogen and the C-4
substituent will be stable.   Uronic acid units, it is suggested, may be
naturally esterified not only by methyl groups but by interaction with
hydroxyl groups of sugar units elsewhere in the same, or another,
molecule.   It is considered probable that lactone or ester formation of
these types will be promoted in the dry dimethyl sulphoxide.   If this is
so, then addition of dimsyl anion would be expected to convert those
esterified and 4-O-substituted units very extensively into hexenuronates
units.   There would be simultaneous cleavage of the link to the 4-O-
substituent which, if small, would be lost during dialysis or other
procedures which follow methylation.

The Haworth methylation procedure is slow but more rapid at
the early stages than later when it reaches a near plateau.   Why is this
so?   Slow methylations during later stages of the Haworth procedure were
attributed to the decreased solubility of the by then increasingly fully
methylated xylan.   Consider a theoretical xylan (one, however, having
features present in most, if not all, non-endospermic xylans) namely
xylan molecules having some D-glucuronic acid units and 4-O-methyl-D-

Fig. 1.   Hakomori methylation.

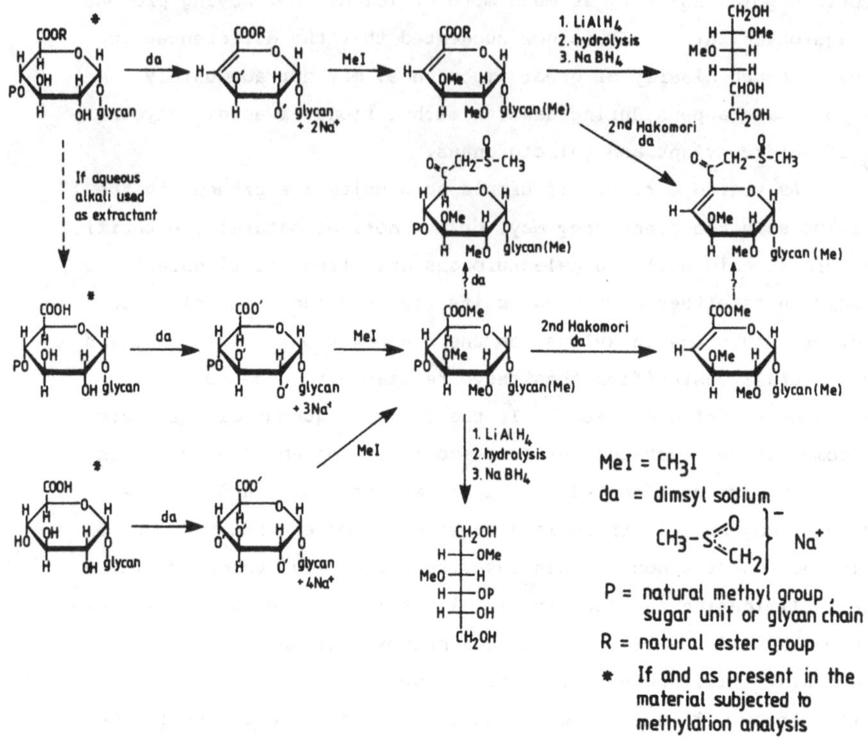

glucuronic acid units on the 0-2 position of some of the main-chain
xylose units.   In the early stages of methylation those uronic acid
units which lack a 4-0-methyl group will quickly acquire one (Fig. 2).
Under the strongly aqueous alkaline conditions the uronic acid units
would be expected to be saponified.   In spite of this, it is suggested
that during the addition of the dimethyl sulphate there will not only be
methylation but some, albeit transient, esterification.   Such
esterification would be expected to be more marked in the region of the
temporary interface between the added reagent and the alkaline solution
of increasingly methylated xylan.   If this is so, there would once more
be competition between saponification and β-elimination.   At any one
moment only a low proportion of the uronic acid units would be esterified
but, over the protracted period of all 'good' methylations, all such units
would be transiently re-esterified until they eventually underwent
β-elimination.   The 'better' the methylation the more extensive would

β-elimination be.   Possibly during extended methylations the
hexenuronate units (known to be acid labile - although reports of their
precise lability vary) would be hydrolysed by the base (Lindberg et al.
1973; Lindberg & Lönngren 1972).

Fig. 2.   Haworth methylation.

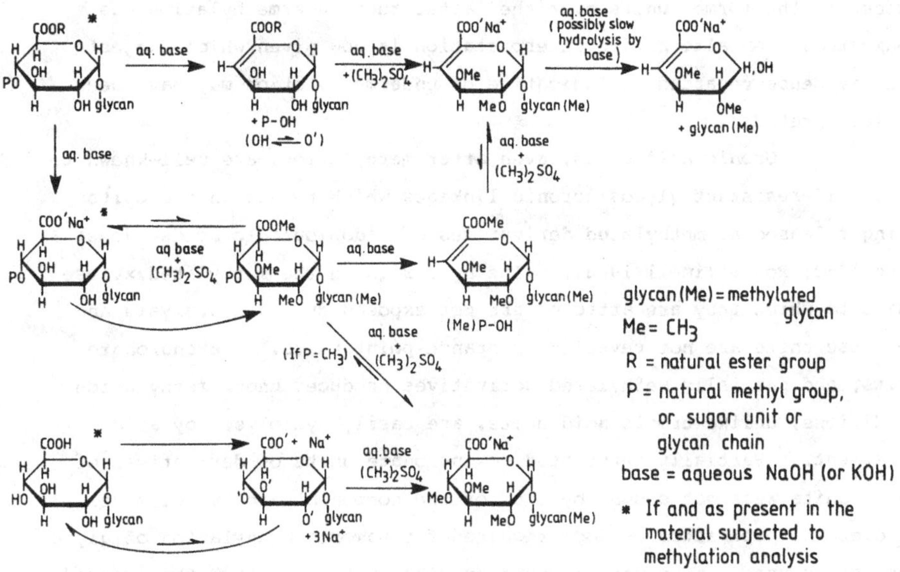

During the course of a methylation, attempts to establish
whether the polysaccharide was nearly fully methylated were commonly
made in three ways.   One involved determination of the methoxyl
content and comparison with the estimated methoxyl content of a fully
methylated glycan of an imagined composition.   Quite apart from errors
in the imagination, the methods formerly used to determine methoxyl
content were not adequately accurate - partly because of the difficulties
in obtaining dry weights of viscous syrups.   It is also possible to
examine methylated materials in the IR for evidence of free hydroxyls
but this method also has uncertainties.   Another, and invaluable,
method became available with the introduction of paper chromatography in

the late 1940's.   It was then possible to monitor the progress of a
methylation by examining hydrolysates of hemicelluloses subjected to an
increasing number of methylation treatments.   In a fully methylated
xylan, for example, there should be a correspondence between the
proportion of units then identified as having carried side-units or side-
chains and the proportion of identifiable terminal units deriving from
such side-chains or side-units (allowance having been made for the
single non-reducing end unit of the main chain).   When there was an
excess of the former units over the latter then undermethylation was
indicated.   An alternative interpretation is now given which suggests
that evidence regarded as indicative of undermethylation may have been
misinterpreted.

　　　　　Uronic acid units, even after methylation, are well-known to
have acid-resistant glycosiduronic linkages which result in the units
being released as methylated derivatives of aldobiouronic acids (Jones &
Wise 1952; Roy & Timell 1968).   The hydroxyls on the methylated xylose
units to which they are attached are not exposed during hydrolysis and
so these units are not revealed as branch-point units.   Hexenuronate
units, and partially methylated derivatives produced under methylation
conditions, unlike uronic acid units, are easily hydrolysed by acid
treatment.   Partially methylated hexenuronate units or derivatives of
these units were not sought by p.c. of the components in hydrolysates.
If over the protracted periods required for Haworth methylation of glycans
the hexenuronate units were removed by alkali (Fig. 2) then the exposed
hydroxyl groups would be methylated and the xylose units to which they
had been attached would cease to be regarded as indicative of undermethylation.   Then, and only then, would the methylation appear to be
complete when comparisons were made between proportions of materials
corresponding to branch-point units and end units.   After hexenuronate
units were formed, and before they were cleaved, the presence of such,
routinely undetected, hexenuronate units would have been equated as
indicative of undermethylation.   Normally the Haworth procedure was not
regarded as having led to complete methylation and, as mentioned,
methylation was often completed by a single Purdie treatment.   It was
frequently remarked 'at the bench' that a Purdie methylation had 'gone
acid' and indeed this did happen.   Possibly hexenuronate or uronic acid
units were in some way responsible for the acidity and, if so, may also
have been responsible for autohydrolysis of the remaining hexenuronate

units and for methylation thereafter of the exposed hydroxyls.   The
proportions of branch-point units and of terminal side units present in
hydrolysates would then agree and the methylation be regarded as
complete.   Varied reports of the presence and absence of methylated
derivatives of uronic acids (as their aldobiouronic acid derivatives
which, as mentioned, hydrolyse with difficulty) may be partly explained
by different exposures to methylation conditions and of losses consequent
on β-elimination.

In summary, it is suggested that the length of Haworth
methylations is partly due to the time it takes for a transient
esterification —— β-elimination reaction to be followed by loss of
hexenuronate units and for the then exposed hydroxyl groups to be
methylated.   Complete loss of all such units would be equated with
complete methylation;   retention of methylated uronic acid units, which
have not undergone β-elimination would not, however, be translated as
indicative of undermethylation.   The presence of attached hexenuronate
units was very likely to be misinterpreted as indicative of undermethylation.

Methylation by the Hakomori procedure (Fig. 1), although
superficially similar, is very different from that by the Haworth
procedure.   If the uronic acid units are not esterified (due to the lack
of an ester group in the natural state, or to its removal by aqueous
alkali) then during the first part of the two-stage process, namely
addition of dimsyl anion, there will be no hexenuronate formation.
During the second part (addition of methyl iodide) all hydroxyl groups
and the carboxylic groups will be etherified and esterified, respectively,
and some units will undergo β-elimination before the base is destroyed by
the excess of methyl iodide added.   If hemicelluloses are extracted by
aqueous alkali then natural esters will be destroyed.   Such destruction
can be avoided by carrying out extraction with dimethyl sulphoxide which
dissolves many glycans (Bouveng & Lindberg 1965).

Hemicelluloses in mesophyll cells of Italian ryegrass leaves
were found to have uronic acid-, and arabinose-, rich xylan molecules
(S.P. Davies & K.C.B. Wilkie, unpublished).   The material in the
mesophyll cells was methylated directly by suspending them in dimethyl
sulphoxide and treating them with dimsyl anion.   β-Elimination was
concluded to have taken place of 4-O-substituted uronate units.   During
addition of methyl iodide the 4-O-positions (and all others) would be

methyl etherified.   <u>Before</u> the base was destroyed by the methyl iodide,
further β-elimination would take place.   When, contrary to normal
practice, a second addition of dimsyl anion was made there was then
total β-elimination of all uronic acid units.   After the second addition
of base either ethyl iodide or trideuteriomethyl iodide was added to
provide distinctive substituents on still free hydroxyls.   <u>No</u> further
hydroxyls were etherified.   The hemicellulose had been completely
methylated <u>but</u> the hexenuronate units had evidently remained attached to
the methylated (xylan) main chain.   On acid hydrolysis a very high
proportion of hydroxyl groups was exposed.   The results accorded with
the structural feature in Fig. 3.   Fig. 1 outlines various alternative
reaction routes under Hakomori methylation conditions.

Fig. 3.   Structural feature in a xylan from Italian
ryegrass leaf mesophyll rich in L-arabinose and in
D-glucuronic acid units or 4-<u>O</u>-methyl-D-glucuronic
acid units or both.

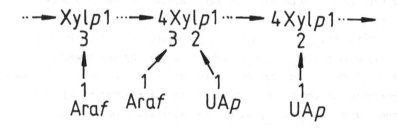

$$Xylp = D\text{-xylopyranose}$$
$$Araf = L\text{-arabinofuranose}$$
$$UАp = D\text{-glucopyranuronic acid}$$
$$or\ 4\text{-methyl ether}$$

9.   Hydrolysis of methylated hemicellulose or methylated glycan is
effected in a variety of ways according to whether the methylated
derivatives are wanted as methyl glycosides or as free sugars.
Frequently irrespective of which are wanted, initial depolymerisation is
effected by heating the fully methylated material with anhydrous

methanolic hydrogen chloride (most simply made by adding acetyl
chloride to dry methanol).   Methyl glycosides of fully methyl
etherified glycosides (e.g. the methyl 2,3,5-tri-$\underline{O}$-methyl-L-
arabinofuranosides) may be partly lost due to their volatility unless
precautions are taken (e.g. use of a sturdy, well-protected, sealed tube).
Other procedures for depolymerisation yielding free sugars involve
heating the methylated material with trifluoroacetic acid, or with formic
acid.   In each case the procedure is followed by treatment with hot
dilute mineral acid to complete hydrolysis.   Neutralisation of
hydrolysates may involve the addition of, for example, silver carbonate
(for HCl) or barium carbonate or hydroxide (for sulphuric acid).   In
each case the precipitate formed may coprecipitate, as salts, compounds
such as uronic acid derivatives containing carboxylic groups, or
compounds may be selectively, or generally, adsorbed on the inorganic
precipitate.   A procedure which can avoid such losses is to destroy
excess of hydrochloric acid with diazomethane in ether (but there is of
course an explosive, and poison, hazard from the diazomethane).
Hydrolysis may or may not be complete and there may be degradative
losses.   It is desirable that hydrolysis be checked by repeating the
same, and other, procedures to obtain reproducible results.
Reproducibility of results is not easily obtained even under ostensibly
the same conditions if glycans have uronic acid units.
10.   Separation of methylated components in a mixture prior to the late
1940's was by classical techniques, such as distillation under reduced
pressure, of materials which were difficult to handle in such ways.
In spite of the formidable difficulties brilliant studies were carried
out during that period and major, and many minor, aspects of structure
detected.   When p.c. was introduced in the late 1940's, and widely
adopted soon after for the study of sugars, and methylated sugars, it
became much easier to identify, quantify and separate components in
mixtures.   Over the past $\underline{ca}$. 20 years the methods of choice, where the
apparatus and chemicals are available, are g.c. and m.s. to separate and
identify known and unknown compounds.

It was difficult or impossible to detect and identify very
small proportions of methylated sugars (and their derivatives) by p.c.
as can now be done by g.c..   It is stressed that any earlier failure to
report the presence of a structural feature on the basis of p.c. (and
column chromatographic) studies did not prove the absence of such a

feature.   The same materials would have to be re-examined by the
present more sensitive methods before a positive or negative conclusion
could be reached.

It is important to consider the differences between what
could safely be concluded on the basis of single, and indeed normal,
studies of a hydrolysate of a methylated material compared to what could
be established by repeated studies of a material (possibly methylated on
different occasions in various ways) using combinations of swift and
sensitive techniques and equipment (e.g. Hakomori methylation, and
g.c.-m.s.).   A single study of any one material by any one method may
reveal the presence of minor components, but are they genuine or are they
'noise' or due to contamination?   Repeated studies of hemicelluloses
from Italian ryegrass leaf and stem, and from leaf mesophyll, made it
clear that some g.c. 'noise' was due to components which could no
longer be ignored because their presence was regularly indicated.   As
mentioned, minor components may be of major significance if they derive
from particularly important sites in polymers; they could have derived
from branch-points, or could affect the conformation of molecules.
Even more important they may represent more (possibly much more)
abundant material largely destroyed or lost.   The 'minor' components
could give guidance on what should be monitored when improving methods
to enable the glycans containing these features to be recovered and
studied more adequately.

11.   Preparation and identification of derivatives of methylated sugars
formerly involved structural proof or correlation of the physical
properties of derivatives with values reported in the literature
(commonly m.p.'s of derivatives which could be relatively easily
crystallised).   Some crystals were sticky until washed, and a few on re-
examination by g.c. have been found to be contaminated by a coat of
syrup i.e. of other components not separated, and therefore not known to
be present, when the earlier studies were carried out.   G.c. separation
of components allied to m.s. provides an excellent indication of the
identity of the various components in mixtures and has made qualitative
studies more sensitive and quantitative studies easier and more accurate
than when p.c. and the m.p's. of sugar derivatives were primarily used to
establish identity.

The results obtained by procedures such as those outlined
and by ancillary or similar studies were commonly used to present

structural features in, or structures of, hemicellulosic glycans.   Most
early structural studies of polysaccharides were carried out by
scientists trained as organic chemists.   Organic chemists do not
normally carry out structural studies on material known to be either
impure, or worse, of such a nature that the very meaning of the term
pure is unclear.   The word 'structure', unless qualified, will conjure
up in the average chemist's mind (and probably even more in the minds of
those less highly trained in chemistry) the idea that a published
hemicellulosic structure is, like most other organochemical structures,
a unique statement informing the reader that in the molecules of the
material studied this atom is linked in the way shown to that one and so
on.   The sugar units of each polysaccharide molecule do, of course, have
unique structures.   In hemicellulosic glycans the sequence of such sugar
units is, however, normally uncertain except in the case of linear
homoglycans having only one type of unit linked glycosidically in one
way as in amylose and cellulose.   Under such circumstances there are no
ambiguities about the primary structure (the covalently-bonded structure)
other than about d.p. and the distribution of molecules of different d.p.'s.

Most of the structural features in hemicellulosic glycans,
have stood up to, and almost certainly will continue to stand up to, the
rigours of further study.   Uncertainty about some is difficult to
dispel.   One may not have isolated quite the same molecular population
even from the same plantstuff and so a feature reported earlier but not
confirmed later could be present in the plant's total hemicellulosic
population but absent from the fraction (sub-population) isolated and
studied on a later occasion.

What are presented as hemicellulosic structures are better
described as 'structural assemblages' (Wilkie 1983) which present
possible permutations of structural features conveying information in an
easily assimilated form of chemical shorthand which requires to be
translated back into its original form to communicate the true
information.   Structural assemblages require to be periodically
reconsidered, and, in effect, to be mentally dismembered into their
constituent parts, and then reconstitued from these, and any other parts
subsequently discovered.   If a jigsaw puzzle having an abstract, and
unknown, design fell to pieces we might reassemble the parts to give a
satisfactory fit, indeed even an attractive picture, but it would
probably not be the right picture.   If several such puzzles were

dropped at the same time then the uncertainty over the verisimilitude of
each reassembled picture would be even greater. An analagous
situation applies in studies of the hemicelluloses and of glycans
derived from them particularly if, during fractionation or chemical
treatments, some of the pieces of each hemicellulose, or glycan, puzzle
have been lost, or altered, or have been concluded to belong to an
accompanying one.

Chemists, and others, studying the hemicelluloses, pectic
substances and gums, except in moments of euphoria or haste to publish,
accept that most 'structures', other than the simplest, are possible
structural models yet to be proved right or wrong.   Other, more soundly
based, models may be proposed and these should cause earlier ones to be
reconsidered and occasionally discarded.   In practice, many models which
are less soundly based than others continue to be quoted on terms of
equality.   Certainly simple models are easier to remember than complex
ones and those hedged with qualifications about their significance are
likely to be viewed with less favour than others where such comments,
although still relevant, were not specifically, and inconveniently made.
Once a 'structure' is embedded in the literature, and much quoted, it is
singularly resistant to removal.

Little attention has, so far, been paid above to a major and,
literally, vital reason for differences in the hemicelluloses of plants
of one species, namely anatomical differences and developmental changes
in the plant.   For reasons explained earlier (See 8.) it was formerly
impractical to carry out many studies of the structures of hemicelluloses
of plants at different stages of growth by Haworth methylation procedures.
The studies were too slow. Most studies by chemists were, until quite
recent years, on plants of ill-defined age, and not infrequently on
ill-defined parts of these plants.   Most annual and perennial plants,
including grasses and legumes, upon which studies of hemicelluloses
were carried out extensively, were collected at or near the time of
normal harvest. Now that structural studies can be carried out more
speedily it is possible to examine the hemicelluloses of plants at
different stages of growth and to repeat studies until reproducible
results are obtained, or the reasons for non-reproducibility accounted
for.   Repeated studies are essential if an objective is to make
'fundamental' statements about the hemicelluloses, or glycans, of a
plant, or to make comparisons between the hemicelluloses of a plant at

different stages of growth, or between those from different parts of the plant, or between those from plants which are taxonomically closely related.   Although repetitive structural studies on the hemicelluloses of developing plant were formerly impractical, simple quantitative and qualitative studies were frequently carried out by determining the sugars in hydrolysates of unfractionated hemicelluloses extracted from plants and parts of plants at different stages of growth.   These showed that there are considerable variations in the sugar composition of the total hemicelluloses from young and old plants (Fig. 4) (Buchala et al. 1971; Buchala & Wilkie 1973; Buchala & Wilkie 1974).

The hemicelluloses of the grasses and cereals will be briefly commented on; fuller details and references are reported elsewhere (Wilkie 1979).   The numerous and varied studies set some points already made, and to be made, in perspective.   The remarks are not limited to the Gramineae but are relevant to hemicelluloses from plants of other families, although the particular sugars, and plant parts, may there differ.   In the non-endospermic parts of grasses (leaves and stem) the dominant glycan is a xylan.   The sugar compositions were determined of xylans from many plantstuffs and of the unfractionated hemicelluloses from which they had been derived.   There were higher proportions of non-xylose units in the latter than could be accounted for by the xylans isolated.   This was not surprising in view of the criteria guiding fractionation procedures.   It was, however, necessary to fill a blank in knowledge by speculating on the nature(s) of the glycan(s) from which quantitatively unaccounted sugars had originated.   Such speculation was based on consideration of glycans of frequently doubtfully 'proved' structure from other, sometimes taxonomically quite remote, plants.   It was possible that L-arabinose and D-galactose not quantitatively accounted for by 'typical' grass and cereal xylans derived from a pectic type homoarabinan (similar to one reported from peanuts) or from an arabinogalactan (similar to one from larch).

The following type of statement on the xylans from grasses is widely accepted: "Esparto grass is unusual in that it yields a homoxylan (i.e. a xylan composed only of D-xylopyranose units) (Chanda et al. 1950).   There have been many studies on xylans from wheat leaf and straw (Aspinall & Mahomed 1954; Aspinall & Meek 1956; Erenthal et al. 1954) and they have been shown to have low proportions of L-arabino-furanosyl units (attached glycosidically to 3-0-positions of main chain

Fig. 4.   Total hemicellulose from bamboo, <u>Arundinaria japonica</u> and of materials from it soluble and insoluble in water (Wilkie 1983).

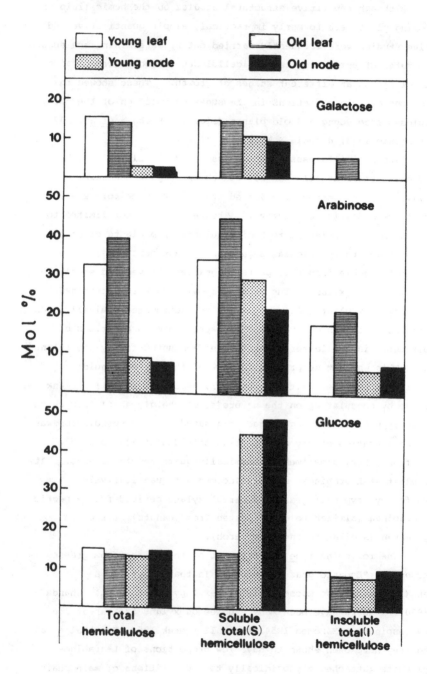

D-xylose units), and a low or lower proportion of D-glucuronopyranuronic
acid units, or of units of its 4-methyl ether, or of both, attached to
2-0-positions of other D-xylose units.   Xylans of many other grasses
and cereals are similar to those from wheat.   The xylan from maize
(corn) hull displays unusual structural features which account, amongst
other things, for the presence of D-, and some L-, galactoses (Erenthal
et al. 1954; Roudier 1959; Whistler & Corbett 1955; Montgomery et al.
1957)."   Such types of statement are now, I believe, misleading.
[Another xylan was isolated from esparto which had a higher proportion
of arabinofuranosyl units than found in the xylans of wheat, barley,
oats and other grasses and cereals (Aspinall et al. 1953).   It was
commented that the arabinose in esparto hemicellulose did not derive from
a pectic-type arabinan (Aspinall et al. 1953)].

        Replicate, and varied, structural studies have been carried
out on the hemicelluloses, and on fractionated hemicelluloses, from
Italian ryegrass (Lolium multiflorum) (J.S. Al-Hakkak & K.C.B. Wilkie,
unpublished).   Highly substituted xylans were concluded to be present
having all of the features reliably reported in the non-endospermic xylans
of any grass.   Most hemicelluloses isolated for earlier structural
studies had been obtained from fully grown plants - probably at about the
time of normal harvest.   Had younger plants been used as sources of
hemicelluloses for studies, and had different fractionation techniques
consequently been employed, then conclusions about the nature of the
hemicelluloses would have been very different.

        Companion studies on Italian ryegrass leaf mesophyll cell-
wall hemicelluloses and on those from the mesophyll of other grasses
(S.P. Davies & K.C.B. Wilkie, unpublished) show that their hemicellulosic
compositions are similar to one another but very different from those in
the more abundant cell-wall material in the leaves and stems.   In young
plants the proportion of mesophyll and other thin-walled cells is higher
than it is in older plants.   In the latter it is, accordingly, to be
expected that the contribution made by the hemicelluloses of such cells
will be diluted, and interpretation confused, by the increasing domination
of the hemicelluloses from other types of cell.   It may be premature to
account for all hemicelluloses in terms of the few types of cell and
tissue so far studied.

        The hemicelluloses of Italian ryegrass have features which
show that some xylan molecules are very highly substituted.   They have

structural features previously reported present in the xylans from maize.
These infrequently encountered features are often implied to be rare;
it is true that they have been rarely reported in the xylans of grasses
other than maize.   Structural features in the total hemicelluloses of
Italian ryegrass are compatible with the presence of several xylans
ranging from a family having arabinose to xylose in the ratio of ca. 1
to 10 and no, or little uronic acid to, at the other extremity, a xylan
having the two sugars in the ratio of ca. 7 to 10 and a corresponding
proportion of xylose units substituted by D-glucuronic acid or its 4-
methyl ether.   The xylan from the mesophyll is similar to the latter and
has a very high proportion of uronic acid substituents.   Some of the
D-xylose units of the mesophyll xylan have both an arbinose and a uronic
acid substituent.   Such substitution is indicated in the arabinose-, and
uronic acid-, rich materials of the leaf hemicelluloses.   The highly
substituted xylans have $\underline{O}$-β-D-xylopyranosyl-(1→2)-L-arabinofuranosyl,
$\underline{O}$-α-D-galactopyranosyl-(1→3)-L-arabinofuranosyl side chains, and
possibly $\underline{O}$-α-D-, and -L-, galactopyranosyl-(1→4)-β-D-xylopyranosyl-(1→2)-
L-arabinofuranosyl side chains.

      The highly substituted xylans derive from the thin-walled cells
(mesophyll cells) and the xylans with low degree of substitution from the
epidermis and vascular tissues.   Other hemicelluloses present include
(1→3) and (1→4) heterolinked D-glucan, rhamnogalacturonan, highly
substituted (1→5)-arabinoglycan branched on the 2-$\underline{O}$-position, (1→4)-
galactan, and (1→3) and (1→6) linked galactan which may be an arabinogalactan.   There is evidence indicative of low proportions of a 6-$\underline{O}$-
branched xyloglucan and of a mannoglycan (possibly a glucomannan).   The
amount of a fructan present fluctuates depending on the time of the
collection of the plant.

      It is improbable that the polydiversity, polydispersity and
polymolecularity (Reid & Wilkie 1969) of the hemicululosic populations
of any anatomically complex plant part will be adequately described in
terms of glycans obtained by fractionation.   The studies on Italian
ryegrass indicate that variations between the populations of
hemicelluloses at different stages of growth of a plant primarily
reflect variations in the population and types of cells in the plantstuff
and, superimposed on these, variations caused by chemical and handling
procedures.   There is nothing to indicate that the variations noted
between the xylans, and other hemicellulosic glycans, from plants which

are taxonomically closely related are species-specific other than in terms relating to the anatomy of the plant and its stage of development.

It is suggested that the following points be considered when planning studies and when interpreting the results of present and earlier studies of hemicelluloses, and of fractions, including allegedly pure glycans, obtained from them.   Many points raised below cannot yet be answered.   Where this is the case conclusions should be viewed with some reservation.

a. How representative was the material isolated of the range and types of glycan molecules in the plantstuff?

b. Was some of the material degraded during isolation or during structural analysis?

c. Is the structure primarily an averaged assemblage of all the aspects of structure deduced present somewhere in any of the molecules, or is it possible to regard the material as a distinct glycan rightly and meaningfully called pure?

d. Could hemicellulosic glycans have unique structures?   At present there is little evidence that complex hemicellulosic glycans do, but it should be noted that the facts that the molecular populations are mixed, and that chemical degradation takes place, would make it singularly difficult to detect uniformity of structure in molecular sub-populations.

e. If the glycan is, say, a xylan, and was obtained by fractionation of a hemicellulose, is the material recovered representative of all the material that should be termed xylan, or is it only a selected part of the xylan molecular population?   Does a 'pure' xylan (or other glycan) contain molecules of all the d.p.'s of the glycan in the total population of hemicellulosic molecules, and do the molecules have all of the structural features associated with any of the molecules called, in this case, xylans?   Are there several distinct xylans meriting separation and study or are they all part of a wide molecular spectrum?

f. Where there are side-units or side-chains, are they randomly positioned along a main chain, or are they in some way in an ordered arrangement;   e.g. are they at regular positions along the main chain, or in regular or in random clumps?

g. Where the glycan core is not linear is it of the 'herring-bone' or multi-branched (bush) type?

h. Is there a main chain of one type of sugar unit or, if more than
    one type of unit is deduced present in the core (including the
    main chain), are these units alternating, or in groups, or in
    irregular sequence?

i. Does the main chain carry pendant chains and, if so, are these
    chains internally glycosidically linked in the same way as in
    the main chain?

j. Are there non-carbohydrate moieties (lignin, protein, peptide,
    acetyl groups etc.) covalently linked before, and lost, unsought,
    or modified during, isolation or study of hemicelluloses?

k. Are what are regarded as polysaccharides in fact detached parts
    of a supermolecular complex possibly comprising several
    polysaccharides (and other components) as, in isolated state,
    we know them?

l. How certain is it that the 'bits and pieces', detected by
    methylation or structural analytical methods, derived from the
    particular glycan to which they are claimed to belong?   [Where
    there have been many studies of highly, and differently,
    fractionated materials, and the sub-fractions (including 'pure'
    glycans) consistently display the same features, then there is
    reasonable evidence that these features are indeed part of the
    molecules with which they are apparently associated - but they
    are not necessarily linked in the way presented.]

m. Has consideration been given to what was detectable using the
    equipment available and the techniques employed at the time and
    place where the work was carried out?   Some recent studies still
    involve the use of methods employed in the late 1940's.   It is
    an elementary fact that by g.c. and by m.s. examination of
    products of methylation analysis many minor, and some major,
    sugar derivatives which would have escaped identification by paper
    chromatography are now detectable much more readily.   Absence of
    comment about a structural feature should be viewed against the
    background of what it was reasonable to expect could be detected
    when and where the work reported was carried out.

n. Can conclusions about the hemicelluloses or glycans be
    extrapolated with safety to predict the nature of the
    hemicellulose in other plants or in the same plant at a different
    stage of growth?

Fig. 5.   The Hemicellulose Circle.

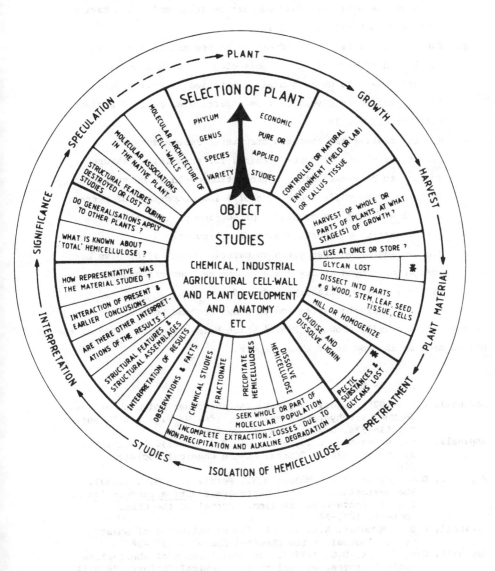

o. Is it possible to conclude the anatomical origin of distinctive
    parts of the hemicellulosic population (e.g. molecules having
    certain structural features)?

p. Can structural features deduced to be present in a complex
    population of hemicellulosic molecules (e.g. in a total
    hemicellulose) be unambiguously identified as having derived
    from a particular glycan or from a particular part of a plant,
    cell, or cell-wall?

q. How informative are averaged structural assemblages of the
    structures of any glycans? Where, for example, a feature
    appears to be present in low proportion could it be that its
    presence is actually indicative of a low proportion of certain
    molecules having a much higher proportion of that feature and
    its absence from otherwise similar molecules?

r. Has consideration been given to material which might have been
    lost or destroyed at any stage of the various processes
    involved in its extraction and study?

In conclusion attention is drawn to 'the hemicellulose
circle' (Wilkie 1983) shown in Fig. 5 which brings together some of the
points made in the course of this chapter.

REFERENCES

Aspinall, G.O. (1970). Pectins, plant gums and other polysaccharides.
    In The Carbohydrates, vol. IIB, eds. W. Pigman & D. Horton,
    pp. 515-536. New York: Academic Press.
Aspinall, G.O., Greenwood, C.T., Sturgeon, R.J. (1961). Degradation of
    xylans by alkali. Journal of the Chemical Society,
    3667-74.
Aspinall, G.O., Hirst, E.L., Moody, R.W., Percival, E.G.V. (1953).
    The hemicelluloses of esparto grass (Stipa tenacissima L.).
    The arabinose-rich fraction. Journal of the Chemical
    Society, 1631-34.
Aspinall, G.O. & Mahomed, R.S. (1954). The constitution of wheat straw
    xylan. Journal of the Chemical Society, 1731-36.
Aspinall, G.O. & Meek, E.G. (1956). The constitution of wheat-straw
    hemicelluloses. Journal of the Chemical Society, 3830-34.
Bailey, R.W. (1973). Structural carbohydrates. In Chemistry and
    Biochemistry of Herbage, vol. 1, eds. G.W. Butler & R.W.
    Bailey, pp. 157-211. New York: Academic Press.
Blake, J.D., Murphy, P.T., Richards, G.N. (1971). Isolation and A/B
    classification of hemicelluloses. Biochemical Journal, 20,
    656-64.
Blake, J.D. & Richards, G.N. (1971). Evidence for molecular aggregation
    in hemicelluloses. Carbohydrate Research, 18, 11-21.
Bouveng, H.O. & Lindberg, B. (1965). Acetylated wood polysaccharides.
    Methods in Carbohydrate Chemistry, 5, 147-50.

Buchala, A.J., Fraser, C.G., Wilkie, K.C.B. (1971). Quantitative studies on the polysaccharides in the non-endospermic tissues of the oat plant in relation to growth. Phytochemistry, 10, 1285-91.

Buchala, A.J., Fraser, C.G., Wilkie, K.C.B. (1972). Extraction of hemicellulose from oat tissues during the process of delignification. Phytochemistry, 11, 1249-54.

Buchala, A.J. & Wilkie, K.C.B. (1973). Total hemicelluloses from wheat at different stages of growth. Phytochemistry, 12, 499-505.

Buchala, A.J. & Wilkie, K.C.B. (1974). Total hemicelluloses from Hordeum vulgare plants at different stages of maturity. Phytochemistry, 13, 1347-51.

Chanda, S.K., Hirst, E.L., Jones, J.K.N. (1950). The constitution of xylan from esparto grass (Stipa tenacissima L.). Journal of the Chemical Society, 1289-97.

Conrad, H.E. (1972). Methylation of carbohydrates with methylsulfinyl anion and methyl iodide in dimethyl sulfoxide. Methods in Carbohydrate Chemistry, 7, 361-64.

Erenthal, I., Montgomery, R., Smith, F. (1954). The carbohydrates of the Gramineae. II. The constitution of the hemicelluloses of wheat straw and corn cobs. Journal of the American Chemical Society, 76, 5509-14.

Gaillard, B.D.E. (1961). Separation of linear from branched polysaccharides by precipitation as iodine complexes. Nature, 191, 1295-96.

Hakomori, S. (1964). Rapid permethylation of glycolipids and polysaccharides, catalysed by methylsulfinyl carbanion in dimethyl sulfoxide. Journal of Biochemistry (Tokyo), 55, 205-208.

Harwood, V.D. (1952). The action of acid chlorite on wheat straw. Tappi, 35, 549-55.

Haworth, W.N. (1915). A new method of preparation of alkylated methylated sugars. Journal of the Chemical Society, 107, 8-16.

Hirst, E.L. & Percival, E. (1965). Methylation of polysaccharides and fractionation of methylated products. Methods in Carbohydrate Chemistry, 5, 287-96.

Jayme, G. & Lang, F. (1963). Cellulose solvents. Methods in Carbohydrate Chemistry, 3, 75-83.

Jones, J.K.N. & Stoodley, R.J. (1965). Fractionation using copper complexes. Methods in Carbohydrate Chemistry, 5, 36-38.

Jones, J.K.N. & Wise, L.E. (1952). The hemicelluloses present in aspen wood. Journal of the Chemical Society, 3389-93.

Lindberg, B. & Lönngren, J. (1972). Specific degradation of polysaccharides containing uronic acid residues. Methods in Carbohydrate Chemistry, 7, 142-48.

Lindberg, B., Lönngren, J., Thompson, J.L. (1973). Degradation of polysaccharides containing uronic acid residues. Carbohydrate Research, 28, 351-57.

Meier, H. (1965). Fractionation by precipitation with barium hydroxide. Methods in Carbohydrate Chemistry, 5, 45-46.

Montgomery, R. & Smith, F. (1955). The carbohydrates of the Gramineae VII. The constitution of a water-soluble hemicellulose of the endosperm of wheat. Journal of the American Chemical Society, 77, 3325-28.

Montgomery, R., Smith, F., Srivastava, H.C. (1957). Structures of corn
    hull hemicelluloses Part IV. Partial hydrolysis and
    identification of 3-0-α-D-xylopyranosyl-L-arabinose and
    4-0-β-D-galactopyranosyl-L-arabinose. Journal of the
    American Chemical Society, 79, 698-700.
Morrison, I.M. (1973). Isolation and analysis of lignin-carbohydrate
    complexes from Lolium multiflorum. Phytochemistry, 12,
    1979-84.
Morrison, I.M. (1974). Lignin-carbohydrate complexes from Lolium perenne.
    Phytochemistry, 13, 1161-65.
Neilson, M.J. & Richards, G.N. (1978). The fate of soluble lignin-
    carbohydrate complex produced in bovine rumen. Journal of
    the Science of Food and Agriculture, 29, 513-19.
Purdie, T. & Irvine, J.C. (1903). The alkylation of sugars. Journal of
    the Chemical Society, 83, 1021-37.
Reid, J.S.G. & Wilkie, K.C.B. (1969). Polysaccharides of the oat plant
    in relationship to growth. Phytochemistry, 8, 2045-51.
Roudier, A. (1959). The linkage of residual D-glucuronic acid and 4-0-
    methyl-D-glucuronic acid in the xylan of wheat straw.
    Comptes Rendues de l'Academie des Sciences, Paris, 248,
    1432-35.
Roudier, A.J. & Eberhard, L. (1965). Recherche sur les hemicelluloses
    du bois de Pin Maritime des Landes. IV. Polyosides extraits
    de ce bois par l'eau bouillante. Constitution d'une
    arabinane presente parmi ceux-ci. Bulletin de la Societe
    Chimique de France, 460-64.
Roy, N. & Timell, T.E. (1968). The acid hydrolysis of glycosides. XI.
    Effect of pH on the hydrolysis of acid disaccharides.
    Carbohydrate Research, 7, 17-20.
Sandford, P.A. & Conrad, E.E. (1966). The structure of the Acetobacter
    aerogenes A3 (S1) polysaccharide. 1. A re-examination using
    improved procedure for methylation analysis. Biochemistry,
    5, 1508-17.
Schulze, E. (1891). Zur Kenntniss der chemischen Zusammensetzung der
    pflanzlichen Zellmembranen. Berichte der Deutsche Chemische
    Gesellschaft, 24, 2277-87.
Shimizu, K. (1981). β-Elimination of 2-0-(4-0-methyl-D-gluco-
    pyranosyluronic acid)-D-xylose with methylsulphinyl
    carbanion and hydrolysis of the hex-4-endopyranosiduronic
    linkage. Carbohydrate Research, 92, 219-24.
Stephen, A.M. (1983). Other plant polysaccharides. In The Polysaccharides,
    vol. 2, ed. G.O. Aspinall, pp. 97-193. New York: Academic
    Press.
Timell, T.E. (1964). Wood hemicelluloses. Advances in Carbohydrate
    Chemistry, 19, 247-302.
Timell, T.E. (1965). Wood hemicelluloses. Advances in Carbohydrate
    Chemistry, 20, 409-83.
Whistler, R.L. & BeMiller, J.N. (1958). Alkaline degradation of
    polysaccharides. Advances in Carbohydrate Chemistry, 13,
    289-329.
Whistler, R.L. & BeMiller, J.N. (1963). Holocelluloses from annual
    plants. Methods in Carbohydrate Chemistry, 3, 21-22.
Whistler, R.L. & Corbett, W.M. (1955). Ologosaccharides from partial
    hydrolysis of corn fiber hemicellulose. Journal of the
    American Chemical Society, 77, 6328-30.

Whistler, R.L. & Feather, M.S. (1965). Hemicellulose extraction from
    annual plants with alkaline solutions. Methods in
    Carbohydrate Chemistry, 5, 144-45.
Whistler, R.L. & Richards, E.L. (1970). Hemicelluloses. In The
    Carbohydrates, vol. IIA, eds. W. Pigman & D. Horton,
    pp. 447-469. New York: Academic Press.
Whistler, R.L. & Sannella, J.L. (1965). Fractional precipitation with
    ethanol. Methods in Carbohydrate Chemistry, 5, 34-36.
Wilkie, K.C.B. (1979). The hemicelluloses of grasses and cereals.
    Advances in Carbohydrate Chemistry and Biochemistry, 36,
    215-64.
Wilkie, K.C.B. (1983). Hemicellulose. Chemtech, 13, 306-19.
Wise, L.E., Murphy, M., D'Addieco, A.A. (1946). Chlorite holocellulose,
    its fractionation and bearing on summative wood analysis
    and on studies on the hemicelluloses. Paper Trade Journal,
    122, 35-43.
Worth, H.G.J. (1967). The chemistry and biochemistry of pectic
    substances. Chemical Reviews, 67, 465-73.
Zitko, V., Bishop, C.T. (1965). Fractionation of pectins from
    sunflowers, sugar beet, apples and citrus fruit. Canadian
    Journal of Chemistry, 43, 3206-14.

2   DEVELOPMENTS IN THE ISOLATION AND ANALYSIS OF CELL WALLS
    FROM EDIBLE PLANTS

R.R. Selvendran
AFRC Food Research Institute, Norwich, NR4 7UA, England

B.J.H. Stevens
AFRC Food Research Institute, Norwich, NR4 7UA, England

M.A. O'Neill
AFRC Food Research Institute, Norwich, NR4 7UA, England

Abstract.   Improved methods are described for the isolation
and purification of cell-wall material from edible plants.
These methods include the complete disruption of the tissue
structure by wet ball-milling and the use of solvents, such
as Na deoxycholate, phenol-acetic acid-water and aqueous
dimethyl sulphoxide, which have a high affinity for
intracellular compounds.

The extraction of pectic substances with water, chelating
agents and alkali is discussed.   Extraction with cyclo-
hexanediamine tetra acetate (CDTA) at 20°C, followed by
extraction with Na_2CO_3 at 1°C and then at 20-22°C, is
recommended.

Hemicelluloses can be sequentially extracted with M KOH at
1°C, then at 20-22°C, followed by extraction with 4 M KOH
at 20-22°C.

The methods of characterisation of individual polymers are
also reviewed.   These include: a) determination of monomeric
composition, b) the use of methylation and GC-MS techniques,
both before and after mild or partial acid or enzymic
hydrolysis, for linkage analysis and sequencing of oligosaccharide fragments.   Examples are given of the application
of these methods to the study of polysaccharides and
complexes of polysaccharides with proteins and/or
polyphenols, from potatoes, cabbage, carrots, runner beans
and apples.

Key words: cell wall, cell-wall material, polysaccharides,
polysaccharide complexes, pectic substances, hemicelluloses,
cellulose, oligosaccharides, xyloglucans, hydroxyproline-
rich glycoproteins, proteoglycans, vegetables, fruits,
chromatography, methylation analysis, mass spectrometry.

INTRODUCTION

In recent years there has been considerable interest in the
cell walls of edible plants because they serve as sources of dietary
fibre (Southgate 1976; Selvendran 1983a & 1984) and as determinants of
texture (Knee & Bartley 1981).   Edible plant organs present special

problems to the cell-wall biochemist because they may contain high levels
of starch (potatoes and cereals), intracellular proteins (cereals and
peas), and polyphenols (apples and sugar beet) which make the isolation
of relatively pure cell walls from them difficult.   However, in order
to get unambiguous evidence on the occurrence and structure of cell-wall
glucans, glycoproteins, phenolic esters and polyphenolic material it is
imperative that the cell walls are free of the intracellular compounds
cited above.   The isolated cell walls are extracted with aqueous
inorganic solvents under various conditions to solubilise the wall
polymers (O'Neill & Selvendran 1980a, 1983; Ring & Selvendran 1981;
Selvendran & DuPont 1984).   Alternatively the cell walls can be
fractionated by sequential treatments with specific wall degrading
enzymes and some chemical reagents (Talmadge et al. 1973; Bauer et al.
1973).

Both approaches have advantages and disadvantages and it is
best to use a combination of them.   The solubilised components are
purified by column chromatography (ion-exchange, partition or gel
filtration) usually using two or more chromatographic systems, and the
purified polymers are characterised by a combination of chemical and
enzymic methods.

In our laboratory, improved methods have been developed for
the isolation and characterisation of cell-wall polymers from
parenchymatous and lignified tissues of a range of edible plant organs.
This article concentrates on the methods which we have developed, or
found useful, for our studies.   Attention is, however, drawn to some of
the alternative methods used for the analysis of cell-wall polysaccharides.
Throughout the paper the problems associated with the methods are
indicated and in some instances measures for overcoming them are given.
Such information would help cell wall researchers to assess the relative
merits of the published methods and the factors that have to be borne
in mind when choosing, or devising, the most appropriate method for a
particular plant tissue.

ISOLATION AND PURIFICATION OF CELL WALLS
Isolation of cell walls: Practical considerations

To isolate relatively pure cell walls every effort must be
made to (i) avoid co-precipitation of intracellular compounds with the
cell wall material (CWM), (ii) remove the starch and cytoplasmic proteins

quantitatively, and (iii) minimise cell-wall enzyme activity as much as possible.   In potatoes and oats, for example, the starch to proteins to cell wall ratios are about 15:0.7:1 and 7.7:1.6:1.0, respectively (Selvendran & DuPont 1980 a,b).   Incomplete removal of starch and proteins would, therefore, seriously interfere with the subsequent fractionation and analysis of the cell-wall preparations.

In our studies, coprecipitation effects were avoided by using aqueous inorganic solvents, instead of aqueous (aq.) alcohol.   To minimise the formation of oxidation products of polyphenols, 5 mM sodium metabisulphate was incorporated into the extraction medium.   To achieve complete removal of proteins and starch it was necessary to ensure 'complete' disruption of tissue structure and then use solvents which have high affinity for these compounds.   The first objective was achieved by wet ball-milling the tissue triturated in an aqueous solvent. Proteins were removed by sequential treatment with 1% (w/v) aq. sodium deoxycholate (SDC) or 1.5% (w/v) aq. sodium laurylsulphate (SLS) and phenol: acetic acid: water (PAW) (2:1:1, w/v/v), and starch was removed with dimethyl sulphoxide containing 10% water (90% aq. DMSO).   To minimise enzyme activity low temperature extraction (frozen powder (-20°C) was blended with SDC or SLS at 20°C) and treatment with PAW were used.   As a representative example, the preparation of CWM from potatoes is outlined.

Preparation of CWM from fresh potatoes

The method with the relative amounts of material and extractants used is shown schematically in Fig. 1.   The tubers are skinned, sliced into small pieces, dropped into liquid N_2 and ground to a fine powder.   The frozen powder is usually stored at -20°C until required.   The powder (50g) or freshly cut material is blended with 1% aq. SDC or 1.5% aq. SLS containing 5 mM $Na_2S_2O_5$, in a Waring blender for 2 min and then with an Ultraturrax for a further 3 min; a few drops of octanol are added to minimise frothing.   The triturated material is filtered through nylon cloth and washed with two bed volumes of 0.5% SDC or SLS containing 3 mM $Na_2S_2O_5$; from the filtrate the bulk of the SDC-soluble polymers, which are mainly of intracellular origin, could be isolated.   The resulting product (Residue-R1) is ball-milled in 0.5% SDC or SLS containing 3 mM $Na_2S_2O_5$ for 15 h at 2°C, which gives optimal cell disruption for potatoes.   This treatment gives a homogeneous

product which can be readily separated by centrifugation, after
filtration through a nylon sieve (3 mm pore size) to remove the porcelain
balls.   The supernatant is decanted and retained for the isolation of
cold-water-soluble pectic substances.   The residue is washed with
distilled water by centrifugation and then extracted twice with PAW; a
short treatment in a blender gives a uniform suspension.   The residue
obtained by centrifugation is washed with distilled water to give
residue-R_3, which contains much starch.   The starch is removed from
residue-R_3 by two extractions with 90% aq. DMSO and washed by
centrifugation with distilled water (x6) to give a residue which gives a
negative reaction with I_2/KI.   The material thus obtained contains a
very small amount of adsorbed DMSO.   However, a product completely free
of DMSO is obtained by suspending the pellet in distilled water and
dialysing against distilled water for 15 h.   The residue obtained by
centrifugation is defined as the purified cell wall material; yield 0.6g
dry weight.   It is best to freeze-dry an aliquot of the CWM for analysis
of the constituent sugars and amino acids, and to store the bulk of it in
a frozen state until required for fractionation studies.   For details of
the method, particularly the steps involved in the removal of starch, the
reader is recommended to consult Selvendran (1975a), Ring & Selvendran
(1978), and Selvendran & DuPont (1980b), particularly the last reference.

General comments on the method

In the above method, the following seven points should be
noted; these comments are generally applicable to other tissues as well.
(1) Filtration of the material homogenised in 1% aq. SDC is recommended
because it removes the bulk of the intracellular proteins, polyphenols
etc. and only a small amount (<20%) of the cold-water-soluble pectic
substances.   (2) Wet ball-milling the triturated material is necessary
to disintegrate the cell structure and render the contents accessible to
the solvents.   The average time for ball-milling is about 15 h, but for
certain tissues e.g. apple parenchyma, 7 h is adequate.   The tissues
should not be ball-milled too fine, as the resulting product does not
settle readily on centrifugation.   The optimum time for ball-milling
should be determined for each product, in preliminary experiments.   The
wet ball-milling treatment does not result in detectable breakdown of
cell-wall polymers (Selvendran et al. 1979; Selvendran & DuPont 1984).
The bulk (80%) of the cold-water-soluble pectic substances could be

Fig. 1.   Scheme for the preparation of CWM from fresh
           potatoes.

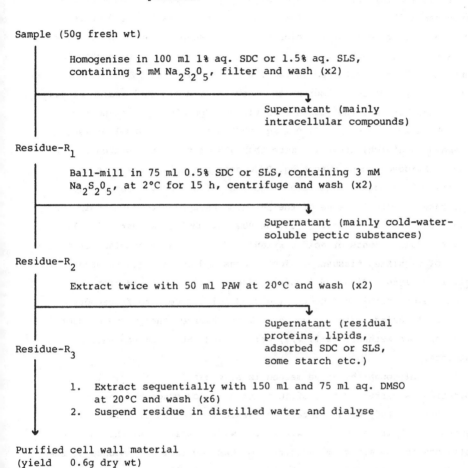

Sample (50g fresh wt)

    Homogenise in 100 ml 1% aq. SDC or 1.5% aq. SLS,
    containing 5 mM $Na_2S_2O_5$, filter and wash (x2)

    Supernatant (mainly
    intracellular compounds)

Residue-R_1

    Ball-mill in 75 ml 0.5% SDC or SLS, containing 3 mM
    $Na_2S_2O_5$, at 2°C for 15 h, centrifuge and wash (x2)

    Supernatant (mainly cold-water-
    soluble pectic substances)

Residue-R_2

    Extract twice with 50 ml PAW at 20°C and wash (x2)

    Supernatant (residual
    proteins, lipids,
    adsorbed SDC or SLS,
    some starch etc.)

Residue-R_3

    1. Extract sequentially with 150 ml and 75 ml aq. DMSO
        at 20°C and wash (x6)
    2. Suspend residue in distilled water and dialyse

Purified cell wall material
(yield   0.6g dry wt)

isolated from the 0.5% SDC extract.   The dialysed extract is
concentrated, and ethanol is added at 20°C to a final concentration of
80%;   after 6 h the precipitated polymers are removed by centrifugation.
The contaminating proteins & starch could be removed enzymatically
(Selvendran & DuPont 1980a & 1984).   The amount of cell-wall
polysaccharides solubilised by the 0.5% SDC treatment when expressed as
a percentage of the carbohydrate content of the purified CWM is as follows:
Potatoes 3.2, apples 4.5, runner beans 6.0, wheat bran <1% and oats 10.0.
(4) PAW treatment removes residual intracellular proteins, some starch,
adsorbed SDC or SLS, lipids and pigments from the residue-R_2.   Very

small amounts (<1%) of cell-wall polysaccharides are solubilised, but
some of the loosely held cell wall enzyme glycoproteins may be extracted
(Selvendran 1975a).   It is useful to note that unless the product is
very rich in intracellular proteins e.g. peas, soybeans and most cereals,
the PAW treatment may be omitted.   However, for the above products it is
essential.   We have obtained intracellular protein-free CWM,
quantitatively from soya flour by the above treatments (Selvendran et al.
1981).   (5) With all the starch-rich tissues (potatoes, oats and rye
flour) which we have examined, 90% aq. DMSO quantitatively solubilised
the remaining starch; this was more than 90% of the total starch in the
tissue.   Evidence to show that aq. DMSO does not solubilise more than
2% of the CWM of potatoes has been reported (Ring & Selvendran 1978).
In the case of cereals, some of the β-D-glucans appear to be solubilised
(Selvendran & DuPont 1980b).   (6) The 90% aq. DMSO treatment solubilised
only very small amounts of acidic xylans and acidic arabinoxylans from
the CWM of lignified tissues of dicotyledons and cereals (Selvendran
et al. 1979, 1980; Ring & Selvendran 1980).   (7) For lipid-rich
products (e.g. certain oil-containing seeds) it is best to freeze-dry
the material, grind it to a fine powder, and extract the ground product
with chloroform:methanol (2:1, v/v) prior to aq. SDC, PAW and aq. DMSO
treatments.

　　　　Although the above method is somewhat long, the final
preparation is virtually free of intracellular compounds and therefore
enables one to study with greater confidence the nature of cell-wall
polymers and their association within the wall matrix.   As this method
enables one to isolate gram quantities of CWM, appreciable amounts of the
individual polymers could be isolated for characterisation and also to
study their properties.   In the next section, the methods used for the
sequential extraction of the CWM and fractionation of the isolated
polymers are discussed, and the final section deals briefly with the
methods used for the characterisation of the carbohydrate polymers.

SEQUENTIAL EXTRACTION AND FRACTIONATION OF CELL-WALL POLYMERS
Sequential extraction of CWM
　　　　The CWM of an organ, from a dicotyledonous plant, containing
both parenchymatous and lignified tissues is first depectinated, and
then delignified, usually, by treatment with sodium chlorite and acetic
acid at 70°C (Wise et al. 1946).   The resulting holocellulose is washed

thoroughly with distilled water and extracted with alkali of increasing strength containing 10 mM sodium borohydride and in some instances 3-4% boric acid. The material insoluble in the above reagents is called α-cellulose. If the organ is relatively free of lignin, as in the case of parenchymatous tissues of vegetables and fruits, the delignification treatment can be omitted. As the chlorite/acetic acid treatment brings about oxidative cleavage of phenolic cross-links, including the isodityrosine cross-links of the hydroxyproline-rich wall glycoproteins ('extensin') and modifies some of the wall protein amino acids (Selvendran et al. 1975; O'Neill & Selvendran 1980; Fry 1982b), deletion of this step has facilitated the detection, isolation and characterisation of polysaccharide-protein-polyphenol complexes from the depectinated CWM of cabbage (Stevens & Selvendran 1984b), carrot (Stevens & Selvendran 1984c) apples (Stevens & Selvendran 1984d) and runner beans (O'Neill & Selvendran, unpublished results).

In this section we shall concentrate mainly on the chemical methods which we have used for the sequential extraction of CWM from parenchymatous tissues of vegetables and fruits. Based on our experience, a scheme is finally proposed which may be of general application for the fractionation of CWM from other parenchymatous and immature tissues.

Extraction of pectic substances: general considerations

Pectic substances are, for their most part, the most readily extractable polymers in the cell wall but usually suffer some degradation in the process. Above pH5, breakdown by transelimination becomes significant at elevated temperatures (Albersheim et al. 1960) and de-esterification will occur in alkaline conditions in the cold and in strongly acid conditions (Doesburg 1965). Below pH 3, hydrolysis will occur of the labile arabinofuranoside linkages by which many of the associated neutral polysaccharides are joined to the main rhamnogalacturonan 'backbone'.

The most commonly used extractants of pectic substances are water (cold or hot), chelating agents (cold or hot) and dilute alkali (cold). From the alcohol-insoluble-residues of most parenchymatous tissues small but significant amounts of pectic substances have been extracted with cold water (Stevens & Selvendran, unpublished results). In our earlier studies we isolated the cold-water-soluble pectic

substances from the 1% SDC or 1.5% SLS extracts used in the initial
stages of purification of the CWM (Ring & Selvendran 1978; O'Neill &
Selvendran 1980; Stevens & Selvendran 1980; 1984 a,c).   However, for
reasons already stated, we now recommend the isolation of the bulk of the
cold-water-soluble pectic substances from the 0.5% SDC extract (Fig. 1).
Hot water is frequently used and until recently we routinely used water,
at pH5 for 2 h at 80°C, for the first stage in fractionating CWM, in
order to separate the more highly esterified pectins from those less
esterified which are extractable with chelating agents   (Ring &
Selvendran 1981; O'Neill & Selvendran 1983).   Even under these conditions,
recent studies with cabbage cell wall polysaccharides have shown that
some degradation can still occur, resulting in polysaccharides rich in
neutral sugars (Stevens & Selvendran 1984a).   In view of these results
it is now debatable whether extraction with hot water serves any useful
purpose.

      The less esterified pectins, particularly those in the middle
lamella, are complexed mainly with Ca++ ions and require the use of
chelating agents to complex with them and thereby release the pectin.
EDTA (Na salt), ammonium oxalate or sodium hexametaphosphate have been
used for this purpose.   Although EDTA is only really effective above
pH7 which would result in degradation, Stoddart et al. (1967), using
EDTA at pH 4-5, obtained nearly the same yield of apple pectin, but over
a longer time, as at pH6.8.   In order to maintain good chelating
ability with minimum degradation we have favoured the use of 1% ammonium
oxalate at pH5 for 2 h at 80°C.   Other workers (Stoddart et al. 1967)
have used 2 to 3% sodium hexametaphosphate at pH 3.7 for up to 4 h at
90-100°C and claimed that only minimal degradation occurred, although at
pH4.2 Aspinall & Cottrell (1970) found de-esterification and
transelimination.   We have avoided the use of hexametaphosphate
because it could not be completely removed from the isolated polysaccharides.

      Cyclohexanediamine tetraacetate (CDTA) has recently been used,
apparently very effectively, giving yields of pectin, at room temperature,
at least as good as those of the more conventional chelating agents at
high temperatures and consequently with reduced degradation (Jarvis et al.
1981; Jarvis 1982).   This procedure has been used to extract pectic
substances from the CWM of onions (Redgwell & Selvendran, unpublished
results).   However, the merits of this procedure have yet to be fully

assessed.

        Cold, dilute alkali has also been used either with, or
following extraction by, chelating agents.   As with the other methods
there is considerable variation, largely dependent on the origin of the
tissue, in the proportion of pectin extracted.   Voragen et al. (1983)
reported that the proportions of uronic acid extracted with cold 0.05 M
NaOH + 0.005 M EDTA were as high as 95% and 97%, of the total uronide
that could be solubilised from the alcohol-insoluble residues (AIR) of
strawberries and cherries, respectively, and 82% for apples (var. Golden
Delicious).   Using this method, with the alkali + EDTA adjusted to pH 10,
we could only solubilise 24% of the total uronic acid from a different
variety of apple CWM, although uronide amounting to 5% of the CWM was
solubilised during preparation of the CWM (Stevens & Selvendran 1984d).

        Extraction with 0.05 M Na_2CO_3 at 1°C after extraction with
CDTA gave fairly good yields of pectins from onions and apples (12.6%
and 8.1% of the CWM respectively) (Redgwell & Selvendran, unpublished
results; Selvendran & Stevens, unpublished results).   Jarvis et al.
(1981) found that, after extraction with chelating agents, 0.1 M Na_2CO_3
was efficient for the extraction of galactan-rich pectic material from
the CWM of potatoes.   Table 1 shows yields of pectins extracted under
various conditions from apples (var. Cox's Orange Pippin).

        With most tissues some pectic substances are also extracted
with the hemicelluloses, but some remains with the α-cellulose residue.
Pectins associated with the α-cellulose have been isolated by the use of
cellulase (Stevens & Selvendran 1980; O'Neill & Selvendran 1983).
Recently the use of n-methyl morpholine oxide (MMNO) has proved useful
(Joseleau et al. 1981; Chambat et al. 1984) although the basic nature
of this reagent is likely to cause degradation as well as de-esterification.

Extraction of hemicelluloses: General comments.

        Hemicelluloses are generally defined as the polysaccharides
that are solubilised from the delignified cell walls (holocellulose)
with increasing concentrations of alkali.   KOH is preferred because the
potassium acetate formed on neutralization with acetic acid is more
soluble than sodium acetate in the aqueous alcohol which is used to
precipitate the hemicellulose fractions.   Alkali solubilises the
polysaccharides: (i) by hydrolysing the ester linkages between
polysaccharides, and also between polysaccharides and non-carbohydrate

Table 1.   Composition of apple pectins extracted from CWM under various conditions.

Method of extraction	Yield % of CWM	Sugar composition (µg/mg)						Uronic acid
		Rha	Ara	Xyl	Man	Gal	Glc	
1.5%SLS at 10°C[a,b]	(5.7)	8	46	4	16	34	33	299
H_2O at 80°C[b] then	6.9	20	219	13	10	39	12	685
1% NH_4 Oxalate at 80°C[b]	6.3	27	159	12	3	37	10	707
0.05M NaOH 0.005M EDTA at 1°C[c]	16.3	8	48	4	3	28	16	560[e]
0.05M CDTA at 20-22°C[d]	12.2	13	102	12	3	32	24	632
then 0.05M Na_2CO_3 at 1°C[d]	5.4	8	78	7	2	18	13	639
then 0.05M Na_2CO_3 at 20-22°C[d]	2.7	19	259	24	4	78	16	227[e]

a Solubilised during the ball-milling stage of purification of the CWM.

b Stevens & Selvendran 1984 d.

c Ruperez, Selvendran & Stevens, unpublished results.

d Selvendran & Stevens, unpublished results.

e The low values for uronic acid may be due to the precipitation of the pectic material which took place during treatment with acid and subsequent hydrolysis.

components such as phenolic acids (e.g. esters of hydroxycinnamic acids), and (ii) by disrupting the hydrogen bonding between the hemicelluloses and the cellulose microfibrils.   Generally, strong alkali (4M KOH) is required to solubilise the bulk of the xyloglucans and glucomannans which are associated with the α-cellulose; the solubility of the glucomannans is enhanced by the presence of 3-4% boric acid.   The borate ion complexes with the cis hydroxyl groups of the mannose residues, thereby increasing their solubility.

Early work by Wise et al. (1947) showed that the extraction of the hemicelluloses from the holocellulose was rapid with concentrations up to 10% KOH but slow above this level.   As a result of this they recommended sequential extraction with 5%, then 24% KOH. These concentrations, or 1M and 4M KOH, were used for extracting hemicelluloses from pears (Jermyn & Isherwood 1956) and runner beans (Selvendran 1975b).   We routinely use these concentrations of alkali because they effect a partial fractionation of the hemicelluloses, but in some cases an intermediate extraction with 10% (or 2M) KOH is useful. Carpita (1983; 1984) has recently shown that partial fractionation of maize hemicelluloses could be obtained by using 1M, 2M and 4M KOH sequentially.   Degradation of the hemicelluloses can be minimised by using oxygen free solutions and by reducing latent aldehydic groups with potassium or sodium borohydride.

With some tissues, e.g. depectinated apple CWM, extraction with 1M KOH at 1°C followed by extraction at room temperature also gave a partial fractionation of the polysaccharides; those extracted at 1°C were rich in xylose, whereas those extracted at room temperature were richer in arabinose (Stevens & Selvendran 1984d).   Similar observations have been made with lupin hypocotyl cell walls (Monro & Bailey 1975). Chaotropic agents such as urea or guanidinium thiocyanate (GTC) have also proved useful for extracting some of the mannose-rich polymers from the depectinated cell walls of apples (Stevens & Selvendran 1984d) and lupin hypocotyls (Monro et al. 1976).

Recommended method for the sequential extraction of CWM from parenchymatous tissues

The recommended method with the relative amounts of material and solvents is shown schematically in Figs. 2 & 3.   Depectination with CDTA at 20°C causes minimum degradation of pectic substances.   Extraction

Fig. 2.   Scheme for the extraction of pectic substances.

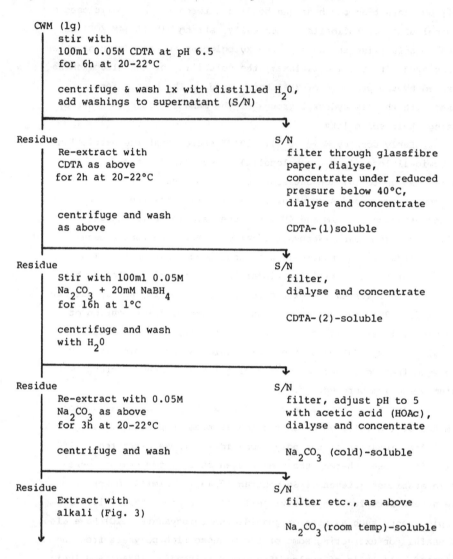

Fig. 3.   Scheme for the extraction of hemicelluloses

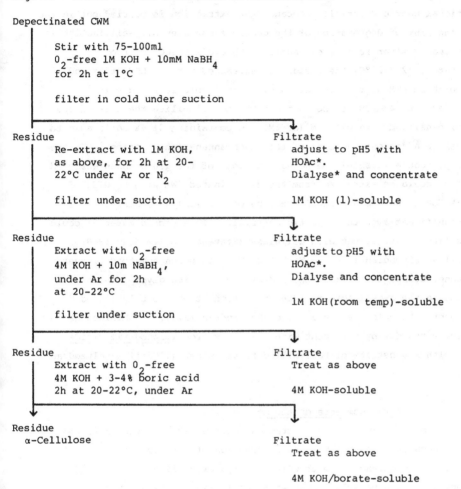

Depectinated CWM

    Stir with 75-100ml
    O_2-free 1M KOH + 10mM $NaBH_4$
    for 2h at 1°C

    filter in cold under suction

Residue
    Re-extract with 1M KOH,
    as above, for 2h at 20-
    22°C under Ar or N_2

    filter under suction

Filtrate
    adjust to pH5 with
    HOAc*.
    Dialyse* and concentrate

    1M KOH (1)-soluble

Residue
    Extract with O_2-free
    4M KOH + 10m $NaBH_4$,
    under Ar for 2h
    at 20-22°C

    filter under suction

Filtrate
    adjust to pH5 with
    HOAc*.
    Dialyse and concentrate

    1M KOH(room temp)-soluble

Residue
    Extract with O_2-free
    4M KOH + 3-4% boric acid
    2h at 20-22°C, under Ar

Filtrate
    Treat as above

    4M KOH-soluble

Residue
    α-Cellulose

Filtrate
    Treat as above

    4M KOH/borate-soluble

*   If a precipitate forms, it should be separated by centrifugation.

of the depectinated cell walls with 0.05M Na_2CO_3 at 1°C for 2h solubilises additional pectic material.   Because the extraction is carried out in the cold the risk of degradation of the pectic substances by β-elimination is minimised.   After further extraction with 0.05M Na_2CO_3 at room temperature (20°-22°C) the residue is extracted with 1M KOH at 1° and then with 1M KOH at room temperature.   The insoluble residue is extracted with 4M KOH at room temperature to solubilise the bulk of the xyloglucans, and then with 4M (or 6M) KOH containing 3%-4% boric acid to solubilise additional amounts of the glucomannan-rich fraction.   In the above procedure it should be noted that some of the glucomannan-enriched fraction could be extracted from the depectinated CWM with 6M GTC. The 4M KOH (+ borate)-insoluble residue of α-cellulose contains small but significant amounts of pectic material.   This pectic material could be isolated by treatment with cellulase (Stevens & Selvendran 1980; O'Neill & Selvendran 1983).   With the CWM from parenchymatous tissues of runner beans, the bulk of the hydroxyproline-rich glycoproteins remains in the α-cellulose residue from which it can be solubilised by treatment with chlorite/acetic acid (Selvendran 1975), or obtained as an insoluble residue by treatment with cellulase from Trichoderma viride grown with a substrate of ball-milled filter paper (O'Neill & Selvendran 1983).

General comments on the method.

1.   For solubilising pectic substances, CDTA at 20°C is the preferred solvent because it causes minimum degradation of the pectins.

2.   Following extraction with CDTA, the hydroxyproline-rich glycoproteins can be solubilised with two 15 min treatments with chlorite/acetic acid (O'Neill & Selvendran 1980).

3.   Extraction with dilute Na_2CO_3 at 1°C de-esterifies the pectins, thus minimising degradation by β-elimination.   The extraction at room temperature probably solubilises the pectic substances that are in association with the hemicellulosic polymers.

4.   1M KOH solubilises the bulk of the hemicellulosic polymers. Further fractionation of these polymers by anion exchange chromatography has shown that they contain mainly polysaccharide - protein - polyphenol complexes (Stevens & Selvendran 1984b & c;   O'Neill & Selvendran, unpublished results).

5.   4M KOH solubilises the bulk of the 'free' xyloglucans and some

glucomannans.

6.   It is preferable to concentrate the hemicellulosic polymers (obtained after dialysis of the neutralised material) by evaporation under reduced pressure and then use the concentrate directly for fractionation by ion-exchange chromatography.   Any precipitated material can be separated by centrifugation and analysed separately.   Freeze-drying the concentrate, or precipitation with alcohol, to isolate the polymers should be avoided except for the purpose of analysis.

Solubilisation of hydroxyproline-rich glycoproteins

The bulk of the Hyp-rich glycoprotein, somewhat modified, was solubilised from the depectinated CWM of runner beans by treatment with chlorite/acetic acid (Selvendran et al. 1975, 1977; Selvendran 1975b). With a shorter treatment, less modified wall glycoproteins have been solubilised and characterised (O'Neill & Selvendran 1980).   Following our earlier work, Hyp-rich glycoproteins have been solubilised from the CWM of suspension-cultured tomato tissue (Mort & Lamport 1977).

The solubilisation of the Hyp-rich glycoprotein is probably due to oxidative degradation of the isodityrosine cross-links in the wall proteins (Fry 1982b).   However, the disruption of phenolic ether (or ester) linkages between the N-terminal amino acid residues and the $-NH_2$ groups of lysine may also play a part.   Firm evidence for the occurrence of the above linkages in the cell-wall proteins of dicotyledons has not yet been obtained.

Solubilisation of pectic substances linked to phenolic acids

Mild acid extraction of the CWM of sugar beet resulted in the solubilisation of an arabinan-rich pectic fraction which had substituted cinnamic acid derivatives linked to it, probably by ester bonds.   Mild acid hydrolysis released, among other compounds, p-coumaric and ferulic acids from the arabinan fraction (Selvendran & Stevens, unpublished results).   Following the procedure described by Fry (1982a, 1983) using Driselase, we have released, from sugar beet CWM, a series of oligo-saccharides containing arabinose associated with p-coumaric and ferulic acids (Stevens & Selvendran, unpublished results).

Fractionation of cell-wall polymers

In the study of cell-wall polymers, one of the first

requirements is their purification and fractionation into more or less
homogenous individual polymers.   That means that further fractionation
of the polymer yields only material having the same specific rotations
and the same ratio of monosaccharide residues.   In Methods in
Carbohydrate Chemistry, Vol. 5, a range of procedures is described for
fractionating neutral and acidic polysaccharides.   Not all the methods
are of preparative value, although they may give information which can
be used to interpret or correct results obtained by other means.

Pectins

In addition to being very heterogeneous in size and
composition, neutral polysaccharides, possibly of pectic origin, are
often also extracted.   Pectins have been initially separated from the
bulk of the neutral polysaccharides by precipitation with ethanol
(Kertesz 1951; Foglietti & Percheron 1968), cupric acetate solution
(Aspinall et al. 1967, 1968) or quaternary ammonium salts such as cetyl
trimethyl ammonium bromide (Cetavlon, CTAB) (Foglietti & Percheron 1968;
Salimath & Tharanathan 1982).   Cetyltrimethyl ammonium hydroxide and
CTAB have been used to purify arabinan of mustard seed (Hirst et al. 1965)
and apple cell walls (Aspinall & Fanous 1984).

With solutions of copper salts the relatively insoluble
copper salts of the pectins are formed, but Cu^{++} ions will also complex
with -OH groups of some netural polysaccharides (see the section on
Hemicelluloses) which could co-precipitate (Stevens & Selvendran 1984a;
Blake & Richards 1971).   Co-precipitation can also occur with the
alcohol precipitation methods.   Under normal conditions precipitation
with the quaternary ammonium salts is specific for acidic polysaccharides.
The precipitated complexes are relatively insoluble, which makes it
possible to precipitate them from very dilute solutions.   Scott (1960;
1965) has described the use and applications of this method.   Isolated
pectins have been separated into fractions of decreasing acidity and
increasing netural sugar content by fractional precipitation with ethanol
(Barrett & Northcote 1965; Foglietti & Percheron 1968) or, after
saponification, with increasing concentrations (up to 0.22M) of sodium
acetate followed by precipitation with ethanol (Aspinall et al. 1970;
Salimath & Tharanathan 1982; Zitko & Bishop 1965).

Further fractionation can also be achieved by ion-exchange
and gel-filtration chromatography.   Before the introduction of the

modified dextran-based ion-exchange materials, DEAE cellulose was used
(and often still is) with elution with phosphate buffers at slightly
acid pH (Neukom et al. 1960; Aspinall et al. 1967).   'Neutral'
polysaccharides, usually with low levels of uronic acid, are not
retained on the columns.   The bulk of the pectins can be fractionated
according to their degree of esterification, the less esterified pectins
generally eluting with the higher ionic strength buffers (Deventer-
Schriemer & Pilnik 1976).   However, some acidic polysaccharides adsorb
strongly onto the cellulose and require the use of strong chaotropic
reagents such as urea, or NaOH up to 0.5M, to release them (Neukom et al.
1960), NaOH causing de-esterification or degradation by β-elimination
(Knee 1970).

The dextran-based ion-exchange materials such as DEAE
Sephadex are in bead forms and permit higher flow rates, but considerable
shrinkage occurs with increasing ionic strength and separation of pectic
polysaccharides is no better than on DEAE cellulose.   DEAE Sephacel is
based on a beaded form of cellulose combining the properties of
cellulose with the advantages of the high flow rates and non-shrinkage
at high ionic strengths.   Recoveries from DEAE Sephadex and DEAE
Sephacel are often poor, sometimes as low as 66% (Aspinall et al. 1968;
Stevens & Selvendran 1984a).

More recently DEAE Sepharose CL-6B, an agarose-based ion-
exchange material with higher exclusion limits than the dextran-based
materials, has given improved separations.   Stepwise elution was
found to give better and more reproducible separations than gradient
elution from the phosphate form (Souty et al. 1981), but in later
applications gradient elution has been used (Barbier & Thibault 1982;
Thibault 1983).   Macroporous acrylamide polymers have recently become
available as ion exchangers with high ion-exchange capacities and
avoiding many of the limitations associated with the polysaccharide-based
materials.   Good separation and recovery of onion cell wall pectic
substances has been obtained with DEAE-Trisacryl-CM (LKB Ltd.)
(Redgwell & Selvendran, unpublished results), and this ion-exchange
material has also been used in the purification of arabinans (Joseleau
et al. 1983).

The pectic polysaccharides extracted with the hemicelluloses
from cabbage, carrot and apple have been fractionated during fractionation
of the hemicellulosic polymers, and structural investigations have

revealed that these are probably complexes of pectins with hemicellulosic
polysaccharides such as arabinoxylans or xyloglucans and proteins and/or
phenolics (Stevens & Selvendran 1984 b,c).   Similar results have been
obtained with runner beans (O'Neill & Selvendran, unpublished results).

Fractionation of hemicelluloses

Some fractionation of hemicelluloses can be obtained by
precipitation with ethanol, copper salts or as barium complexes.   These
methods are described in Methods in Carbohydrate Chemistry, Vol V.
Gaillard et al. (1969) and Gaillard & Thompson (1971) studied the use of
solutions of iodine/potassium iodide in calcium chloride for the
separation of highly branched polysaccharides, especially xylans, from
unbranched polysaccharides or polysaccharides having branches of single
terminal residues.   This property, together with precipitation with
CTAB, has recently been applied to the separation of hemicellulosic
polymers of coffee leaves (Cecy & Correa 1984).

Stepwise alcohol precipitation (10 to 90%) has been used
for the fractionation of galactans (Wood & Siddiqui 1972) and has been
particularly useful in the fractionation of heteroxylans from commercial
wheat bran (Brillouet et al. 1982) and from beeswing wheat bran where the
precipitated fractions were further fractionated by chromatography on
DEAE Sephacel.   Recoveries ranged from 48%, for the fraction
precipitated with 40% ethanol, to 87% for those fractions precipitated
at high ethanol concentrations (Selvendran & DuPont, unpublished results).

Precipitation as copper complexes, either with Fehling's
solution or with cupric acetate, has been used in the purification of
xylans and xyloglucans (e.g. Aspinall & Wilkie 1956; Aspinall et al. 1977;
Ring & Selvendran 1981).   Glucomannans can be purified by precipitation
with barium hydroxide which probably complexes with the vicinal 2,3-cis-
hydroxy groups of the mannose (Meier 1958, 1965).   Because the 3,4-
hydroxy groups of galactose also have this configuration, galactans
containing (1->6)-linked galactose or polysaccharides with a high
proportion of terminal galactose, can also be precipitated.   At higher
concentrations of barium hydroxide 4-0-methyl glucuronoxylans can be
precipitated.

Anion-exchange chromatography is the most widely used method
of fractionating hemicellulosic polysaccharides.   As with pectins, DEAE
cellulose is often used, but xylans with low degrees of substitution

often require elution with urea or NaOH which can release polymeric compounds, rich in xylose and glucose, from the cellulose (Stevens & Selvendran, unpublished results), so the use of cellulose is not recommended unless the use of these eluants can be avoided.   Phosphate, borate and acetate are probably the most widely used counter-ions, but carbonate and acetate have also been used.   As with the pectic polymers, DEAE Sephadex and DEAE Sephacel have more recently been used; DEAE Sephadex to obtain primary separation of 'neutral' and acidic poly-saccharides, with further fractionation on DEAE Sephacel, although DEAE Sephacel has also been used for the primary separation.   DEAE Sephacel (acetate form) has been used to fractionate the 1M KOH-soluble polysaccharides of carrot CWM into a range of complexes including pectic-xylans and pectic-arabinans/galactans.   Some fractions were probably also complexed with proteins and/or phenolics (Stevens & Selvendran 1984c).   DEAE cellulose or DEAE Sephacel in the borate form has been useful in the purification of xyloglucans (Siddiqui & Wood 1971; O'Neill & Selvendran 1983).   From cabbage CWM two xyloglucan fractions were obtained, one of which was rich in mannose (Stevens & Selvendran 1984b).   Although recoveries of hemicellulosic polysaccharides from DEAE Sephacel are generally better than those of the pectic polymers, low recoveries are sometimes obtained, as above with some wheat-bran acidic arabino-xylans.   DEAE Sepharose CL-6B has recently been used in the fractionation of hemicellulosic polymers of carrot (Aspinall et al. 1983), and in the purification of xyloglucans from apple cell walls the borate form provided a better separation and slightly better recoveries than DEAE Sephacel (borate) (Ruperez, Selvendran & Stevens, unpublished results).

Recently hydroxylapatite has been used in the fractionation of proteoglycans from runner-bean cell walls.   After an initial fractionation on DEAE Sephadex (chloride) fractions were chromatographed on hydroxylapatite.   The polysaccharides which were bound, and eluted with phosphate buffer, were of relatively low carbohydrate contents compared with those of the unretained fractions (O'Neill & Selvendran, unpublished results).

It is difficult to give any firm recommendation for methods using anion-exchange chromatography, but DEAE Sepharose CL-6B ($\overline{PO_4^{---}}$ form) and DEAE Trisacryl-CM (PO_4^{---} or Cl^- form) give better recoveries and separations than the earlier materials.

Isolation & fractionation of CWM from lignified tissues

The preparation of CWM from lignified tissues involves extraction of the subdivided fibres with 1% SDC, followed by wet ball-milling the triturated material in distilled water for 36-48 h.   The extended ball-milling period is required to finely subdivide the fibres so that the cell-wall polymers can be quantitatively extracted. Treatments with PAW and aq. DMSO are usually not necessary.

Because very little, if any, pectic material can be solubilised, the depectination stage is omitted.   The bulk of the hemicellulosic polymers from the CWM of lignified tissues (e.g. parchment layers of mature runner-bean pods) could only be solubilised after delignification.   Delignification is usually carried out by treatment with sodium chlorite and acetic acid at 70°C, the duration depending on the extent of lignification;   several hours (4) are required for heavily lignified tissues (Green 1963), but only 1 to 2 h for grass cell walls (Morrison 1974).   Delignification of the CWM from parchment layers of mature runner-bean pods for 1.5 h was found to result in incomplete removal of lignin, but 3 h was adequate.   Delignification does not usually remove substantial amounts of hemicellulosic polymers (Buchala et al. 1972), although some of the glycoproteins and acidic xylans are solubilised.   From the holocellulose of parchment layers a small proportion (20%) of the acetylated xylans could be solubilised with absolute DMSO at 20°C.   Further extraction of the residue with 1M and 4M KOH solubilised the bulk of the remaining acidic xylans to leave the α-cellulose residue, which, on hydrolysis, gave >92% glucose (Selvendran, unpublished results).   The degrees of polymerisation of the acidic xylans solubilised by 1M and 4M KOH were 75 and 200 respectively (Selvendran et al. 1977).   Before fractionation of the hemicelluloses by anion-exchange chromatography, it is better to effect partial fractionation by stepwise precipitation with increasing concentrations of ethanol (Reid & Wilkie 1969; Selvendran & DuPont, unpublished results).

CHARACTERISATION OF CELL-WALL POLYMERS
Determination of the monosaccharide composition of cell walls

Neutral monosaccharides.   Essentially quantitative hydrolysis of neutral non-cellulosic polysaccharides can be achieved with 1M H_2SO_4

at 100°C for 2h (Selvendran et al. 1979) or 2M trifluroacetic acid at
120°C for 2 h (Albersheim et al. 1967).   The quantitative release of
glucose from cellulose requires pretreatment with 72% (w/w) H_2SO_4 to
facilitate dissolution of the polymers followed by dilution of the acid
and heating at 100°C (Jermyn 1955).   The released sugars are commonly
separated and quantified by GLC (Dutton 1973), with the alditol acetates
being the derivatives of choice because they give a single peak for each
sugar (Sawardeker et al. 1965).   Sugars found in plant cell walls can
be separated on packed columns containing 3% w/w OV-225 in less than one
hour (Selvendran et al. 1979), although improved resolution and shorter
run times may be obtained with capillary columns (Blakeney et al. 1982).

Uronic acids.   Uronic acids are commonly estimated
colourimetrically with carbazole (Bitter & Muir 1962), which requires
corrections for interference by neutral sugars (Selvendran et al. 1979),
or with m-hydroxydiphenyl (Blumenkrantz & Asboe-Hansen 1973), which is
the current method of choice for sensitivity and minimal interference
from neutral sugars (Selvendran & DuPont 1984).

Attempts to analyse uronic acids by GLC have had limited
success (Blake & Richards 1970).   However, Ford (1982) has recently
described a semi-micro method for the analysis of galacturonic acid in
pectin using silylated aldono-1,4-lactones which is claimed to be suitable
for routine analysis of polygalacturonic acid in plant material.

Recently Voragen et al. (1982) have described the
separation of galacturonic, mannuronic and glucuronic acid by HPLC on
strong anion-exchange columns.   However, the columns suffer from low
stability and attempts to use alternative columns (amino bonded or C18
reverse phase) were unsuccessful (Voragen et al. 1982).

Lignin.   For an assessment of the methods used for
estimation of lignin from vegetables and cereals, see Selvendran &
DuPont (1984).

Methylation analysis
Introduction.   Methylation analysis is one of the most
important methods for determining the type(s) of glycosidic linkages in
a polysaccharide (Lindberg 1972).   Early procedures often required
repeated treatments to obtain complete methylation (Bouveng & Lindberg
1960) and have largely been replaced by the method developed by Hakomori
(1964) where the polysaccharide, in dimethylsulphoxide, is treated with

the strong base sodium methylsulphinyl carbanion and alkylated with
methyliodide.   After isolation the methylated polysaccharide can be
analysed, after work up, as the partially methylated alditol acetates
(Lindberg 1972).   These derivatives are easily examined by GLC, and due
to their low masses and simple fragmentation patterns, are suitable for
GC-MS (Lindberg 1972; Jansson et al. 1976).

The separation of partially methylated alditol acetates by
GLC can be carried out on packed columns containing OV-225 or ECNSS-M
(Lindberg 1972) Fig. 4.   In favourable cases the use of both columns
may facilitate identification of co-eluting derivatives.   This can be
illustrated with 2,3-di-$\underline{O}$-Me-xylitol and 2,3,4,6-tetra-$\underline{O}$-Me-galactitol
which co-elute on OV-225 but are separated on ECNSS-M (Lindberg 1972).
With complex mixtures of partially methylated alditol acetates, the use
of capillary columns offer improved resolution (Lomax & Conchie 1982),
although a range of columns may still be required (Geyer et al. 1982).
Figure 5 illustrates the improved resolution and shorter run time
obtained on an OV-225 WCOT capillary column.

In principle, methylation analysis of complex heteroglycans
should give equimolar proportions of non-reducing terminal groups and
branch points.   Excess branch points, which indicate undermethylation,
may arise from the poor solubility of the polysaccharide in
dimethylsulphoxide.   This may be overcome by heating the sample to 60°C
or by the addition of tetramethyl urea (Narui et al. 1982).   Improved
etherification may also be achieved by increasing the period of
exposure to base (Jansson et al. 1976) or by using potassium methylsulphinyl carbanion (Valent et al. 1980).   The presence of branched
glucosyl residues in methylation analysis of whole cell walls should be
interpretated with caution due to partial methylation of cellulose (Ring
& Selvendran 1978; O'Neill & Selvendran 1980b).   It must, however, be
noted that certain cell-wall polysaccharides such as arabinoxylans (Ring
& Selvendran 1980) and arabinans (Stevens & Selvendran 1984a) give penta-
$\underline{O}$-acetyl-pentitol derivatives after repeated methylations and suggest the
presence of doubly branched sugars.

Methylated sugars, especially the highly-methylated
derivatives that are very volatile, may be lost during acid hydrolysis
and work-up procedures (Lindberg 1972).   Recently, Blakeney et al.
(1983) have described the use of 1-methylimidazole as a catalyst for
acetylation of alditols in the presence of borate ions and the potential

Figs. 4 & 5.   Separation of the partially-methylated alditol acetates from the IM KOH soluble polyglycan of runner-bean cell walls.

(4) Packed column (3m x 2mm) containing 3% OV-225 and (5) 25m x 0.2mm WCOT OV-225 capillary column.   Peaks correspond to:

1. $2,3,5-Me_3-Araf$;   2. $2,3,4-Me_3-Xyl$;   3. $2,3,4-Me_3-Arap$;

4. $3,5-Me_2-Ara$;   5. $2,5-Me_2-Ara$;   6. $2,3-Me_2-Ara$;   7. $2,3-Me_2-Xyl$ and $2,3,4,6-Me_4-Gal$;   8. $2,3,6-Me_3-Gal$;   9. $2,3,6-Me_3-Glc$;

10. $2,3-Me_2-Gal$.

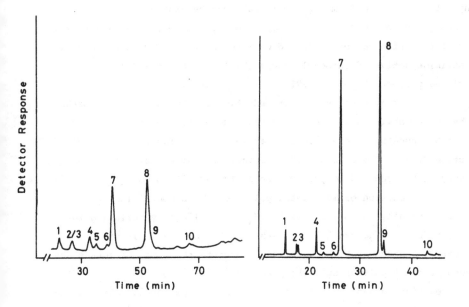

Fig. 4   Fig. 5

for loss of volatile components is reduced (Harris et al. 1984).
Stevens & Selvendran (1984b) have reported an underestimation of
terminal xylose in cabbage xyloglucan, which has also been observed in
the methylation analysis of isoprimeverose (α-D-Xylp-(1->6)-D-Glcp)
which only gave 60% recovery of 2,3,4-tri-O-Me xylitol (O'Neill &
Selvendran, unpublished results).   Whether this derivative readily
degrades on acid hydrolysis or is lost during work-up procedures is
unclear.

Application of methylation analysis to acidic polysaccharides.
Methylation analyses of polysaccharides containing uronic acid residues
are facilitated by the reduction of the acidic sugars to their neutral
homologues.   By performing the reduction with a deuterated reagent, the
methyl ethers derived from the uronic acid can be distinguished by
combined gas chromatography - mass spectrometry (GC-MS) (Lindberg 1972).
Carboxyl reduction can be performed on water soluble polysaccharides via
carbodiimide-activated carboxyl groups (Taylor & Conrad 1972) or the
methylated polymer (Lindberg 1972).

In many cases efficient reduction of the carbodiimide method
can be achieved with a single treatment.   However, the procedure may
have to be repeated several times to achieve a satisfactory degree of
reduction (Lindberg & Jansson 1978; Aspinall & Fanous 1984), and the
recovery of the reduced polymer may be as low as 60% (Gowda et al. 1983).

Reduction of methylated polysaccharides containing uronic
acids is generally performed with $LiAlH_4$ (or $LiAl^2H_4$) by refluxing in
tetrahydrofuran (Lindberg 1972) or ether/dichloromethane mixtures
(Lindberg & Jansson 1978).   With this procedure there may be losses due
to adsorption of polymer onto the precipitate formed on destruction of
excess reducing reagent.   This can be avoided by performing the
reduction with LiB^2H_4, but it is essential that all carboxyl groups are
fully esterified as the reagent is not effective at reducing the free
acid.   A comparison of the reduction of methylated pectins with $LiAl^2H_4$
and LiB^2H_4 showed both methods to be effective, although there were apparent
losses of derivatives from galacturonic acid (Stevens & Selvendran 1984a).
It is unclear, however, whether these losses arose from incomplete
reduction or degradation by β-elimination of native, esterified
galacturonic acid on treatment with the strong base, or both.   It
should be emphasised that prior to methylation of pectic polymers by the
Hakomori procedure, it is essential to de-esterify the material to

minimise base-catalysed degradation (Aspinall & Fanous 1984).

Simple extensions of methylation analysis.   Partial acid
hydrolysis in combination with methylation analysis is a simple
procedure for determining the point of attachment of acid-labile sugars
(Lindberg & Jansson 1978).   Partial acid hydrolysis of the methylated
xyloglucan from potato showed Araf residues to be linked to C-2 of
xylose; however, small but significant hydrolysis of the glucan
backbone was also observed (Ring & Selvendran 1981).   Similar studies
with beeswing-wheat-bran arabinoxylan showed Araf to be attached to C-2
and C-3 of the xylose, but again some hydrolysis of the backbone was
observed (Ring & Selvendran 1980).

A combination of methylation analysis and partial acid
hydrolysis demonstrated that the oxalate-soluble pectic polysaccharide
from cabbage contained a rhamnogalacturonan backbone to which a highly
branched arabinan was linked, possibly via (1->4)-linked galactose
oligosaccharides, to C-4 of rhamnose (Stevens & Selvendran 1984a).

By performing partial hydrolysis prior to methylation,
undesirable depolymerisation can be minimised.   Treatment of the
xyloglucan from runner bean with 10mM oxalic acid removed almost all the
arabinose and fucose.   Methylation of the degraded polymer showed these
sugars to be attached to C-2 of xylose or galactose but could not
distinguish which (O'Neill & Selvendran 1983).

Using a combination of methylation, partial hydrolysis,
ethylation and complete hydrolysis, the ring size of sugars can be
determined (Darvill et al. 1980a).   This has been applied in
structural studies of rhamnogalacturonan 1 (McNeil et al. 1980) and
glucuronoarabinoxylan (Darvill et al. 1980b) isolated from suspension-
cultured sycamore cell walls.

Useful structural information can be obtained from the
specific degradation of methylated polysaccharides containing esterified
uronic acid residues by β-elimination (Lindberg et al. 1975).   A
single treatment of the methylated polysaccharide, in dimethylsulphoxide,
with sodium methylsulphinyl carbanion results in the complete loss of
hexuronic acid (Aspinall & Rosell 1977).   Elimination of terminal non-
reducing uronic acids can unambiguously demonstrate its point of
attachment.   If, however, the uronic acid is substituted at C-4 and
forms part of a chain the degradation becomes more complex (Lindberg
et al. 1975).   With polysaccharides containing simple chemical repeat

units, the interpretation of such degradations may be relatively
straightforward.   However, with plant cell-wall polysaccharides such as
rhamnogalacturonan-I (McNeil et al. 1980a) or glucuronoarabinoxylans
(Darvill et al. 1980b), unambiguous data is not so readily obtained.

The release of oligosaccharide fragments from polysaccharides and their characterisation.

Mass spectra of methylated oligosaccharide alditols.

Methylated oligosaccharide alditols have proved to be the most suitable
derivatives for MS sequencing (Lonngren & Svensson 1974).   Samples of
sufficient purity can be introduced directly into the mass spectrometer
using a probe inlet whereas complex mixtures require GC-MS (Hamming &
Foster 1972) or LC-MS (McNeil et al. 1982a).   The most commonly used
ionisation technique is electron impact (EI-MS) which gives valuable and
diagnostic ions although molecular or pseudomolecular ions are rarely
observed (Lonngren & Svensson 1974).   Molecular ions may be readily
obtained with chemical ionisation (CI-MS) techniques where the sample
reacts with an ionised gas (Hunt 1974).   However, recent evidence
suggests that CI-MS must be interpreted with caution due to various
elimination processes and adduct formation (McNeil 1983).   Pseudomolecular ions may also be obtained with fast atom bombardment (FAB-MS)
methods although further information from this technique may be limited
(Dell et al. 1983).

Valuable structural information is generally obtained using
EI-MS and this is illustrated in Fig. 6.   The nomenclature corresponds
to that of Kotchetkov & Chizhov (1966) except that the alditol moiety
has been termed 'ald' (Nilsson & Zopf 1982).   For further details on
MS sequencing of oligosaccharides, see the articles by Karkkainen (1971),
Moor & Waight (1975), and Kovacik et al. (1978), and Selvendran (1983).

Partial acid hydrolysis.   Selective partial acid hydrolysis
is commonly employed to generate oligosaccharides (Conrad et al. 1966).
Treatment of the anionic microbial polysaccharide, gellan gum, with
0.5M H_2SO_4 for 4 h at 100°C produced mainly the aldobiouronic acid
Glcp-A(1->4)-Glcp, whereas 0.2M TFA for 2 h at 100°C gave relatively
high yields of acidic tri- and tetrasaccharides:

		Relative Amount
GlcpA-(1->4)-Glcp-(1->4)-Rhap		20
Glcp-(1->4)-GlcpA-(1->4)-Glcp		40
Glcp-(1->4)-GlcpA-(1->4)-Glcp-(1->4)-Rhap		35
GlcpA-(1->4)-Glcp-(1->4)-Rhap-(1->3)-Glcp		5

From these data the polysaccharide chemical repeat unit could be determined (O'Neill et al. 1983).

Fig. 6.   Probe EI-MS of the methylated trisaccharide alditol methyl ester, Glcp-(1->4)-GlcpA-(1->4)-Glcp, obtained from gellan gum.

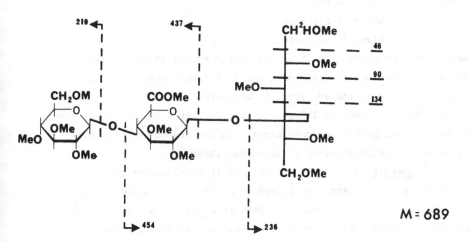

M = 689

Relative abundance of some of the diagnostic ions

Ion (m/z)	Fragment	%	Ion (m/z)	Fragment	%
658	M-31	0.3	437	baA$_1$	4.1
644	M-45	0.9	405	baA$_2$	2.8
643	M-46	0.5	373	baA$_3$	0.8
630	M-59	0.4	296	aldJ$_1$	18.8
629	M-60	1.3	236	ald	95.3
599	M-90	5.3	219	aA$_1$	59.5
567	M-122	1.8	187	aA$_2$	100.0
514	baldJ$_1$	5.4	155	aA$_3$	46.5
454	bald	5.8	134		36.2
			90		52.8

Kovacik et al. (1968) have shown that for glucuronic and
galacturonic acid the intensities of the aA_1 and aA_2 ions differ.   For
the former $aA_1 \ll aA_2$ and the latter $aA_1 \simeq aA_2$.   This has been confirmed by
examination of the M.S. for the aldobiouronic acid from wheat bran
Ring & Selvendran 1980) and gellan gum which gave the ratio 1.0:4.5
(O'Neill et al. 1983), whereas the aldobiouronic acid isolated from
runner-bean pectin (GalpA(1-2)-Rhap) gave 1.0:1.2 (O'Neill &
Selvendran, unpublished results).

Partial acid hydrolysis of wheat bran with 0.2M TFA at 100°C
for 2 h gave acidic di- and trisaccharides in the ratio 1.0:2.0 (Ring &
Selvendran 1980).   Combined GC-MS of the methylated derivatives showed
that the oligosaccharides were:

GlcpA-(1->2)-Xylp

GlcpA-(1->2)-Xylp-(1->4)-Xylp

By performing the methylation with deuterated methyl iodide it was
possible to determine the ratio of GlcpA to 4-0-Me-GlcpA, from the
relative abundance of the aA_1 ions, to be approximately 3.0:1.0 (Ring &
Selvendran 1980; Selvendran 1983b).   A similar study of the aldobiouronic
acid from cabbage pectin was consistent with the absence of methoxyl
substituents on GalpA (Stevens & Selvendran 1984a).

Acetolysis.   During acetolysis (1->6)-linkages are
preferentially cleaved and this procedure has proved of value in the
structural studies of yeast mannans (Stewart et al. 1968) and also as
a means of obtaining oligosaccharides containing acid-labile sugars
(Aspinall 1972).

Acetolysis has been used to partially characterise
xyloglucans from cell walls (Kooiman 1961; Aspinall et al. 1977; Kato &
Matsuda 1980; O'Neill & Selvendran 1983).   Acetolysis of the xyloglucan
from runner-bean cell walls produced a range of oligosaccharides that
were characterised by GC-MS of the methylated oligosaccharide alditols
(O'Neill & Selvendran 1983) as:

	Relative amount
Araf-(1->2)-Xylp	1.0
Galp-(1->2)-Xylp	13.0
Glcp-(1->4)-Glcp	4.0
Fucp-(1->2)-Galp-(1->2)-Xylp	6.0
Glcp-(1->4)-Glcp-(1->4)-Glcp	1.3

It is clear that oligosaccharides containing fucose can be obtained.
However, the above illustrates that acetolysis is not entirely selective
as depolymerisation of the backbone has occurred.

Acetolysis has also been used to investigate the structural
features of lemon-peel pectin (Aspinall et al. 1968).   Three major
acidic oligosaccharides were obtained, one of which, GalpA-(1->2)-Rhap-
(1->2)-Rhap, could not be obtained by partial acid hydrolysis due to the
acid-lability of the Rhap-(1->2)-Rhap linkages.   Unlike soybean pectin
(Aspinall et al. 1967) no neutral oligosaccharides were obtained in
acetolysis of lemon-peel pectin (Aspinall et al. 1968), although both
contained netural sugars.

Partial alkaline hydrolysis.   Partial alkaline hydrolysis
of polysaccharides is limited due to the fact that exposed reducing
groups readily undergo base catalysed degradation (Aspinall 1972).
Alkaline degradation has, however, been used to obtain hydroxyprolyl-
arabinosides from plant cell walls (Lamport 1967).   Treatment of the
hydroxyproline-rich lectin from potato (Allen et al. 1978) also produced
hydroxyprolyl-arabinosides which were shown by direct-insertion - MS
of the methylated glycopeptides (Ashford et al. 1982) to be identical
to the glycopeptides from tobacco (Akiyama & Kato 1976, 1977) and
runner-bean (O'Neill & Selvendran 1980) cell walls.

Methods based on β-elimination.   Degradations based on
β-elimination of methylated acidic polysaccharides using sodium
methylsulphinyl anion in dimethylsulphoxide have limited application in
the production of oligosaccharide fragments due to base-catalysed
degradation of exposed reducing sugars (Lindberg et al. 1975).   In an
alternative procedure the β-elimination degradation is performed with the
organic base 1,5-diazabicyclo [5,4,0] undec-5-ene (DBU) in benzene/acetic
anhydride (Aspinall et al. 1975).   The liberated reducing groups are
protected by acetylation, allowing the isolation of oligosaccharide
derivatives which can be characterised (Aspinall & Chaudhari 1975).
However, due to problems in removing excess reagent (Aspinall 1982) the
method has not been extensively used.

Degradations involving β-elimination can also be performed on
methylated polysaccharides that contain free hydroxyls at specific
positions.   The free hydroxyls are oxidised to keto or aldehydo groups,
and subsequent treatment with base eliminates substituents β to the
oxidised group (Svensson 1978).

Using a combination of β-elimination (sodium methylsulphinyl carbanion) oxidation and a second β-elimination (DBU), the structures of 7 differently linked glycosyl residues attached to position 4 of 2,4-linked rhamnose in sycamore rhamnogalacturonan-1 have been demonstrated (McNeil et al. 1982b).

HPLC of alkylated oligosaccharide alditols.   A major problem in the sequencing of polysaccharides, especially when only small amounts of material are available, is the efficient and rapid separation of oligosaccharides and their derivatives.   Valent et al. (1980) were the first to demonstrate the potential of reverse-phase HPLC for separating alkylated oligosaccharide alditols.   Subsequently, McNeil et al. (1982a) have reviewed the application of HPLC and HPLC-MS for separating and characterising a number of alkylated oligosaccharides from different polysaccharides.

Recently, we have investigated the potential of reverse-phase HPLC (O'Neill & Redgwell, unpublished results) for the characterisation of oligosaccharides from partial acid hydrolysates of kiwi-fruit gum (Redgwell 1983) and enzymatic digests of runner-bean xyloglucan (O'Neill & Selvendran, unpublished results).   After partial depolymerisation of the native polysaccharide(s), the products are initially separated by gel-filtration (Biogel P-2), reduced to oligosaccharide alditols with NaB^2H_4, methylated and separated by reverse-phase HPLC on an ODS column using 60% aq acetonitrile as the mobile phase. Although the resolution obtained is less than that for methylated/ethylated derivatives (McNeil et al. 1982a), it gives sample of sufficient purity for probe MS.   Similar studies (O'Neill, unpublished results) with acidic oligosaccharides derived from gellan gum (O'Neill et al. 1983) have shown that two acidic trisaccharide alditol derivatives that could only be partially resolved by GLC can be well resolved by HPLC (Fig. 7). It is apparent that the presence of esterified carboxyl groups increases retention times when compared with similar neutral derivatives.

NMR of polysaccharides.   The application of ^1H and ^{13}C NMR for the structural investigation of polysaccharides has been reviewed (Perlin & Casu 1982).   Native polysaccharides often give broad signals due to the short relaxation times of polymer protons, although this can be partially overcome by using elevated temperatures and solvents such as dimethylsulphoxide (Perlin & Casu 1982).

Examination of the polysaccharide gellan gum by ^1H-NMR proved

Fig. 7.   Separation of methylated oligosaccharide alditol methyl esters
of oligosaccharides derived from gellan gum.

A. Reverse phase HPLC on Zorbax ODS with 60%aq CH$_3$CN as mobile phase.
B. GLC on a column (45cm x 4mm) containing 1% OV-1.   The peaks
correspond to derivatives of: 1. Glcp-(1->4)-GlcpA-(1->4)-Glcp.
2. GlcpA-(1->4)-Glcp-(1->4)-Rha;   3. Glcp-(1->4)-GlcpA-(1->4)-GlcpA-
(1->4)-Rhap.

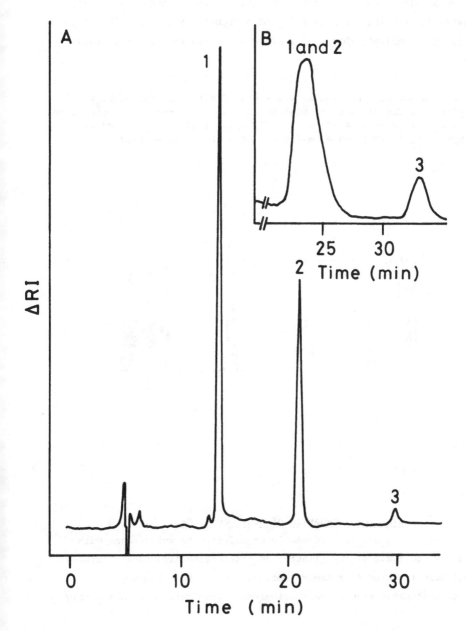

difficult due to the high viscosity of the polymer at low concentrations.
However, the methylated polysaccharide gave four well resolved anomeric
protons (Fig. 8), which was consistent with a tetrasaccharide repeat unit
(O'Neill et al. 1983).   The introduction of fourier transform techniques
(FT-NMR) has also made it possible to obtain ^{1}H-NMR spectra on as little
as 30µg of alkylated oligosaccharide alditols (Robertson et al. 1981).
However, due to ring and methoxyl proton signal overlap (Perlin & Casu
1982), it is unlikely that such studies would reveal other structural
features.

Fig. 8.   A 300 MHz proton n.m.r. spectra (C^2HCl_3, internal $CHCl_3$) of
methylated gellan gum showing signals for anomeric protons corresponding
to three β-linkages at 4.22ppm ($\underline{J}_{1,2}$ 7.67Hz), 4.42ppm ($\underline{J}_{1,2}$7.68Hz),
4.63ppm ($\underline{J}_{1,2}$ 7.68Hz) and an α-linkage at 5.31ppm (unresolved).

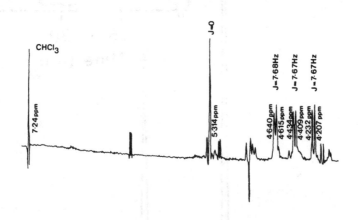

Enzymatic degradation of polysaccharides.

Xyloglucans.   Enzymatic degradation of xyloglucans with
cellulase (EC 3.2.1.4) has illustrated interesting structural features.
Xyloglucans isolated from seeds (Kooiman 1961) have a low degree of
side-chain substitution whereas xyloglucans from cell walls are highly

substituted.   Extensive regions of the backbone are substituted with
single xylosyl residues whereas, often, less extensive regions are
substituted with di- and trisaccharides (Kato & Matsuda 1980; O'Neill &
Selvendran, unpublished results).   It is also apparent that the extent
of depolymerisation is dependent on the xyloglucan.   The xyloglucan
from mung bean, when treated with Trichoderma viride cellulase, gave
fragments Dpn 8-10 accounting for $\simeq$ 70% of the polymer (Kato & Matsuda
1980b), whereas the xyloglucan from runner bean gave fragments of
Dpn 2-5 and 10-12 accounting for 57% and 32% respectively (O'Neill &
Selvendran, unpublished results).

        Mixed linkage β-D-glucans.   Mixed linkage β-glucans account
for a high proportion of the endosperm cell wall of oats and barley.
Recent structural analysis of barley β-glucan using purified β-glucanases
(Woodward & Fincher 1982) has shown that the polysaccharide is composed
of cellotrio- and cellotetraosyl residues separated by a single (1->3)
linkage (Woodward et al. 1983).

        Enzymic degradation of pectins.   Early studies of the
enzymic degradation of pectins were limited due to the uncertain origins
of the pectins used and the use of commercial enzymes (Aspinall et al.
1968) which normally contain a number of other hydrolases (Waibel et al.
1980).   Endopolygalacturonanase degradation of partially-hydrolysed
soy-sauce pectin demonstrated the attachment of xylose to C-3 of (1->4)-
linked GlcpA, a structure that had been previously observed in lemon-
peel pectin (Aspinall et al. 1968).

        Exoglycosidase degradations.   Using α-L-fucosidase
(EC 3.2. 1.51) from beef kidney it was possible to hydrolyse 85% of the
fucose from runner-bean xyloglucan.   However, attempts to remove
selectively terminal galactose residues with α- or β-D-galactosidases
were unsuccessful (O'Neill & Selvendran 1983).   Kato & Matsuda (1980b)
found that the exo-β-D-galactosidase from Charonia lampas could remove
terminal galactose from oligosaccharides derived from mung-bean
xyloglucan, suggesting steric hinderance in the whole polymer may affect
enzyme action.   The α-D-galactosidase from green coffee beans was,
however, capable of removing the single galactose residue attached to
serine in potato lectin (Allen et al. 1978) and runner-bean cell wall
glycoprotein (O'Neill & Selvendran 1980).   Using an α-L-arabino-
furanosidase from Sclerotinia fructigena (Laborda et al. 1973), it was
possible to demonstrate that the terminal arabinose in the hydroxyprolyl-

tetraarabinosides from potato lectin was α-linked (Ashford et al. 1982), which is the same as similar glycopeptides isolated from tobacco cell walls (Akiyama et al. 1980).

A combination of chemical and enzymatic degradations have been used to demonstrate the nature of the protein-polysaccharide linkages in wheat-endosperm arabinogalactan-peptide. Using partial acid and alkaline hydrolysis followed by exoglycosidase digestion, it was found that hydroxyproline was substituted by a β-D Galp residue (Strahm et al. 1981; McNamara & Stone 1981).

CONCLUSIONS

Using improved methods we have isolated relatively pure cell walls from a range of edible plant organs. Sequential extraction of the cell-wall material with aqueous inorganic solvents, followed by fractionation of the mixtures of polymers by column chromatography, has led to the isolation of a range of carbohydrate-containing polymers. The purified polymers have been partially characterised by a combination of chemical and enzymic methods. Thus xyloglucans have been isolated from potatoes (Ring & Selvendran 1981), runner beans (O'Neill & Selvendran 1983), cabbage (Stevens & Selvendran 1984b) and apples (Stevens & Selvendran 1984d); and hydroxyproline-rich glycoproteins/ proteoglycans from runner beans (O'Neill & Selvendran 1980). Also isolated from cabbage and apples (refs as above), carrots (Stevens & Selvendran 1984c) and runner beans (O'Neill & Selvendran, unpublished results) have been small, but significant amounts of polysaccharide-protein and/or polyphenol complexes where the polysaccharide moieties were xyloglucans, arabinoxylans and pectic substances, or a combination of them. In immature cabbage leaves the ratio of 'free' xyloglucan to xyloglucan-pectic-protein complex is about 10:1 (Stevens & Selvendran 1984b). Small amounts of acidic arabinoxylan-protein-polyphenol and acidic arabinoxylan-protein-polyphenol-β-D-glucan complexes have been isolated from beeswing wheat bran and de-hulled oats (Selvendran & DuPont, unpublished results).

Some of these complexes may be cell-wall enzymes which have been immobilised by phenolic and glycosidic linkages and located at specific sites within the wall, and the others may have a role as linking compounds within the cell-wall matrix. In cabbage, as in other tissues (Pusztai et al. 1971; Selvendran & O'Neill 1982), the

hydroxyproline-rich glycoprotein is closely associated with the pectic
substances, in contrast to tissues such as sycamore callus (Heath &
Northcote 1971) and runner-bean parenchyma (Selvendran 1975b; O'Neill &
Selvendran 1983) in which it is associated with the cellulose.   These
studies show that the polysaccharide-protein-polyphenol complexes are
fairly widespread in higher plant cell walls, although the bonding
between the polymer moieties and their functions within the walls is
uncertain.   These are attractive areas for future work.

ACKNOWLEDGEMENTS

   The authors thank Miss Susan DuPont and Mr. R.J. Redgwell
for some of the unpublished results, Dr. P.S. Belton for n.m.r.
spectroscopy, and Mr. J. Eagles for help and advice with GLC-mass
spectrometric analysis.

REFERENCES

Akiyama, Y. & Kato, K. (1976). Agricultural and Biological Chemistry,
     40, 2343-2348.
Akiyama, Y. & Kato, K. (1977). Agricultural and Biological Chemistry,
     41, 79-81.
Akiyama, Y., Mori, M. & Kato, K. (1980). Agricultural and Biological
     Chemistry, 44, 2487-2489.
Albersheim, P., Neukom, H. & Deuel, H. (1960). Archives of Biochemistry
     and Biophysics, 90, 46-51.
Albersheim, P., Nevins, D.J., English, P.D. & Karr, A. (1967).
     Carbohydrate Research, 5, 340-345.
Allen, A.K., Desai, N.N., Neuberger, A. & Creeth, J.M. (1978).
     Biochemical Journal, 171, 665-674.
Ashford, D., Desai, N.N., Allen, A.K., Neuberger, A., O'Neill, M.A. &
     Selvendran, R.R. (1982). Biochemical Journal, 201, 199-208.
Aspinall, G.O. (1972). In Elucidation of organic structures by physical
     and chemical methods, 2nd edition, eds. K.W. Bentley &
     G.W. Kirby, pp. 379-450. London: Wiley-Interscience.
Aspinall, G.O. (1982). In The polysaccharides, Vol. 1, ed. G.O. Aspinall,
     pp. 35-131. New York: Academic Press.
Aspinall, G.O., Begbie, R., Hamilton, A. & Whyte, J.N.C. (1967).
     Journal of the Chemical Society (C), 1065-1070.
Aspinall, G.O. & Cottrell, J.W. (1970). Canadian Journal of Chemistry,
     48, 1283-1289.
Aspinall, G.O. Cottrell, J.W., Molloy, J.A. & Uddin, M. (1970).
     Canadian Journal of Chemistry, 48, 1290-1295.
Aspinall, G.O. & Chaudhari, A.S. (1975). Canadian Journal of Chemistry,
     53, 2189-2193.
Aspinall, G.O., Craig, J.W. & Whyte, J.L. (1968). Carbohydrate Research,
     7, 442-452.
Aspinall, G.O. & Fanous, H.K. (1984). Carbohydrate Polymers, 4, 193-214.
Aspinall, G.O., Fanous, H.K. & Sen, A.K. (1983). In Unconventional sources
     of dietary fibre, ed. I. Furda, pp. 33-48. Washington, D.C.:
     American Chemical Society.

Aspinall, G.O., Hunt, K. & Morrison, I.M. (1967). Journal of the
    Chemical Society (C), 1080-1086.
Aspinall, G.O., Krishnamurthy, T.N., Mitura, W. & Funabashi, M. (1975).
    Canadian Journal of Chemistry, 53, 2182-2188.
Aspinall, G.O. & Rosell, K.G. (1977). Carbohydrate Research, C23-C26.
Aspinall, G.O. & Wilkie, K.C.B. (1956). Journal of the Chemical Society,
    1072-1076.
Barbier, M. & Thibault, J-F. (1982). Phytochemistry, 21, 111-115.
Barrett, A.J. & Northcote, D.H. (1965). Biochemical Journal, 94, 617-627.
Bauer, W.D., Talmadge, K.W., Keegstra, K. & Albersheim, P. (1973). Plant
    Physiology, 51, 174-187.
Bitter, T. & Muir, H.M. (1962). Analytical Biochemistry, 4, 330-334.
Blake, J.D. & Richards, G.N. (1970). Carbohydrate Research, 14, 375-268.
Blake, J.D. & Richards, G.N. (1971). Carbohydrate Research, 17, 253-268.
Blakeney, A.B., Harris, P.J., Henry, R.J., Stone, B.A. & Norris, T.
    (1982). Journal of Chromatography, 249, 180-182.
Blakeney, A.B., Harris, P.J., Henry, R.J. & Stone, B.A. (1983).
    Carbohydrate Research, 113, 291-299.
Blumenkrantz, N. & Asboe-Hansen, G. (1973). Analytical Biochemistry, 54,
    484-489.
Bouveng, H.O. & Lindberg, B. (1960). Advances in Carbohydrate Chemistry,
    15, 53-89.
Brillouet, J-M., Joseleau, J-P., Utille, J-P. & Lelievre, D. (1982).
    Journal of Agricultural and Food Chemistry, 30, 488-495.
Buchala, A.J., Fraser, C.G. & Wilkie, K.C.B. (1972). Phytochemistry, 11,
    1249-1254.
Carpita, N.C. (1983). Plant Physiology, 72, 515-521.
Carpita, N.C. (1984). Phytochemistry, 23, 1089-1093.
Cecy, I.I.T. & Correa, J.B. (1984). Phytochemistry, 23, 1271-1276.
Chambat, G., Barnoud, F. & Joseleau, J-P. (1984). Plant Physiology, 74,
    687-693.
Conrad, H.E., Bamburg, J.R., Epley, J.D. & Kindt, T.J. (1966).
    Biochemistry, 5, 2808-2817.
Darvill, A.G., McNeil, M. & Albersheim, P. (1980a). Carbohydrate
    Research, 86, 309-315.
Darvill, J.E., McNeil, M., Darvill, A.G. & Albersheim, P. (1980b).
    Plant Physiology, 66, 1135-1139.
Dell, A., Morris, H.R., Egge, H., Von Nicolai, H. & Strecker, G. (1983).
    Carbohydrate Research, 115, 41-52.
Deventer-Schriemer, W.H. & Pilnik, W. (1976). Lebensmittel-Wissenschaft+
    Technologie, 9, 42-44.
Doesburg, J.J. (1965). Pectic substances in fresh and preserved fruits
    and vegetables. Wageningen, I.B.V.T.-Communication No. 25.
Foglietti, M.J. & Percheron, F. (1968). Carbohydrate Research, 7, 146-
    155.
Ford, C.W. (1982). Journal of the Science of Food and Agriculture, 33,
    318-324.
Fry, S.C. (1982a). Biochemical Journal, 203, 493-504 (1982).
Fry, S.C. (1982b). Biochemical Journal, 204, 449-455 (1982).
Fry, S.C. (1983). Planta, 157, 111-123.
Gaillard, B.D.E., Thompson, N.S. & Morak, A.J. (1969). Carbohydrate
    Research, 11, 509-519.
Gaillard, B.D.E. & Thompson, N.S. (1971). Carbohydrate Research, 18,
    137-146.
Geyer, R., Geyer, H., Kuhnhardt, S., Mink, W. & Stirm, S. (1982).
    Analytical Biochemistry, 121, 263-274.

Gowda, D.C., Reuter, G. & Schauer, R. (1983). Carbohydrate Research, 113, 113-124.

Green, J.W. (1963). In Methods in carbohydrate chemistry, Vol. 3, ed. R.L. Whistler, pp. 9-21. New York & London: Academic Press.

Hakomori, S. (1964). Journal of Biochemistry (Tokyo), 55, 205-213.

Hamming, M.C. & Foster, N.G. (1972). Interpretation of mass spectra of organic compounds, pp. 138-216. New York: Academic Press.

Harris, P.J., Henry, R.J., Blakeney, A.B. & Stone, B.A. (1984). Carbohydrate Research, 127, 59-73.

Heath, M.F. & Northcote, D.H. (1971). Biochemical Journal, 125, 953-961.

Hirst, E.L., Rees, D.A. & Richardson, N.G. (1965). Biochemical Journal, 95, 453-458.

Hunt, D.F. (1974). In Advances in mass spectrometry, Vol. 6, ed. A.R. West, pp. 517-522. London: Applied Science.

Jansson, P.E., Kenne, L., Liedgren, H., Lindberg, B. & Lonngren, J. (1976). Chemical Communications, University of Stockholm, pp. 1-76.

Jarvis, M.C., Hall, M.A., Threlfall, D.R. & Friend, J. (1981). Planta, 152, 93-100.

Jarvis, M.C. (1982). Planta, 154, 344-346.

Jermyn, M.A. (1955). In Modern methods of plant analysis, Vol. II, eds. K. Paech & M.V. Tracey, pp. 197-225. Berlin: Springer.

Jermyn, M.A. & Isherwood, F.A. (1956). Biochemical Journal, 64, 123-132.

Joseleau, J-P., Chambat, G. & Chumpitazi-Hermoza, B. (1981). Carbohydrate Research, 90, 339-344.

Joseleau, J-P., Chambat, G. & Lanvers, M. (1983). Carbohydrate Research, 122, 107-113.

Karkkainen, J. (1971). Carbohydrate Research, 17, 1-10.

Kato, Y. & Matsuda, K. (1980a). Agricultural and Biological Chemistry, 44, 1751-1758.

Kato, Y. & Matsuda, K. (1980b). Agricultural and Biological Chemistry, 44, 1759-1766.

Kertesz (1951). The pectic substances. New York & London: Interscience Publishers.

Knee, M. (1970). Journal of Experimental Botany, 21, 651-662.

Knee, M. & Bartley, I.M. (1981). In Recent advances in the biochemistry of fruits and vegetables, eds. J. Friend & M.J.C. Rhodes, pp. 133-160. London & New York: Academic Press.

Kochetkov, N.K. & Chizhov, O.S. (1966). Advances in Carbohydrate Chemistry, 21, 39-93.

Kooiman, P. (1961). Recueil des Travaux Chimiques des Pays-Bas, 80, 849-852.

Kovacik, V., Bauer, S., Rosik, J. & Kovac, P. (1968). Carbohydrate Research, 8, 282-290.

Kovacik, V., Mihalov, V., Hirsch, J. & Kovac, P. (1978). Biomedical Mass Spectrometry, 5, 136-145.

Laborda, F., Fielding, A.H. & Byrde, R.J.W. (1973). Journal of General Microbiology, 79, 321-329.

Lamport, D.T.A. (1967). Nature (London), 216, 1322-1324.

Lindberg, B. (1972). In Methods in Enzymology, Vol. 28, part B, ed. V. Ginsburg, pp. 178-195. New York: Academic Press.

Lindberg, B. & Jansson, P.E. (1978). In Methods in Enzymology, Vol. 50, ed. V. Ginsburg, pp. 3-35. New York: Academic Press.

Lindberg, B., Lonngren, J. & Svensson, S. (1975). Advances in Carbohydrate Chemistry and Biochemistry, 31, 185-240.

Lomax, J.A. & Conchie, J. (1982). Journal of Chromatography, 236, 385-394.

Lonngren, J. & Svensson, S. (1974). Advances in Carbohydrate Chemistry
    and Biochemistry, 29, 41-106.
McNamara, M.K. & Stone, B.A. (1981). Lebensmittel-Wissenschaft+
    Technologie, 14, 182-187.
McNeil, M., Darvill, A.G. & Albersheim, P. (1980). Plant Physiology, 66,
    1128-1134.
McNeil, M., Darvill, A.G., Aman, P., Franzen, L-E. & Albersheim, P.
    (1982a). In Methods in Enzymology, Vol. 83, ed. V. Ginsburg,
    pp. 3-45. New York: Academic Press.
McNeil, M., Darvill, A.G. & Albersheim, P. (1982b). Plant Physiology, 70,
    1586-1591.
McNeil, M. (1983). Carbohydrate Research, 123, 31-40.
Meier, H. (1958). Acta Chemica Scandinavica, 12, 144-146.
Meier, H. (1965). In Methods in carbohydrate chemistry, Vol. 5, ed.
        R.L. Whistler, pp. 45-46. New York & London: Academic Press.
Monro, J.A. & Bailey, R.W. (1975). Carbohydrate Research, 41, 153-161.
Monro, J.A. & Bailey, R.W. (1976). Phytochemistry, 15, 175-181.
Moor, J. & Waight, E.S. (1975). Biomedical Mass Spectrometry, 2, 36-45.
Morrison, I.M. (1974). Phytochemistry, 14, 505-508.
Mort, A.J. & Lamport, D.T.A. (1977). Analytical Biochemistry, 82, 289-309.
Narui, T., Takahashi, K., Kobayashi, M. & Shibata, S. (1982).
        Carbohydrate Research, 103, 293-301.
Neukom, H., Deuel, H., Heri, W.J. & Kundig, W. (1960). Helvetica Chimica
        Acta, 43, 64-71.
Nilsson, B. & Zopf, D. (1982). In Methods in Enzymology, Vol. 83, ed.
        V. Ginsburg, pp. 46-58. New York: Academic Press.
O'Neill, M.A. & Selvendran, R.R. (1980a). Biochemical Journal, 187, 53-63.
O'Neill, M.A. & Selvendran, R.R. (1980b). Carbohydrate Research, 79, 115-
        124.
O'Neill, M.A. & Selvendran, R.R. (1983). Carbohydrate Research, 111, 239-
        255.
O'Neill, M.A., Selvendran, R.R. & Morris, V.J. (1983). Carbohydrate
        Research, 124, 123-133.
Perlin, A.S. & Casu, B. (1982). In The Polysaccharides, Vol. 1, ed.
        G.O. Aspinall, pp. 133-193. New York: Academic Press.
Pusztai, A., Begbie, R. & Duncan, I. (1971). Journal of the Science of
        Food and Agriculture, 22, 514-519.
Redgwell, R.J. (1983). Phytochemistry, 22, 951-956.
Ring, S.G. & Selvendran, R.R. (1978). Phytochemistry, 17, 745-752.
Ring, S.G. & Selvendran, R.R. (1980). Phytochemistry, 19, 1723-1730.
Ring, S.G. & Selvendran, R.R. (1981). Phytochemistry, 20, 2511-2519.
Robertson, B.K., Aman, P., Darvill, A.G., McNeil, M. & Albersheim, P.
        (1981). Plant Physiology, 67, 389-400.
Salimath, P.V. & Tharanathan, R.N. (1982). Carbohydrate Research, 106,
        251-257.
Sawardeker, J.S., Sloneker, J.H. & Jeanes, A. (1965). Analytical
        Chemistry, 37, 1602-1604.
Scott, J.E. (1963). In Methods of Biochemical Analysis, Vol. 8, ed.
        D. Glick, pp. 145-197. New York & London: Interscience
        Publishers.
Scott, J.E. (1965). In Methods in Carbohydrate Chemistry, Vol. 5,
        ed. R.L. Whistler, pp. 38-44. New York & London: Academic
        Press.
Selvendran, R.R. (1975a). Phytochemistry, 14, 1011-1017.
Selvendran, R.R. (1975b). Phytochemistry, 14, 2175-2180.
Selvendran, R.R. (1983a). In Dietary fibre, eds. G.G. Birch & K.J. Parker,
        pp. 95-147. London & New York: Applied Science Publishers.

Selvendran, R.R. (1983b). In Recent developments in mass spectrometry in biochemistry, medicine and environmental research, 8, ed. A. Frigerio, pp. 159-176. Amsterdam: Elsevier Scientific Publishing Co.

Selvendran, R.R. (1984). American Journal of Clinical Nutrition, 39, 320-337.

Selvendran, R.R., Davies, A.M.C. & Tidder, E. (1975). Phytochemistry, 14, 2169-2174.

Selvendran, R.R. & O'Neill, M.A. (1982). In Encyclopedia of plant physiology, New Series, Vol. 13A, Plant Carbohydrates I, eds. F.A. Loewus & W. Tanner, pp. 515-583. Berlin, Heidelberg, New York: Springer-Verlag.

Selvendran, R.R., Ring, S.G. & March, J.F. (1977). In Cell wall biochemistry, eds. B. Solheim & J. Raa, pp. 115-117, Bergen Univesitets-forlaget.

Selvendran, R.R., March, J.F. & Ring, S.G. (1979). Analytical Biochemistry, 96, 282-292.

Selvendran, R.R., Ring, S.G. & Du Pont, M.S. (1979). Chemistry and Industry (London), 225-230.

Selvendran, R.R. & DuPont, M.S. (1980a). Journal of the Science of Food and Agriculture, 31, 1173-1182.

Selvendran, R.R. & Du Pont, M.S. (1980b). Cereal Chemistry, 57, 278-283.

Selvendran, R.R., Ring, S.G., O'Neill, M.A. & Du Pont, M.S. (1980). Chemistry and Industry (London), 855-888.

Selvendran, R.R., Ring, S.G. & Du Pont, M.S. (1981). In The analysis of dietary fibre in food, ed. W.P.T. James & O. Theander, pp. 95-121. New York & Basel: Marcel Dekker Inc.

Selvendran, R.R. & Du Pont, M.S. (1984). In Developments in food analysis techniques - 3, ed. R.D. King, pp. 1-68. London & New York: Elsevier Applied Science Publishers.

Siddiqui, I.R. & Wood, P.J. (1971). Carbohydrate Research, 17, 97-108.

Southgate, D.A.T. (1976). In Fiber in human nutrition, eds. G.A. Spiller & R.S. Amen, pp. 31-72. New York: Plenum Press.

Souty, M., Thibault, J.F., Navarro-Garcia, G. Lopez-Roca, J.M. & Breuils, L. (1981). Science des Aliments, 1, 67-80.

Stewart, T.S., Mendershausen, P.B. & Ballou, C.E. (1968). Biochemistry, 7, 1843-1854.

Stevens, B.J.H. & Selvendran, R.R. (1980). Journal of the Science of Food and Agriculture, 31, 1257-1267.

Stevens, B.J.H. & Selvendran, R.R. (1984a). Phytochemistry, 23, 107-115.

Stevens, B.J.H. & Selvendran, R.R. (1984b). Phytochemistry, 23, 337-347.

Stevens, B.J.H. & Selvendran, R.R. (1984c). Carbohydrate Research, 128, 321-333.

Stevens, B.J.H. & Selvendran, R.R. (1984d). Carbohydrate Research (in press).

Stoddart, R.W., Barrett, A.J. & Northcote, D.H. (1967). Biochemical Journal, 102, 194-204.

Strahm, A., Amado, R. & Neukom, H. (1981). Phytochemistry, 20, 1061-1063.

Svensson, S. (1978). In Methods in Enzymology, Vol. 50, ed. V. Ginsburg, pp. 33-38. New York: Academic Press.

Talmadge, K.W., Keegstra, K., Bauer, W.D. & Albersheim, P. (1973). Plant Physiology, 51, 158-173.

Taylor, R.L. & Conrad, H.E. (1972). Biochemistry, 11, 1383-1388.

Thibault, J.F. (1983). Phytochemistry, 22, 1567-1571.

Valent, B., Darvill, A.G., McNeil, M., Robertson, B.K. & Albersheim, P. (1980). Carbohydrate Research, 79, 165-192.

Voragen, A.G.J., Schols, H.A., DeVries, J.A. & Pilnik, W. (1982).
    Journal of Chromatography, 244, 327-336.
Voragen, A.G.J., Timmes, J.P.J., Linssen, J.P.H., Schols, H.A. & Pilnik,
    W. (1983). Zeitschrift fur Lebensmittel-Untersuchung und-
    Forschung, 177, 251-256.
Waibel, R., Amado, R. & Neukom, H. (1980). Journal of Chromatography,
    197, 86-91.
Wise, L.E., Murphy, M. & D'Addieco, A.A. (1946). Paper Trade Journal,
    122, 35-43.
Wise, L.E. & Ratliff, E.K. (1947). Analytical Chemistry, 19, 459-462.
Wood, P.J., Siddiqui, I.R. (1972). Carbohydrate Research, 22, 212-220.
Woodward, J.R. & Fincher, G.B. (1982). European Journal of Biochemistry,
    121, 663-670.
Woodward, J.R., Fincher, G.B. & Stone, B.A. (1983). Carbohydrate
    Polymers, 3, 207-225.
Zitko, V. & Bishop, C.T. (1965). Canadian Journal of Chemistry, 43,
    3206-3214.

3 INTERACTION OF ENZYMES INVOLVED IN CELL-WALL
   HETEROPOLYSACCHARIDE BIOSYNTHESIS

K.W. Waldron
Department of Botany, University of Glasgow,
Glasgow G12 8QQ, U.K.

C.T. Brett
Department of Botany, University of Glasgow,
Glasgow G12 8QQ, U.K.

Abstract. Biosynthesis of cell wall heteropolysaccharides
requires the interaction of several enzymes. "Precise"
and "imprecise" models for the interactions of these enzymes
are proposed, and the sites of synthesis, initiation
mechanisms and termination mechanisms are discussed.
Evidence concerning the interactions of enzymes in the
synthesis of xyloglucan, glucuronoxylan, arabinoglucuronoxylan, rhamnogalacturononan, glucomannan,
galactoglucomannan, arabinogalactan and maize root mucilage
is reviewed. Important areas for future research are
outlined.

Key words: Plant cell wall; heteropolysaccharide
biosynthesis; enzyme interactions; xyloglucan;
glucuronoxylan; arabinoglucuronoxylan; rhamnogalacturonan;
glucomannan; galactoglucomannan; arabinogalactan; maize
root mucilage.

INTRODUCTION

The higher plant cell wall is a structure of great complexity
at the molecular level. The carbohydrate portion of the wall contains
at least nine types of monosaccharides or monosaccharide derivatives,
and these carbohydrate residues are linked together as polysaccharides by
a variety of different glycosidic bonds (Table 1). Additional complexity
is provided by other wall components (water, protein, lignin and other
phenolics, ions, lipids, etc.), and all these materials are linked by a
host of non-covalent bonds, supplemented by some covalent bonds and
perhaps also by concatenation (Lamport & Epstein 1983). In addition,
polysaccharides of a given type are thought to be heterogenous in size
and perhaps also in their detailed structure.

Since this complexity is still far from being unravelled,
those working on wall biosynthesis have no precise knowledge of the
structure of the molecules whose formation they are studying. However,

as knowledge of wall structure advances, it becomes possible to ask more
searching questions about biosynthesis. It is now clear that the
biosynthesis of any one cell-wall polysaccharide requires the action of
more than one enzyme, since even the simplest homopolysaccharide appears
to require separate enzymes for initiation and elongation. This review
concentrates on the cell-wall heteropolysaccharides, whose biosynthesis
probably requires three or more enzymes. It aims to summarise what
is known about the interactions of these enzymes and to suggest questions
which might be asked in planning future research.

GENERAL CONSIDERATIONS
Precision of synthesis
Cell-wall heteropolysaccharides consist of a backbone chain
of glycosyl residues, to which side-chains of one or more glycosyl units
may be attached. These side-chains may themselves contain branch points
at which further side-chains are attached. Any of the glycosyl residues
may be attachment points for non-glycosyl material (e.g. methyl groups,
acetyl groups, protein and phenolics).

It is not clear at present with what degree of precision
these heteropolysaccharides are synthesised. Two extreme situations
might be envisaged. If the synthesis were absolutely precise, every
molecule would be formed with exactly the same primary structure, as is
the case with proteins which have undergone no post-translational
modification. If it were absolutely imprecise, no two molecules would
be alike, even with respect to the types of glycosyl residues present.

Neither of these extremes seems likely at present. It has
not proved possible to purify polysaccharides to homogeneity (Wilkie
1985), and hence absolute precision seems not to occur. On the other
hand, some degree of precision is provided by the specificities of the
biosynthetic enzymes, so that it is at least possible to distinguish
between types of polysaccharides on the basis of the glycosyl residues
present and the linkages between them.

The degree of precision of synthesis is thought to lie
somewhere between these two extremes. It is generally accepted that
the overall size of cell-wall heteropolysaccharides is not precisely
determined; the size is generally much greater than the diameter of any
one protein molecule, so that size cannot easily be determined by the
specificity of enzyme action, and there is no evidence for any template

Table 1.   Glycosidic bonds found in plant cell wall
polysaccharides.   The list is not exhaustive.   For details,
see Aspinall (1980) and Darvill et al. (1980).
Abbreviations: RG: Rhamnogalacturonan;   AG Arabinogalactan.

Bond	Polysaccharides in which the bond is found
Ara α(1→2) Ara	Arabinan
Ara α(1→3) Ara	Arabinan: RG I
Ara α(1→5) Ara	Arabinan: RG I; AG I
Ara α(1→3) Gal	AG I
Ara (1→6) Gal	AG II
Ara (1→4) Rha	RG I
Ara (1→2) Xyl	Arabinoxylan
Ara α(1→3) Xyl	Arabinoxylan
Fuc (1→2) Gal	Xyloglucan
Gal β(1→5) Ara	Arabinoxylan
Gal β(1→3) Gal	AG II
Gal β(1→4) gal	AG I; galactan; RG I
Gal β(1→6) Gal	AG II; galactan
Gal α(1→6) Man	Galactomannan; galactoglucomannan
Gal (1→4) Rha	RG I
Gal β(1→2) Xyl	Xyloglucan
Gal β(1→4) Xyl	Xylan
GalU α(1→4) GalU	RG I
GalU α(1→2) Rha	RG I
Glc β(1→3) Glc	Callose; β(1→3), β(1→4)-glucan
Glc β(1→4) Glc	Cellulose; β(1→3), β(1→4)-glucan; glucomannan; galactoglucomannan; xyloglucan
Glc β(1→4) Man	Glucomannan; galactoglucomannan
GlcU β(1→6) Gal	AG II
GlcU β(1→2) Man	Glucuronomannan
GlcU α(1→2) Xyl	Glucuronoxylan
Man β(1→4) Glc	Glucomannan; galactoglucomannan
Man α(1→4) GlcU	Glucuronomannan
Man β(1→4) Man	Glucomannan; galactoglucomannan
Rha (1→4) GalU	RG I
Rha (1→2) Rha	RG I
Xyl β(1→2) Ara	Xylan
Xyl (1→4) Gal	Xylan
Xyl α(1→6) Glc	Xyloglucan
Xyl β(1→6) Man	Glucuronomannan
Xyl β(1→4) Xyl	Xylan

mechanism for defining size (only for a homopolysaccharide, cellulose,
has a template been postulated (Marx-Figini 1976).   Given this
variation in size, two rather more realistic models can be put forward
as the poles of current thought on the precision of synthesis.

1. Imprecise synthesis.   In this model, the biosynthetic enzymes are
sufficiently specific to define the glycosyl residues present and their
linkages.   However, they are not sufficiently specific to give any
regularity to the primary structure (Fig. 1a).   Hence side-chains will
be of variable length and will be attached irregularly along the
backbone.   The backbone can be synthesised independently of the side-
chains.   If the backbone contains more than one type of glycosyl
residue, they are irregularly arranged.   Non-glycosyl components are
present at some potential attachment points, but not at others.   The
molar ratio of the glycosyl residues present is expected to vary
depending on the relative availability of suitable substrates, and this
will be true also of the amounts of non-glycosyl substituents present.
Structural variation may also occur as a result of variations in the
relative amounts of the biosynthetic enzymes available at the site of
synthesis and as a result of changes in their state of activation or
inhibition.

2. Precise synthesis.   In this model, the enzymes which synthesise the
polysaccharides are absolutely specific in all respects except that they
are unable to determine the overall size of the molecule (Fig. 1b).
This is most easily envisaged as occurring by the synthesis of identical,
repeating subunits of oligosaccharide size, linked in a precise manner,
but whose total number in the molecule is not precisely defined.   This
degree of precision could be achieved by a requirement for the enzymes
to add glycosyl residues (and other components) in a precise sequence;
this sequence would repeat itself through a number of cycles during the
synthesis of the polysaccharide.   Such precision could be achieved
provided the diameter of the subunits was of the same order as the
diameter of the biosynthetic enzymes or enzyme complex.   A mechanism of
this type is seen in the formation of the peptidoglycan of bacterial cell
walls (Shockman & Barrett 1983).   The subunit oligosaccharides might be
comparable in complexity to the N-linked oligosaccharide chains of glycoproteins, whose glycosyl residues are linked in a precise structure and

Fig. 1.   Heteropolysaccharides undergoing biosynthesis by
(a) imprecise and (b) precise mechanisms.   A,B and C
represent different monosaccharides.   Arrows represent
glycosyl transfers in progress at one particular moment;
UDP-sugars are depicted, since they are the most common
sugar-nucleotide precursors, but this should not be taken
to imply either that all the precursors are uridine
derivatives or that no intermediates, such as polyprenyl-
phosphate-sugars, are involved in the glycosyl transfer.
The overall direction of synthesis of the backbone is from
right to left.

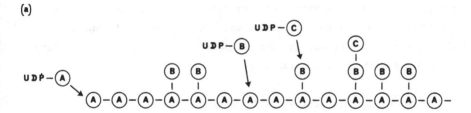

can number up to about 20 residues (Elbein 1979). One way in which such
a precise mechanism could be achieved is by construction of each subunit
on a lipid (or protein) carrier, followed by addition of the complete
subunit to the growing chain, as occurs in bacterial peptidoglycan and
the Salmonella O-antigen (Tonn & Gander 1979). However, there is little
evidence for the participation of lipid intermediates in heteropolysaccharide biosynthesis in plants (Maclachlan 1985), and the participation
of lipid or protein intermediates is in no way necessary for this degree
of precision to be achieved. Characteristic features of this "precise"
mechanism are the following: (a) the backbone of the molecule cannot
grow without concomitant addition of side-chains and non-glycosyl
substituents; (b) the structure of the molecule will be independent of
the availability of substrates, the relative amounts of the biosynthetic
enzymes present, and their degree of activation or inhibition; (c) the
rate of synthesis of the polysaccharide will be limited to the speed of
the slowest enzyme, which may in turn be limited by substrate availability
or by enzyme inhibition or availability.

Later in this review, evidence concerning the degree of
precision of synthesis will be reviewed. The terms "precise" and
"imprecise" will be used as defined above. It should be noted that
intermediate situations are possible, with some parts of the molecule
being synthesised with a greater degree of precision than others. It
should also be noted that evidence concerning precision of synthesis
cannot necessarily be obtained from data concerning the regularity or
otherwise of structure. In particular, precise mechanisms of synthesis
could fail to result in regularity of structure for any of the following
reasons: (a) structural modification subsequent to synthesis, e.g.
action of cell-wall hydrolases or transglycosylases, which might act in
a random manner; (b) variation between cell types, if structural studies
are carried out on samples from a heterogeneous cell population; (c)
variations at different stages of the cell cycle; (d) random
degradation during isolation or analysis of the polysaccharide (see
Wilkie (1985) for a more detailed discussion of some of these points).
Imprecise mechanisms, on the other hand, might lead to apparent
regularities of structure if over-zealous attempts are made to isolate
"pure" polysaccharide fractions.

Sites of synthesis

Current evidence suggests that cell-wall heteropolysaccharides are synthesised within the cell rather than outside the plasma membrane. It seems probable that the major site of synthesis is the Golgi apparatus (Robinson & Kristen 1982).   The endoplasmic reticulum may also be involved (Bolwell & Northcote 1983), but this organelle probably only plays a major part in those cases where the heteropolysaccharide is synthesised as a glycoprotein, such as in the synthesis of mucilage by maize root tips (Green & Northcote 1978; Bowles & Northcote 1974).

The glycosyltransferases are membrane-bound and, by analogy with those involved in glycoprotein synthesis, may form multi-enzyme complexes (Ivatt 1981; Schwartz & Roden 1975).   Alternatively, they may be present in different parts of the Golgi apparatus, so that, for instance, addition of some side-chains may occur at the maturing face or in the Golgi vesicles, after the backbone has been completed. Techniques for subfractionation of the plant Golgi apparatus are only now becoming sufficiently advanced to permit exploration of this possibility.

It is probable that some covalent bonds between polysaccharides are formed within the cell wall (Stoddart & Northcote 1967; Albersheim 1976).   This bond formation may or may not be balanced by concomitant bond breakage.   Whether or not this bond formation in the cell wall is regarded as "synthesis" is a matter of definition, but it is important to note that cell-wall polysaccharides cannot necessarily be regarded as having reached a final, stable form by the time they leave the endo-membrane system.

Initiation reactions

Special enzymes are thought to be required for hetero-polysaccharide initiation, since the glycosyltransferases responsible for backbone elongation appear to require a stretch of pre-formed backbone onto which to add further sugars. By analogy with starch and glycogen, the initiation steps may involve addition of sugars to a protein "primer" (Tandecarz & Cardini 1978).   Other types of primer are also possible (c.f. the possibility of inositol as a primer for callose synthesis (Kemp & Loughman 1974)).   The greater the precision of synthesis of the backbone, the more likely it is that special initiation mechanisms will be necessary.

Termination reactions

Little attention has been paid to possible termination mechanisms for heteropolysaccharide synthesis.   It is tacitly assumed that synthesis ceases when the polysaccharide becomes physically separated from the biosynthetic enzymes during its passage through the endomembrane system.   However, such a crude termination mechanism would provide the cell with little scope for controlling separately the range of sizes of different polysaccharides.   Hence there may be termination reactions, such as the addition of a particular glycosyl residue at the end of the backbone, which would prevent its further elongation.   Such a mechanism could control the average size of the molecules, even if the precise size were not specified.

EVIDENCE CONCERNING ENZYME INTERACTIONS

Only limited information is available concerning enzyme interactions.   This is partly due to certain characteristics of these enzymes which make them difficult to study.   They are membrane-bound, and in most cases it has proved very difficult to solubilise them with detergents.   Hence detailed kinetic studies are difficult, especially since the enzymes often do not exhibit Michaelis–Menten kinetics when analysed in the membrane-bound form.   Since solubilisation is difficult, they are generally studied in an impure form, which means that there is no certainty that only one enzyme is present which will utilise a particular sugar-nucleotide.   This problem is compounded by the uncertainty as to the precise nature of the polysaccharides present in the membrane preparation, since it is one or more of these polysaccharides that acts as a sugar acceptor for the glycosyl transferase.   All these characteristics make the enzymes hard to study.   The following evidence has, however, been obtained.

Xyloglucan

Xyloglucan biosynthesis has been studied using enzyme preparations from pea seedlings (Villemez & Hinman 1975; Ray 1980) and suspension-cultured soybean cells (Hayashi & Matsuda 1981 a,b; Hayashi et al. 1984).   In both these tissues, the presence of UDP-glucose in in vitro enzyme preparations stimulated the incorporation of xylose from UDP-xylose into xyloglucan.   This is to be expected, since xylose is present as a side-chain on a backbone of $\beta(1\rightarrow4)$-linked glucose residues

in this molecule, and the UDP-glucose was presumably needed to form the
backbone.   In the pea system, Ray (1980) reported that the formation
of the glucan backbone was independent of the presence of UDP-xylose.
It is possible that, in the absence of UDP-xylose, the pea system formed
a glucan unrelated to xyloglucan.   However, UDP-glucose stimulated
xylose transfer even when it was present in a pre-incubation rather than
in the same incubation as UDP-xylose.   Hence xyloglucan synthesis in
peas appears to occur by an imprecise mechanism, with the backbone being
formed independently of the side-chains.

In the soybean system, the picture is different.   Matsuda's
group have developed an assay for measuring xylose and glucose transfer
specifically into xyloglucan, as opposed to other glucose-containing and
xylose-containing polysaccharides (Hayashi & Matsuda 1981 a).   They
found that at certain concentrations, UDP-xylose prolonged the
incorporation of glucose from UDP-glucose into xyloglucan.   There was
no evidence that glucose was added to xyloglucan without concomitant
addition of xylose.   Hence the addition of xylose appears to be
essential for elongation of the glucan backbone, indicating a precise
mechanism for the addition of these two sugars.

The newly-incorporated xylose and glucose residues were
found in pentasaccharide or heptasaccharide fragments after partial
digestion of the soybean product.   These fragments could be identified
as part of the xyloglucan structure (Hayashi et al. 1981).   The
pentasaccharide could be converted to the heptasaccharide in a pulse-
chase experiment (Hayashi et al. 1984), indicating that the addition of
the sixth sugar residue, a glucose, was a rate-limiting step in this
in vitro system.

Xyloglucan biosynthesis in soybean suspension-cultured cells
(though apparently not in pea seedlings) may therefore be an example of
a precise biosynthetic system.   This conclusion is supported by the
regularity of structure seen in this type of molecule, since it appears
to be made up of well-defined heptasaccharide and nonasaccharide
subunits (Maclachlan 1985).   The existence of the nonasaccharide
subunit implies, however, that the precision does not extend to the two
additional sugar residues found in the nonasaccharide.   The two sugars
concerned are galactose and fucose; none of the investigators who have
achieved sustained in vitro biosynthesis of xyloglucan have either
added galactose or fucose donors to their in vitro systems or reported

addition of these sugars from endogenous donors.   Hence the addition of
fucose and galactose must not be required for continued growth of the
molecule, and their addition must be by an imprecise mechanism.

Glucuronoxylan and arabinoglucuronoxylan

Investigations into the biosynthesis of these heteropolysaccharides have been complicated by the presence, in the particulate
preparations used, of enzymes which interconvert the sugar-nucleotide
substrates.   The important enzymes in this respect are UDP-glucuronic
acid decarboxylase and UDP-xylose 4-epimerase.   Thus it is likely that
UDP-xylose and UDP-arabinose will be formed in situ from UDP-glucuronic
acid and that UDP-xylose and UDP-arabinose will be interconverted.
However, there are several reports of incorporation of xylose into
β(1→4)-linked xylan chains (Odzuck & Kauss 1972; Dalessandro &
Northcote 1981), and at least in some cases this appears to occur
without the addition of other sugars, indicating synthesis of the xylan
backbone independently of the side-chains (Odzuck & Kauss 1972).
However, the data in general are insufficient to establish this for
certain.

In pea membrane preparations, Waldron & Brett (1983)
demonstrated that sustained incorporation of glucuronic acid into
glucuronoxylan from UDP-glucuronic acid was dependent on the presence of
UDP-xylose.   This is to be expected for the addition of side-chains to
a xylan backbone.   It is not known whether the product contains
arabinose side-chains; these might have been incorporated from UDP-
arabinose derived from the added UDP-xylose.   In the absence of UDP-
xylose, incorporation of glucuronic acid ceased within fifteen minutes;
recent results indicate that the product synthesised from UDP-
glucuronic acid in the absence of UDP-xylose is quantitatively different
to that formed in its presence (Waldron & Brett, unpublished results).
Hence there is no information available as to whether glucuronic acid
can be added to pre-formed xylan chains in this system.   Nor is
information available on this point from the work of Kauss (1967)
concerning a glucuronyltransferase from Zea mays; this enzyme was
probably working in the presence of a xylosyltransferase which was
utilising UDP-xylose formed in the incubation from the added UDP-
glucuronic acid.

Hence it is not clear whether these polysaccharides are

formed by a precise or an imprecise mechanism, as far as the sugar units
are concerned.   Structural studies do not provide further light, since
there is disagreement as to the amount of regularity in the positioning
of the side-chains (Comtat & Joseleau 1981; Rosell & Svensson 1975;
Toman et al. 1981; Kubackova et al. 1979).   It is clear, however, that
addition of the non-carbohydrate components of these polysaccharides is
not required for the synthesis of the carbohydrate portions.   The non-
carbohydrate portions include the 4-0-methyl groups attached to
glucuronic acid and acetyl groups attached to xylose.   It has been shown
that the methyl groups are added to preformed polysaccharide (Kauss &
Hassid 1967), and sustained synthesis of glucuronoxylan was observed in
the absence of any added methyl or acetyl donors (Waldron & Brett 1983).
Hence these groups are added by an imprecise mechanism.

There is preliminary evidence that protein is attached to
some of the nascent glucuronoxylan.   Part of the polymeric product
obtained from UDP-glucuronic acid, using a solubilised enzyme system
derived from that studied by Waldron & Brett (1983), is water-soluble,
and the sustained synthesis of this water-soluble product is dependent
on added UDP-xylose, as is the case with the water-insoluble glucuronoxylan.
The molecular weight of this water-soluble product, as judged by gel
exclusion chromatography on Sepharose CL-6B, was decreased by
incubation with proteinase (Farmer, Waldron & Brett, unpublished results).
Hence part of the glucuronoxylan product may be attached to protein,
which could be acting as a primer.

Solubilisation of the glycosyltransferases involved in
glucuronoxylan biosynthesis offers considerable potential for further
studies on their interaction.   The glucuronyltransferase can be
solubilised by 10% Triton X-100, and since prolongation of its action by
addition of UDP-xylose can be demonstrated in solution, it is likely that
the xylosyltransferase is also solubilised, together with either a
priming system or preformed polysaccharide acceptors (Waldron & Brett,
unpublished results).   Since the efficiency of the system is not
impaired by solubilisation it is likely that the whole biosynthetic
system is solubilised as one unit, which may indicate that a multi-
enzyme complex exists in this case.

Rhamnogalacturonans

Homogalacturonans and at least two distinct types of rhamno-

galacturonans appear to be present in primary cell walls (Chambat &
Joseleau 1980; Darvill et al. 1980).   Some work has been carried out on
incorporation of galacturonic acid into pectic polymers of this type
(Villemez et al. 1965).   However, since no other glycosyl donors were
included in the incubation systems used, it is likely that synthesis of
homogalacturonans was observed.   Hence no definite information is
available concerning the glycosyltransferases involved in rhamnogalacturonan synthesis;   it is not known whether UDP-galacturonic acid
is the galacturonic acid donor, and no rhamnosyltransferase has been
reported in this context.

Incorporation of methyl groups into pectin (as methyl esters
of galacturonic acid) is known to occur by transfer from S-adenosylmethionine onto the preformed polymer, without any concomitant addition
of galacturonic acid (Kauss et al. 1969).   This is clearly an example
of imprecise synthesis as far as the methyl groups are concerned, but it
is not known whether the product was a homogalacturonan or a rhamnogalacturonan.

The rhamnogalacturonans are the most complex cell wall
heteropolysaccharides so far discovered (McNeil et al. 1982; Darvill
et al. 1980).   It has been proposed that their complexity reflects a
role in information transfer in the plant, in addition to any structural
role they may possess (Nothnagel et al. 1983; Spellman et al. 1983).
If so, their biosynthesis may involve mechanisms of greater sophistication
than those required for the synthesis of other cell wall heteropolysaccharides.   The involvement of precise biosynthetic mechanisms seems
likely.

Glucomannans and galactoglucomannans

Considerable work has been carried out on the interaction of
a glucosyltransferase and a mannosyltransferase in synthesising
glucomannan (Villemez 1971; Heller & Villemez 1972 a,b; Brett 1981).
GDP-mannose is required for sustained incorporation of glucose from GDP-
glucose at a high rate, but glucose transfer can continue at a lowered
rate in the absence of GDP-mannose if the GDP-glucose concentration is
high enough (Brett 1981).   Mannose transfer from GDP-mannose can occur
in the absence of added GDP-glucose.   However, the situation is
complicated by the possible presence of epimerases interconverting the
two substrates, and in addition each sugar-nucleotide exerts stimulatory

or inhibitory effects on the enzyme which utilises the other sugar-
nucleotide (Villemez 1971).   Furthermore, there is uncertainty as to
whether glucomannan is the only product, or whether separate glucans and
mannans are made.   As a result of this complexity in the in vitro
systems used, it is not clear whether glucomannan synthesis occurs by a
precise or imprecise mechanism, though the latter would seem more likely.
Structural studies indicate that the glucose and mannose residues are
irregularly arranged along the backbone, which also points to an
imprecise mechanism (Aspinall 1980).   The existence of the stimulatory
and inhibitory effects of the sugar-nucleotides points, however, to a
control mechanism of some sophistication in vivo, even if the synthesis
is imprecise according to our definition.

No information appears to be available concerning
incorporation of galactose residues into galactoglucomannan.

Arabinogalactans

Cell walls contain at least two types of arabinogalactans
(Aspinall 1980).   There is evidence that UDP-arabinose and UDP-galactose
are the sugar donors (Ericson & Elbein 1980).   However, no information
is available concerning the interaction of the enzymes.

Maize root mucilage

This is a complex heteropolysaccharide which is secreted by
the root tip, and which is related to the pectins of the same tissue.
There is clear evidence that the polysaccharide is synthesised as a
glycoprotein (Green & Northcote 1978), and the protein may be acting as
a primer.   Glycolipid intermediates also appear to be involved (Green &
Northcote 1979).   The mucilage is not, however, a cell-wall polysaccharide, and cannot necessarily be expected to be synthesised in the
same way as cell-wall heteropolysaccharides.   Hence it cannot be
assumed from this evidence that lipid intermediates and protein primers
are involved in cell-wall heteropolysaccharide biosynthesis.

CONCLUSIONS AND PROSPECTS FOR THE FUTURE

From the above summary of the evidence, it is clear that our
ignorance about enzyme interactions in this field greatly exceeds the
limited knowledge that has been gleaned so far.   No clear answer can be
given concerning the degree of precision of synthesis of any cell-wall

heteropolysaccharide.   There are indications, however, that the synthesis
is more precise in some cases (e.g. xyloglucan and rhamnogalacturonan)
than in others.   If so, this probably reflects differences in the
functions of the various polysaccharides.   Some functions, such as
recognition or the control of cell extension, may require precise
structure and hence precise synthesis.   Other functions, such as
structural roles, may depend on the overall physical properties of the
molecules and may not require such a high degree of precision.

A number of outstanding challenges face those working in
this field.   The first is the in vitro biosynthesis of a complete cell-
wall heteropolysaccharide.   While this might be possible using impure,
membrane-bound enzymes, a proper understanding of the process will
require the identification, solubilisation and purification of all the
enzymes involved in the synthesis of a particular polysaccharide.   The
purified enzymes would then be recombined, either in the solubilised
form or as a reconstituted membrane of defined composition, and
incubated with the necessary precursors.

A second challenge, very different in nature, is posed by
the great variety of different cell types present in plants.   Since
these cell types are often distinguishable on the basis of their cell
wall morphology, corresponding differences in chemical composition will
occur.   Not all these differences in wall composition will involve
major differences in wall polymer chemistry, but enough examples are
known of characteristic changes in wall polysaccharide biochemistry
accompanying cell differentiation to make it likely that many
differentiated cell types will contain characteristic wall polymers, or
characteristic modifications of those polymers.   Unravelling the
changes in enzyme interactions that bring about these structural
differences will be a major task.   Since biochemists often lose sight
of the great variety of cell types, a list is given to underline the
size of the task (Table 2).   The list is not exhaustive, nor does it
attempt to explore the additional dimension of differences between taxa.

A third challenge lies in discovering the control processes
which govern the differentiation events which give rise to these cell
types.   Northcote and his colleagues have played a most important
pioneering role in this area (Northcote 1985).   Their work points the
way to the elucidation of the mechanisms for the control of the genes
which govern the whole process of wall polymer biosynthesis.

Table 2.   Cell types in higher plants.   The list is not exhaustive,
and there are very many intermediate types (e.g. "Gynoecium parts"
includes style, stigma, ovary, ovules with embryo sac, placentation,
integuments and nucellus), and the various categories frequently
overlap.   For details, see Fahn (1967) and Esau (1977).

Parenchyma: Chlorenchyma, aerenchyma, storage parenchyma (e.g.
endosperm); lobed, folded and armed parenchyma.

Collenchyma: Prismatic, elongate, angular, lamellar and lacunar
collenchyma.

Sclerenchyma: Xylary and extraxylary fibres, fibre-sclereids, sclereids
(brachysclereids, macrosclereids, osteosclereids, astro-
sclereids, filiform sclereids, trichosclereids, etc.; see
also Bailey (1961)).

Cambium and meristems: Procambium, vascular cambium, phellogen, apical
meristems, intercalary meristems.

Secretory structures: Trichomes and glands, hairs, juice sacs, glandular
hairs, scales, stinging hairs, colleters, nectaries,
osmophors, hydathodes, secretory cells, mucilage cells,
secretory spaces, raphid cells, resin ducts, lactifers
(simple and compound, articulated and non-articulated),
tapetum, aleuron(e) layer, stigmatic surfaces, scutellar
epithelium, digestive-absorptive glands in carnivorous plants.

Epidermis and multiple epidermis, including velamen, rhizodermis and
exodermis: Subsidiary cells, guard cells, fibre cells,
sclereids, lithocysts, cystoliths, myrosin secretory cells,
trichomes, glands, silica cells, cork cells, bulliform
cells, root hairs.

Phloem (primary and secondary): Tracheary elements (tracheids and vessel
members), fibres, fibre-tracheids, libriform fibres, axial
and ray parenchyma, tyloses, secretory cells, resin ducts.
"Sapwood", "heartwood", "reaction wood", storage parenchyma
and tracheids.

Phloem (primary and secondary): External, internal and included phloem
(concentric or foraminate); sieve elements (sieve cells,
sieve-tube members), companion cells, albuminous cells,
axial phloem parenchyma, ray parenchyma, fusiform parenchyma,
procumbent and erect parenchyma; fibres (lignified and non-
lignified), fibre-sclereids, storage parenchyma, lactifers,
idioblasts, crystalliferous parenchyma strands.

Periderm: Phellem, phellogen, phelloderm, polyderm, phelloids, lenticels.

Mesophyll: Homogeneous, palisade, spongy, aerenchyma, lobed, internally
ridged.

Floral and fruit tissues: Petals, sepals, androecium, and gynoecium parts,
nectaries, pollen (exine, intine, medine), fleshy and dry
pericarps, testa.

Others: Endodermis (with and without casparian strip), starch sheath,
pulvini, scales, spines, growth layers, transfusion tissue,
root cap cells, contractile cells, adhesive pads.

REFERENCES

Albersheim, P. (1976). The primary cell wall. In Plant Biochemistry, eds. J. Bonner & J.E. Varner, 3rd ed., pp. 225-74. New York: Academic Press.

Aspinall, G.O. (1980). Chemistry of plant cell wall polysaccharides. In The Biochemistry of Plants, vol. 3, ed. J. Preiss, pp. 589-616. New York: Academic Press.

Bailey, I.W. (1961). Comparative anatomy of the leaf-bearing Cactaceae. II. Structure and distribution of sclerenchyma in phloem of Pereskia, Pereskiopsis and Quiabentia. Arnold Arboretum Journal, 42, 144-50.

Bolwell, G.P. & Northcote, D.H. (1983). Arabinan synthase and xylan synthase activities of Phaseolus vulgaris. Subcellular location and possible mechanism of action. Biochemical Journal, 210, 497-507.

Bowles, D.J. & Northcote, D.H. (1974). The sites of synthesis and transport of extracellular polysaccharides in the root tissues of maize. Biochemical Journal, 130, 1133-45.

Brett, C.T. (1981). Polysaccharide synthesis from GDP-glucose in pea epicotyl slices. Journal of Experimental Botany, 32, 1067-77.

Chambat, G. & Joseleau, J-P. (1980). Isolation and characterisation of a homogalacturonan in the primary walls of Rosa cell cultures in vitro. Carbohydrate Research, 85, C10.

Comtat, J. & Joseleau, J-P. (1981). Mode of action of a xylanase and its significance for the structural investigation of the branched L-arabino-D-glucurono-D-xylan from redwood (Sequoia sempervirens). Carbohydrate Research, 95, 101-12.

Dalessandro, G. & Northcote, D.H. (1981). Xylan synthetase in differentiated xylem cells of sycamore trees (Acer pseudoplatanus). Planta, 151, 53-60.

Darvill, A., McNeil, M., Albersheim, P. & Delmer, D.P. (1980). The primary walls of flowering plants. In The Biochemistry of Plants, vol. 1, ed. Tolbert, N.E., pp. 92-162. New York: Academic Press.

Elbein, A.D. (1979). The role of lipid-linked saccharides in the biosynthesis of complex carbohydrates. Annual Reviews of Plant Physiology, 30, 239-72.

Ericson, M.C. & Elbein, A.D. (1980). Biosynthesis of cell wall polysaccharides and glycoproteins. In The Biochemistry of Plants, vol. 3, ed. J. Preiss, pp. 589-616. New York: Academic Press.

Esau, K. (1977). Anatomy of Seed Plants. 2nd edition. New York: Wiley.

Fahn, A. (1967). Plant Anatomy. Oxford: Pergamon Press.

Green, J.R. & Northcote, D.H. (1978). The structure and function of glycoproteins synthesised during slime-polysaccharide production by membranes of the root-cap cells of maize (Zea mays). Biochemical Journal, 170, 599-608.

Green, J.R. & Northcote, D.H. (1979). Polyprenyl phosphate sugars synthesised during slime-polysaccharide production by membranes of the root-cap cells of maize (Zea mays). Biochemical Journal, 178, 661-8.

Hayashi, T., Kato, Y. & Matsuda, K. (1981). Biosynthesis of xyloglucan in suspension-cultured soybean cells. I. Xyloglucan from suspension-cultured soybean cells. Plant and Cell Physiology, 21, 1405-18.

Hayashi, T. & Matsuda, K. (1981a). Biosynthesis of xyloglucan in
     suspension-cultured soybean cells. Evidence that the enzyme
     system of xyloglucan biosynthesis does not contain β(1-4)glucan
     4-β-D-glucosyltransferase activity. Plant and Cell Physiology,
     22, 1571-84.
Hayashi, T. & Matsuda, K. (1981b). Biosynthesis of xyloglucan in
     suspension-cultured soybean cells. Occurrence and some
     properties of xyloglucan 4-β-D-glucosyltransferase and
     6-α-D-xylosyltransferase. Journal of Biological Chemistry,
     256, 11117-22.
Hayashi, T., Nakajima, T. & Matsuda, K. (1984). Biosynthes of xyloglucan
     in suspension-cultured soybean cells. Processing of oligo-
     saccharide building blocks. Agricultural and Biological
     Chemistry, 48, 1023-7.
Heller, J.S. & Villemez, C.L. (1972a). Interaction of soluble glucosyl-
     and mannosyl-transferase enzyme activities in the synthesis
     of a glucomannan. Biochemical Journal, 129, 645-55.
Heller, J.S. & Villemez, C.L. (1972b). Solubilisation of a mannose-
     polymerising enzyme from Phaseolus aureus. Biochemical
     Journal, 128, 243-252.
Ivatt, R.J. (1981). Regulation of glycoprotein biosynthesis by
     formation of specific glycosyltransferase complexes.
     Proceedings of the National Academy of Science, U.S.A., 78,
     4021-5.
Kauss, H. (1967). Biosynthesis of the glucuronic acid unit of
     hemicellulose B from UDP-glucuronic acid. Biochimica et
     Biophysica Acta, 148, 572-4.
Kauss, H. & Hassid, W.Z. (1967). Biosynthesis of the 4-0-methyl-D-
     glucuronic acid unit of hemicellulose B by transmethylation
     from S-adenosyl-L-methionine. Journal of Biological
     Chemistry, 242, 1680-4.
Kauss, H., Swanson, A.L., Arnold, R. & Odzuck, W. (1969). Biosynthesis
     of pectic substances. Localisation of enzymes and products
     in a lipid-membrane complex. Biochimica et Biophysica Acta,
     192, 55-61.
Kemp, J. & Loughman, B.C. (1974). Cyclitol glucosides and their role in
     the synthesis of a glucan from uridine diphosphate glucose
     in Phaseolus aureus. Biochemical Journal, 142, 153-9.
Kubackova, M. Karaczonyi, S., Bilisics, L. & Toman, R. (1979). On the
     specificity and mode of action of a xylanase from Trametes
     hirsuta (Wulf) Pilat. Carbohydrate Research, 76, 177-88.
Lamport, D.T.A. & Epstein, L. (1983). A new model for the primary cell
     wall: a concatenated extensin-cellulose network. Current
     Topics in Plant Biochemistry and Physiology, 2, 73-83.
Maclachlan, G. (1985). Are lipid-linked glycosides required for plant
     polysaccharide biosynthesis? In Biochemistry of Plant Cell
     Walls, eds. C.T. Brett & J.R. Hillman, pp. 199-220.
     Cambridge: Cambridge University Press.
McNeil, M., Darvill, A.G. & Albersheim, P. (1982). Structure of plant
     cell walls. 12. Identification of seven differently linked
     glycosyl residues attached to the 04 of the 2,4-linked
     L-rhamnose residue of rhamnogalacturonan I. Plant Physiology,
     70, 1586-91.
Marx-Figini, M. (1976). Comparative study of cell wall and bacterial
     cellulose biosynthesis and its structural aspects. Applied
     Polymer Symposia, 28, 637-43.

Northcote, D.H. (1985). Control of cell wall formation during growth.
    In Biochemistry of Plant Cell Walls, eds. C.T. Brett &
    J.R. Hillman, pp. 177-197. Cambridge: Cambridge University Press.
Nothnagel, E.A., McNeil, M., Albersheim, P. & Dell, A. (1983). Host-
    pathogen interactions. 22. A galacturonic-acid oligo-
    saccharide from plant cell walls elicits phytoalexins.
    Plant Physiology, 71, 916-26.
Odzuck, W. & Kauss, H. (1972). Biosynthesis of pure araban and xylan.
    Phytochemistry, 11, 2489-94.
Ray, P.M. (1980). Cooperative action of the glucan synthetase and UDP-
    xylose xylosyltransferase of Golgi membranes in the synthesis
    of a xyloglucan-like polysaccharide. Biochimica et
    Biophysica Acta, 629, 431-4.
Robinson, D.G. & Kristen, U. (1982). Membrane flow via the Golgi
    apparatus of higher plant cells. International Review of
    Cytology, 77, 89-127.
Rosell, K-G. & Svensson, S. (1975). Studies on the distribution of
    4-0-methylglucuronic acid residues in birch xylan.
    Carbohydrate Research, 42, 297-304.
Schwartz, N.B. & Roden, L. (1975). Biosynthesis of chondroitin sulphate.
    Solubilisation of chondroitin sulphate glycosyltransferases
    and partial purification of uridine diphosphate-D-galactose:
    D-xylose galactosyltransferase.   Journal of Biological
    Chemistry, 250, 5200-7.
Shockman, G.D. & Barrett, J.F. (1983). Structure, function and assembly
    of cell walls of gram-positive bacteria. Annual Reviews of
    Microbiology, 37, 501-27.
Spellman, R.W., McNeil, M., Darvill, A.G., Albersheim, A. & Dell, A.
    (1983). Characterisation of a structurally complex hepta-
    saccharide isolated from the pectic polysaccharide,
    rhamnogalacturonan II. Carbohydrate Research, 122, 131-53.
Stoddart, R.W. & Northcote, D.H. (1967). Metabolic relationships of
    isolated fractions of the pectic substances of actively
    growing sycamore cells. Biochemical Journal, 105, 45-59.
Tandecarz, J.S. & Cardini, C.E. (1978). A two-step enzymatic formation
    of a glucoprotein in potato tuber. Biochimica et Biophysica
    Acta, 543, 423-9.
Toman, R., Kohn, R., Malovikova, A. & Rosik, J. (1981). Distribution of
    4-0-methyl-D-glucuronic acid units on xylan of the bark of
    white willow (Salix alba L.). Collection of Czechoslovak
    Chemical Communications, 46, 1405-12.
Tonn, S.J. & Gander, J.E. (1979). Biosynthesis of polysaccharides of
    prokaryotes. Annual Reviews of Microbiology, 33, 169-99.
Villemez, C.L. (1971). Rate studies of polysaccharide biosynthesis
    from guanosine diphosphate α-D-glucose and guanosine
    diphosphate α-D-mannose. Biochemical Journal, 121, 151-7.
Villemez, C.L. & Hinman (1975). UDP-glucose-stimulated formation of
    xylose-containing polysaccharides. Plant Physiology, 56,
    (Supplement), 15.
Villemez, C.L., Liu, T.Y. & Hassid, W.Z. (1965). Biosynthesis of the
    polygalacturonic acid chain of pectin by a particulate
    enzyme preparation from Phaseolus aureus seedlings.
    Proceedings of the National Academy of Sciences, U.S.A., 54,
    1626-32.
Waldron, K.W. & Brett, C.T. (1983). A glucuronyltransferase involved in
    glucuronoxylan synthesis in pea (Pisum sativum) epicotyls.
    Biochemical Journal, 213, 115-22.

Wilkie, K.C.B. (1985). New perspectives on non-cellulosic cell-wall polysaccharides (hemicelluloses and pectic substances) of land plants. In Biochemistry of Plant Cell Walls, eds. C.T. Brett & J.R. Hillman, pp. 1-37. Cambridge: Cambridge University Press.

4   THE HELICOIDAL CONCEPT IN PLANT CELL WALL ULTRASTRUCTURE
AND MORPHOGENESIS

A.C. Neville
Zoology Department, University of Bristol,
Woodland Road, Bristol, BS8 1UG, England

S. Levy
Zoology Department, University of Bristol,
Woodland Road, Bristol, BS8 1UG, England

Abstract.   This Chapter reviews the application to plant
cell walls of helicoidal structure - a concept already
familiar in animal skeletal studies.   Helicoids are
'plywood' laminates: in each component layer microfibrils
lie in parallel, in the plane of the layer.   Consecutive
layers are set at a small angle, so that the 'grain'
changes direction through one sense of rotation, like the
steps of a spiral staircase.   In suitably oblique sections,
electron microscopy reveals characteristic arced patterns.
These are widely distributed throughout the plant taxa
(algae, stoneworts, fungi, moss, horsetail, fern,
monocotyledons, dicotyledons and conifers), and are found
in several cell types (parenchyma, collenchyma,
sclerenchyma, tracheids and sclereids).   Internode walls
of _Nitella_ are helicoidal, making it a useful model of
higher plant walls.   Helicoidal regions also occur between
the classical S_1, S_2 and S_3 layers of hardwoods and
softwoods.   Technical details are given as to why arced
patterns are often missed.   Existing hypotheses (ordered
granule, microtubule, multinet growth, terminal complex),
cannot explain helicoidal morphogenesis.   Hence a novel
origin is proposed - self-assembly through a cholesteric
liquid-crystal stage.   Computer graphics are used to plot
the influence of strain gradients upon patterns seen in
sections of helicoidal walls.   This is one possible origin
of herring-bone patterns.

Key words: cellulose orientation, microfibril patterns,
Helicoidal cell walls, self-assembly.

INTRODUCTION: THE HELICOIDAL CONCEPT

Context

The common goal of many botanists is to understand the three-
dimensional architecture of the fibrous component in plant cell walls.
This has traditionally been studied in isolation from similar research
on skeletal systems in animals.   We believe, however, that a comparative
approach may be helpful.   Specifically, the helicoidal interpretation
of crab exoskeleton by Bouligand (1965) has already been applied to

numerous other animal structural materials.   A growing number of
workers are using the same concept to interpret the fine structure of
plant cell walls.   It is our belief that this may promote new thoughts
about wall biosynthesis and morphogenesis.   This hopefully explains a
contribution to a plant cell wall symposium from a Zoology department.

Definitions

The helicoidal model was devised in order to explain a type
of pattern seen in electron micrographs of sections from crab
exoskeleton.   This consists of nests of arcs arranged in ranks, as in
Fig. 1A.

Fig. 1A.   Computer graphics presentation of a section
through a helicoid cut obliquely through its component
layers of microfibrils.   The ranks of arced patterning
are made up of individual segments of microfibrils, and
are diagnostic of a helicoidal structure.

Fig. 1B.   Diagram of a wedge of helicoid, with arced
patterning exposed on the oblique face.   Each 180 degrees
(half turn) of component planes of parallel microfibrils
generates a single row of arcs.   The helicoid is anti-
clockwise (as in Nitella cell wall);   if it were clockwise,
the arcs would point in the opposite direction.

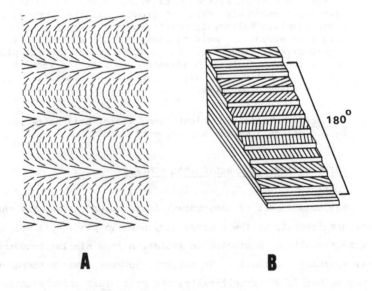

A   B

The explanation (Bouligand 1965) is that the pattern arises
in oblique sections through a special laminated structure known as a
helicoid.   Fibres lie in the plane of each component lamina and are also
arranged in parallel.   The laminae form a stack with the fibres in
consecutive layers set at a small (often constant) angle to each other.
The whole forms a sort of plywood, with the grain constantly changing
direction through one sense of rotation, like the steps of a spiral
staircase.

Oblique sections of such a structure generate the arced
pattern as in Fig. 1B.   Each row of arcs represents a total turn of
180° by the component layers, as marked on the diagram.   Values for the
width of such arcs in biological materials average between 0.1 to 1.0µm,
although much wider examples are known in certain cases.   Arc width
varies with the angle between the plane of sectioning and the component
laminae.

Two points need emphasis.   Firstly, the arc is an illusion
made up of short segments of parallel fibres from each component lamina.
However, although it is in one sense an artifact, the arced pattern can
nevertheless be used as an indicator of helicoidal structure.   Secondly,
careful distinction must be made between a helicoidal structure (drawn
out schematically in projection in Fig. 2A) and a helical structure
(drawn in Fig. 2B).   This is a common source of confusion in current
literature.   Helicoids and helices both occur in plant cell walls.

As will be shown below, helicoids are prevalent in a wide
range of plants.   Layers with helical fibres (often wrongly described
as spiral) are also found, for instance in the wall of the alga
Glaucocystis and in the classical S_1, S_2 and S_3 layers of wood cells.

Objectives

Some examples of arced patterns in electron micrographs of
helicoidal support structures from animals are shown in Figs. 3 - 6.
In addition, similar patterns have been seen in examples of structural
materials from all major animal phyla;   for a list see Neville (1975).
Helicoids represent the most common type of regular extracellular
architecture in animals.   We suspect that they may be equally abundant
in plant cell walls.   Our aims are therefore to review the evidence for
the occurrence and prevalence of helicoids in plant cell wall structure,
together with its implications for current hypotheses of morphogenesis.

We will also consider effects of strain reorientation of microfibrils, caused by growth forces upon helicoidal walls, and mention relevant physicochemical and functional aspects.

Fig. 2. Diagrams to show the difference between a helicoid and a helix: both occur in plant cell walls.

Fig. 2A. Projection view of a sequence of individual microfibrils, arranged as an anti-clockwise helicoid.

Fig. 2B. Drawing of a helix.

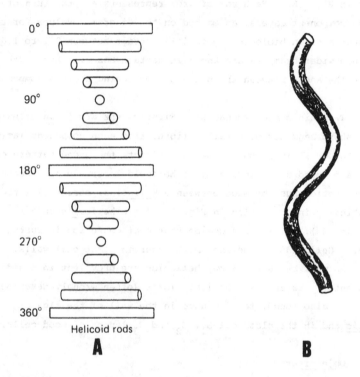

Helicoid rods

A

B

EVIDENCE FOR HELICOIDS IN PLANT CELL WALLS

Helicoidal structure was independently proposed for three different plant cell walls by three groups: by Roland et al. (1977) for Phaseolus hypocotyl, by Peng & Jaffe (1976) for Pelvetia eggs, and by Neville et al. (1976) for oospores of Chara.

Figs. 3 - 6. Electron micrographs of oblique sections through natural helicoids from animal materials.

Fig. 3. Exoskeletal cuticle of an adult termite.

Fig. 4. Oothecal protein from a praying mantis (Sphodromantis tenuidentata) from Neville & Luke (1971). This has been extracted in the form of a mobile cholesteric liquid crystal from the glands which have produced it. The protein was fixed before sectioning.

Fig. 5. Silk moth eggshell (Hyalophora cecropia) from Smith et al. (1971).

Fig. 6. Cod fish eggshell (Gadus morrhua) from Grierson & Neville (1981).

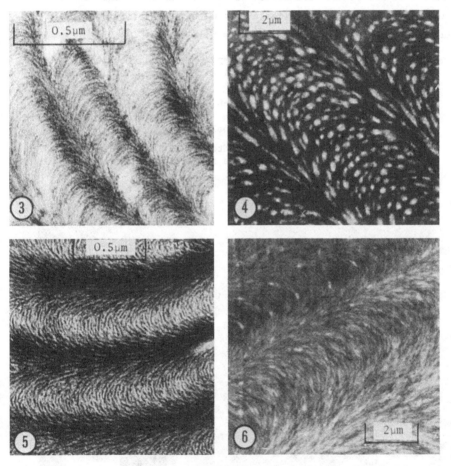

Nitella internode cells - a type example

We have subsequently begun work on the internode wall of
Nitella opaca (Neville & Levy 1984), inspired by an electron
micrograph published by Probine & Barber (1966).   This they interpreted
as having crossed polylamellate structure with the microfibril direction
in adjacent layers crossing orthogonally.   We suspected that it was
really helicoidal.   Oblique electron microscope sections of cell wall
show, when suitably stained, the characteristic arced pattern (Fig. 7A).
We interpret this as a helicoidal array of cellulose microfibrils set in
a matrix.   It has an anti-clockwise twist, i.e. the same sense of twist
as a left-handed corkscrew.   Nitella cell wall is therefore more
regularly structured than previous studies have shown.

Fig. 7.   Two electron micrographs of oblique sections
through Nitella internode cell wall to show the difference
produced by staining technique.

Fig. 7A.   stained in block with uranyl acetate, followed
by lead citrate on the section.   This reveals arced
patterning characteristic of helicoidal structure.

Fig. 7B.   Identical treatment to Fig. 7A but stained with
uranyl acetate on the sections rather than in the block.
No arced patterning is seen.

A   B

Nitella was chosen for detailed study for several reasons.

1. The internode cells are very large (up to 5 cm long), making it easy to control the plane of sectioning for electron microscopy. This is vital to the interpretation of the two-dimensional micrographs as three-dimensional architecture.

2. The cells elongate uniformly along their length (Green 1954).

3. The absence of neighbouring cells (except at the nodes) permits unrestrained growth.

4. It is known that cellulose microfibrils are deposited by apposition next to the plasma membrane (Ray 1967).

5. The cells are accessible to chemicals (e.g. colchicine) directly from the aquatic environment. We plan to take advantage of this in future experiments.

6. Nitella has helicoidal architecture like that of the cell walls in many higher plants. We therefore feel justified in using it as a model for morphogenetic studies. Much work has been done on other aquatic algae (e.g. Oocystis, Glaucocystis, Valonia and Chaetomorpha) but these do not have helicoidal morphology. Nitella therefore seems to be a better model.

Occurrence in plant taxa

As judged from the occurrence of arced patterns in electron micrographs, helicoids are found in those major plant taxa which have been examined (algae, stoneworts, fungi, moss, horsetail, fern, mono-cotyledons, dicotyledons, conifers). A selection of examples is listed in Table 1. A full list of almost fifty examples is given in Neville & Levy (1984). In the original literature, many of these have been wrongly interpreted, or sometimes not at all. We have therefore reinterpreted the micrographs of other workers. Most of the examples still need checking by the tilting test (see below) to confirm the presence of helicoidal structure.

Occurrence in various cell types

Helicoids have been found in a wide variety of cell types (parenchyma, collenchyma, sclerenchyma, tracheids and sclereids). Examples are illustrated in Fig. 8. They may occur in both primary and secondary walls, although in primary walls they may be transient. They are also found in the spherical walls of spores, oospores and zoospores.

ityead:

Table 1.   Selection of plants from a range of taxa for which electron micrographs show arced patterns.   These we interpret as deriving from helicoidally oriented microfibrils.   Asterisked examples have been checked by the goniometric tilting test described in the text.

GROUP	SPECIES	REFERENCE
Algae (Chlorophyceae)	Cylindrocapsa geminella	Hoffman & Hofmann (1975)
	Pithophora oedogonia	Pearlmutter & Lembi (1980)
Algae (Phaeophyceae)	Pelvetia fastigiata	Peng & Jaffe (1976)
Charophyta	Chara vulgaris*	Neville et al. (1976)
	Nitella opaca*	Neville & Levy (1984)
Fungi	Glomus epigaeus*	Bonfante-Fasolo & Vian (1984)
Bryophyta (Musci)	Rhacopilum tomentosum	Schneppe et al. (1978)
Sphenopsida	Equisetum hyemale	Sassen et al. (1981)
Filicopsida	Ceratopteris thalictroides	Sassen et al. (1981)
Spermatophyta (Angiospermae)	Zea mays	Satiat-Jeunemaître (1981)
	Juncus effusus*	Roland (1981)
	Vigna radiata*	Roland et al. (1982)
	Tilia platyphyllos*	Roland & Mosiniak (1983)
Spermatophyta (Gymnospermae)	Pseudotsuga menziesii	Parameswaran & Liese (1981)

Fig. 8.   Drawings of types of plant cells in whose
walls electron microscopy has revealed arced patterning
characteristic of helicoidal structure.

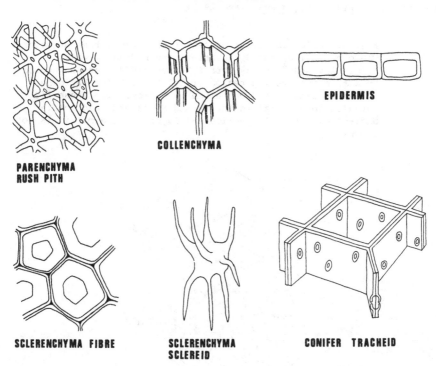

COLLENCHYMA

EPIDERMIS

**PARENCHYMA
RUSH PITH**

SCLERENCHYMA FIBRE

**SCLERENCHYMA
SCLEREID**

CONIFER   TRACHEID

Occurrence in wood

        We suggested that helicoids could form part of the structure
of cell walls in wood (Neville et al. 1976).   Since then several
convincing electron micrographs have been published, and a more
thorough search of the literature has revealed further data, which
although mostly not interpreted as such by the original authors,
support the idea that helicoids often form an integral part of wood
ultrastructure.   A unifying model for wood cell walls has recently been
proposed by Roland & Mosiniak (1983).

        Several papers indicate the existence of a small number of
transition layers between the classical S_1 and S_2, and between the S_2 and
S_3 layers of secondary walls (Fig. 9).   These were based initially on
replicas and layer stripping.   Transition layers were recorded in Pinus
radiata and Eucalyptus elaeophora (Wardrop & Harada 1965);   Picea

jezoensis (Harada 1965); latewood tracheids of Pinus palustris
(Dunning 1968).   Extra layers have also been inferred from birefringence
observations on Pinus virginiana (Tang 1973).

Fig. 9.   Diagrammatic representation of wood fine structure.
The classical S_1 and S_2 layers of the secondary wall,
originally described by Kerr & Bailey (1934), are connected
by a thin intervening region of helicoid.   The building
units are lath-shaped cellulose microfibrils.   These lie
mutually in parallel in the S_1 and S_2 layers, and rotate
anti-clockwise in the helicoidal layers.   The matrix is not
shown.

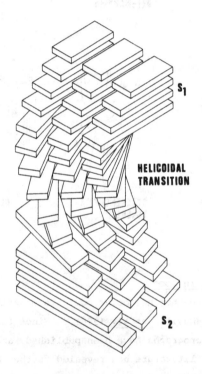

Subsequently, transmission electron microscopy has revealed
that the transitions are helicoidal.   Examples are xylem parenchyma in
Populus tremuloides (Chafe & Chauret 1974); wood vessels, fibres and
parenchyma in Tilia platyphyllos (Roland 1981; Roland & Mosiniak 1983);
sieve wall in Pinus strobus (Chafe & Doohan 1972); needle tracheids in
Pinus sylvestris (Liese 1965); xylem parenchyma in Cryptomeria japonica

(Chafe 1974);   tracheids in <u>Picea</u> <u>abies</u> (Parameswaran & Liese 1982).
So the main secondary (S) layers (which are helically wound around the
cell) are connected by helicoidal "universal plywood", as in Fig. 9.
Wood is a natural plywood.   It may be noted that helicoids are found
not only in softwoods but also in hardwoods (e.g. <u>Eucalyptus</u>, <u>Populus</u>
and <u>Tilia</u>).   In addition to being found in secondary walls of vessels
and fibres of hardwoods, and tracheids of softwoods, helicoids also
occur in parenchyma, which forms a significant volume of wood.

TECHNIQUES FOR REVEALING HELICOIDS

If helicoids are really so widespread in different plant
groups and cell types as the previous sections suggest, we may play
devil's advocate and ask why arced patterns are not seen in even more
electron micrographs?   There are several possible reasons.

1.   Even using the same stain, on the same material, the staining method
can either reveal or conceal an arced pattern in electron microscope
sections.   Thus, if stained with uranyl acetate in the block, followed
by lead citrate on the sections, arcs are clearly seen in <u>Nitella</u>
(Fig. 7A).   However, if both stains are applied on the sections, no
patterning is seen (Fig. 7B).

2.   The magnification may be too low to resolve arced patterning, but
may nevertheless reveal banding corresponding to 180 degree repeats in
the helicoid.

3.   Sections cut vertically to the component layers of a helicoid may
not show arced patterns.   The false impression may be given of a crossed
polylamellate wall with microfibrils alternately in TS and LS.   However,
when tilted with a goniometric stage about an axis in the plane and
parallel to the edge of the section, arced patterns appear.   The sense
of the arcs reverses with direction of tilt (Fig. 10) and this forms a
<u>critical test</u> for identifying a helicoidal system from a two-dimensional
section.   (In an obliquely cut section tilting is not necessary for
arced patterns to be seen).   Examples of tilted tests are asterisked in
Table 1.

4.   In primary cell walls, helicoidal structure may originally be
present, but may become dissipated by growth elongation forces (Roland
<u>et al</u>. 1982).   A sequence of sections taken at progressive stages of
cell growth may thus show arced patterns, then herring-bone patterning,
followed by an irregular microfibrillar appearance.

5.   In preparation of surface replicas, harsh chemical extraction
procedures are often employed, and this may lead to disarrangement of a
helicoidal array of microfibrils.

6.   Fibre X-ray diffraction of cell walls can give the false impression
that microfibrils are randomly distributed, if the beam is perpendicular
to the wall surface, because a ring diagram of reflections is then seen.

7.   Similarly, polarized light analysis of a helicoid in surface view
can give the misimpression that it has isotropic structure, because all
the component microfibril directions cancel each other.

Fig. 10.   Photographs of a model helicoid at various angles
of tilt.   Glass microscope slides have Letratone [C]
representing component planes of microfibrils.   Tilting
in opposite directions reverses the arcs, showing that they
really are patterns.

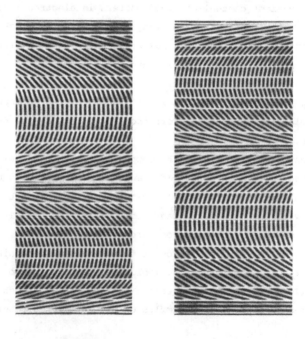

POSSIBLE IMPLICATIONS FOR WALL MORPHOGENESIS

It seems self evident that the cell-wall morphology must be fully understood before any speculation on morphogenetic mechanisms is attempted. We believe that, for the walls already mentioned, we have now reached that point. These walls (i.e. helicoids) seem to be structurally more complex than previously thought. Yet paradoxically that may give a clue as to how they are made.

This section will deal with mechanisms by which microfibrils are thought to be <u>originally positioned</u>. Any changes in these positions brought about by growth strains are considered in the next section. Bearing in mind that many cell walls are decidedly <u>not</u> helicoidal, the last section clearly shows that many are. We may then ask whether existing morphogenetic hypotheses can explain helicoid formation?   If not, do helicoids prompt any novel morphogenetic suggestions?

Ordered granule hypothesis

In his ordered granule hypothesis, Preston (1964) suggested that cellulose microfibril directions might be determined by the closely packed order of cellulose synthetase granules on the outer plasma membrane surface.   This mechanism could well account for certain specific angular changes between layers of microfibrils (e.g. 60°, 90°, 120°).   But it seems unlikely to contain sufficient information to produce the numerous (and systematic) changes in direction in a helicoid. In extreme cases these can change direction as often as every 6 degrees, e.g. in <u>Boergesenia</u> (calculated from Mizuta & Wada 1982).

Microtubule hypothesis

There has been much support for the microtubule hypothesis (Heath 1974).   Microtubules become oriented in specific directions just within the plasma membrane, and are thought to control the direction of extracellular cellulose microfibrils, by somehow acting across the membrane.   Thus microtubules would be expected to lie parallel to the most recently deposited layer of microfibrils.   Lloyd (1983) summarises compelling evidence that microtubules and cellulose microfibrils often lie in parallel, on either side of the cell plasma membrane.   Hardham <u>et al</u>. (1980) found a second group of microtubules oriented at an angle to the first group, in leaf primordia of <u>Graptopetalum paraguayense</u>.   They suggested that this second group predicted the direction of the next layer of microfibrils to be

synthesized.   Similar evidence has been found in elongating guard cells
of Phleum pratense (Palevitz 1981).   It seems unlikely that such a
mechanism could produce a helicoid, which would require revolution in
direction of microtubules, with associated depolymerization and
repolymerization between each of the numerous angular shifts.   This
point has already been realized by Robinson & Herzog (1977) in their
discussion of walls of Oocystis solitaria.   These normally have
orthogonally oriented layers of cellulose microfibrils, but they can be
made to deposit helicoidal walls by treatment with a fungicide, methyl-
benzimidazole-2yl-carbamate (MBC).

Multinet growth hypothesis

During the past few decades the most prominent hypothesis
for wall morphogenesis has been the multinet growth hypothesis of
Roelofsen & Houwink (1953).   This proposed that new cellulose
microfibrils are initially laid down more or less perpendicularly
(transversely) to the direction of cell elongation.   They subsequently
become reoriented by growth forces so as to end up more or less parallel
to the length of the cell.   Thus, the architecture neither begins nor
ends as a helicoidal structure.   Our results on Nitella show that
helicoidal structure is being produced at different stages of cell
elongation.   There is clearly no need for microfibrils to be laid down
transversely to permit cell elongation.   The multinet growth hypothesis
cannot be applied to helicoidal walls.

Terminal complex hypothesis

The freeze-fracture technique has revealed that microfibrils
may be associated with complementary structures in the cell membrane;
these are terminal complexes in the outer (E) face of the membrane, and
rosettes in the inner (P) face of the membrane.   Leading on from the
Singer & Nicolson (1972) fluid mosaic model for membrane structure and
lability, Mueller & Brown (1982a,b) proposed that microfibril
direction is controlled by these complexes moving in the plane of the
membrane, and depositing (and also orienting) microfibrils in their
wakes.   Again, this seems unlikely to give rise to a helicoid, as the
mechanism would need constantly to be changing direction.

Self-assembly hypothesis

None of the above hypotheses seem applicable to helicoid formation. (We do of course appreciate that they were not conceived with helicoids in mind, and that they may well explain morphogenesis of other types of cell wall). All of the hypotheses seem to delegate the problem of microfibril orientation control to the cortical region of the cytoplasm or to the plasma membrane. This does not, however, offer a solution to the direction change mechanism. Perhaps the main significance of the helicoidal model is that it leads in general to new ways of thinking about cell wall morphogenesis, and in particular to an extracellular approach involving self-assembly.

The architecture of helicoids closely resembles that of cholesteric liquid crystals. These represent a state of matter with properties intermediate between those of liquid and solid: they combine viscosity and flow together with some molecular order (revealed by birefringence and X-ray diffraction). Because of their mobility and also because they can self-assemble, they are prime candidates for the spontaneous formation of helicoids, which could later become stabilized. Cholesteric liquid crystalline structure (Fig. 6), and its self-assembly, has been demonstrated for the eggcase protein of praying mantids (Neville & Luke 1971), and the field of helicoidal self-assembling proteins which this created, has recently been reviewed (Neville 1981). Self-assembly was suggested for the formation of helicoids in insect cuticle (Neville & Luke 1969), for moth eggshell (Smith et al. 1971), and for fish eggshell (Grierson & Neville 1981).

That helicoidal plant cell walls might arise by self-assembly was first suggested by Neville et al. (1976). We envisage the spontaneous positioning of new microfibrils by self-assembly on to the inner surface of the previously deposited wall. The angle between consecutive layers of microfibrils would be determined by interaction of electrical forces. These could involve asymmetrical patterns of charges on the microfibril surfaces, and ionic species in the zone of apposition (periplasm). The idea is consistent with the demonstration by radioactive tracer studies that cellulose microfibril deposition occurs next to the cell plasma membrane (Ray 1967). By analogy with synthetic polymer systems, the percentage of cellulose (e.g. 17% in Nitella cell walls; Probine & Preston, 1961) lies in the appropriate range for self-assembling fibre-in-matrix composites.

We suggest that the cell is capable of three sorts of command.   Firstly, a general instruction to make helicoidal wall by self-assembly.   Secondly, a more specific instruction to orient several consecutive layers of microfibrils in one particular direction only;   this could perhaps involve microtubules.   Thirdly, an order to change from one specific direction to another via intervening helicoid (e.g. in wood, Fig. 9).

PHYSICOCHEMICAL ASPECTS OF SELF-ASSEMBLY

Self-ordering systems in biological materials are more economical in energy requirements than non-self-ordering structures, because they do not involve enzymatic control or the hydrolysis of energy-rich phosphate bonds in ATP.   They are particularly appropriate for building skeletal structures, most of which lie outside cells. Just as plant cell wall extensibility can be controlled from the cell within - for instance by auxin-regulated extracellular pH control (Rayle 1973), so self-assembly could be under similar remote control. It could be regulated from the cell by extracellular variation of such simple factors as concentration of hydrogen or other ions.   Cholesteric liquid crystals are in fact known to respond to only small alterations of pH and other ions, and also to concentration changes in small optically active molecules, to give a disproportionately large change in architecture (for a review see Neville 1981).

Hemicellulose has been extracted from cell walls of mung bean hypocotyls, and self-assembled in vitro into birefringent liquid crystalline nodules.   When fixed and sectioned, these showed the arced patterns characteristic of a helicoid (Roland et al. 1977).   Changes in pH and other ionic concentrations do in fact affect the capacity of the hemicellulose subunits to self-assemble (Vian 1982).   In cell walls it could be the hemicellulose matrix which self-assembles, moving the long cellulose microfibrils (to which they are bound in parallel by hydrogen bonds), into a helicoid.   It is interesting to note that hydroxypropyl cellulose can also self-assemble into a cholesteric liquid crystal (Werbowyj & Gray 1976).   It has the cholesteric properties of high angles of optical rotation, dispersing with wavelength, as shown for helicoidal insect cuticle (Neville & Caveney 1969).

It might be permissible to make some physicochemical predictions about the molecules involved.   We may note that both

hemicellulose and hydroxypropyl cellulose are long molecules with
sidechains.   It could be expected that the elements responsible for
helicoid formation in cell walls, are rod-shaped, asymmetrical,
optically active molecules.   Being long, they would probably have a
high radius of gyration in solution, and therefore a high viscosity
would be expected.   Consideration of the types of molecule which form
helicoidal composites (microfibrils set in a matrix), shows that many
sorts of macromolecular combinations occur (Table 2).   Detailed
chemistry seems likely to be overshadowed by molecular shape and surface
properties.

Table 2.   Examples of the chemical variety of helicoidal systems.
The chemistry of both the fibrous and the matrix parts of these
mechanical composites is variable.

HELICOIDAL COMPOSITE	EXAMPLE	REFERENCE
Protein fibres in protein matrix	Fish eggshell	Grierson & Neville (1981)
	Moth eggshell	Smith et al. (1971)
	Mantis eggcase	Neville & Luke (1971)
Protein fibres in polysaccharide matrix	Eelworm cyst	Shepherd et al. (1972)
	Chicken cornea	Coulombre & Coulombre (1975)
Polysaccharide fibres in protein matrix	Insect cuticle	Neville (1975)
	Crab cuticle	Bouligand (1965)
Polysaccharide fibres in polysaccharide matrix	Plant cell wall	Neville et al. (1976)

STRAIN REORIENTATION OF HELICOIDS
        The initial microfibrillar framework in many plant cell walls
becomes reoriented by strains set up during cell enlargement.   In this
section we attempt to suggest how growth strains may influence the
patterns seen in electron micrographs of wall sections.   It will emerge
that our suggestions owe much to the thinking of previous workers; but
our resulting patterns are different because we are considering strains
acting on a helicoid, rather than upon the framework implicated in the
multinet growth hypothesis.
        Thus we adopt several existing observations and concepts as
follows.

1.   New microfibrils are added only at the inner wall surface, as was established by radioactive labelling (Ray 1967) and by polarized light analysis of <u>Nitella</u> (Gertel & Green 1977).

2.   Once deposited, microfibrils are passively reoriented according to their strain environment, which changes with time.   This has also been confirmed by polarizing observations (Gertel & Green 1977).

3.   Cellulose microfibrils are inelastic, and so are supposed to slide with respect to each other through the matrix, rather than individually extend.

4.   Probine & Preston (1961) deduced that there is a strain gradient acting across the wall, with the greatest strains (and therefore largest reorientation of microfibrils) located in the outermost 10 to 20% of wall thickness.   We have replotted this strain gradient and show it in relation to an originally unstrained representation of a helicoid (Fig. 11).   The strain gradient has been applied to the multinet growth concept by Preston (1982), in a re-examination of passive reorientation.

Fig. 11.   Graph of strain gradient across cell wall, replotted from Probine & Preston (1961).   An unstrained helicoid is shown below, represented as in Fig. 2A. The outermost microfibrils will change direction by the greatest amount (as in Fig. 12) because they are subjected to the highest strains.

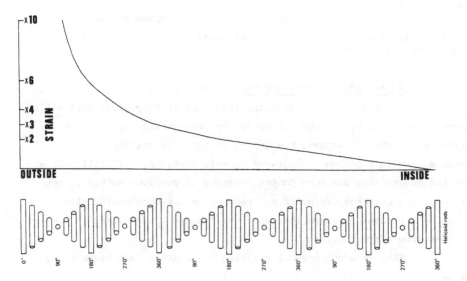

Inspired by Preston's paper, we have applied his mathematical
approach to a helicoid (as opposed to multinet structure), and have
developed a computer programme to plot in detail the effects that a
strain gradient would be expected to have upon the patterns seen in
oblique sections subsequently cut through it.   Some results, plotted
by computer graphics, are shown in Fig. 12.   To make the results more
realistic, we have chosen initial angles (between successive microfibrillar planes) to match typical angles we have calculated for various
species, from electron micrographs in the literature.   Thus, Fig. 12 A-C
(7°) relates to Boergesenia forbesii (Mizuta & Wada 1982); Fig. 12 D-F
(22°) relates to Cymodocea rotundata (Doohan & Newcomb 1976); and Fig.
12 G-I (38°) relates to Equisetum hyemale (Emons & Wolters-Arts 1983).
Both the register of arcs, and their appearance, vary with direction of
sectioning.   Sections cut at right angles to each other (i.e. parallel
to the main strain axis versus across strain axis) have resulting patterns
which are out of phase by $\emptyset$ = 90°.   In each of the examples, an
unstrained pattern is shown first, followed by a section parallel to
strain axis, and finally a section cut across the strain axis.   Fig. 12
shows that the high strains in the outer part of a wall can convert the
arced pattern of an unstrained helicoid into a herring-boned type of
pattern.   The latter is seen in "crossed-polylamellate" walls.   We
suggest that strain acting on a helicoid is one possible origin of
herring-boned pattern, as was deduced empirically by Sargent (1978).
Reis et al. (1982) have presented electron micrograph evidence for the
dissipation of arced patterns in primary wall of mung-bean hypocotyls
(Vigna radiata).   During progressive stages of swelling induced by
ethylene (triggered by treatment with ethrel), the arcs changed first
to herring bones, and then dispersed.   Other papers also include
electron micrographs showing tendency to herring bone appearance in the
outermost layers, whereas the inner layers show undistorted arcs.
Examples are Pisum sativum (Appelbaum & Burg 1971; Lang et al. 1982),
Pelvetia fastigiata (Peng & Jaffe 1976), Cymodocea rotundata (Doohan &
Newcomb 1976), and Pithophora oedogonia (Pearlmutter & Lembi 1978).

FUNCTIONAL ASPECTS

Accurate knowledge of microfibril orientations is important,
as it affects our understanding of the mechanical properties of plant
walls and of plant materials (e.g. wood and textiles).   X-ray

Fig. 12.   Computer graphics plots of patterns predicted in
sections of strained helicoidal cell walls.   The strain
axis is vertical.   The inside of the cell wall (youngest
and therefore least strained) is on the left of each section;
the outside (oldest and most strained) part is on the right
of each section.

A,B,C represent sections of helicoid with an initial angle
of 7 degrees between component planes of microfibrils, such
as occurs in <u>Boergesenia</u> (Mizuta & Wada 1982).   A is
unstrained.   B is sectioned parallel to the strain axis.
C is sectioned perpendicularly to the strain axis.

D,E,F represent sections of helicoid with an initial angle
of 22 degrees, such as occurs in <u>Cymodocea</u> (Doohan & Newcomb
1976).   D is unstrained, E is cut parallel to the strain axis.
F is cut perpendicularly to the strain axis.

G,H,I represent sections of helicoid with an initial angle
of 38 degrees, such as occurs in <u>Equisetum</u> (Emons & Wolters-
Arts, 1983).   G is unstrained.   H is cut parallel to the
strain axis.   I is cut perpendicularly to the strain axis.

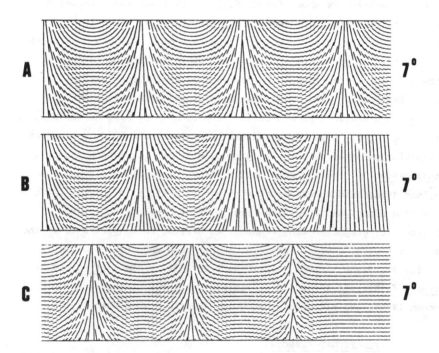

A 7°

B 7°

C 7°

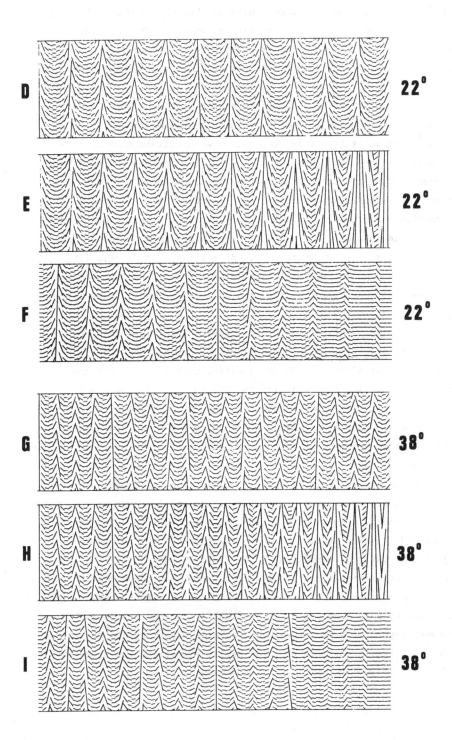

diffraction studies on a helicoid, with the beam passing perpendicularly through its component layers, give ring diagrams which can be misleading. They obscure the true degree of crystallinity of the cellulose microfibrils, because the technique merely gives a composite picture of all the many directions present.   High crystallinity is mechanically important, and accounts for the high Young's modulus for cellulose ($100GNm^{-2}$).   By contrast, the low modulus of the matrix (0.2 to $6.9GNm^{-2}$) allows it to act in combination with cellulose as an effective composite material like fibreglass.   Helicoidal structure represents an improvement upon the basic composite material principle of rods in a matrix.   As a building element it is more versatile than having all of the cellulose microfibrils oriented in parallel.   The latter would tend to split axially and fail to resist shear forces.   A helicoidal ply combines a balance of strength in all directions, with flexibility.   In long cells, the walls form mechanically well-adapted tubular helicoids, with some preference for orientation along the cell axis to resist bending forces.

Vian (1982) makes the interesting suggestion that the outermost tissues of plants possess more layers and more order in their walls than inner tissues.   We would agree, noting especially the occurrence of peripheral helicoids in epidermis, collenchyma and sclerenchyma.   This would strengthen organs by making use of the principle of tubular furniture.   The highest strains are on the outer parts of the structures, so this is where the cells which are specialized for mechanical support should be.   (Rush pith appears to be an exception, being central and yet helicoidal).   During plant development, collenchyma represents the first formed supporting tissue and is known in some cases to be helicoidal.   This would make it strong enough to withstand the bending and twisting strains to which young plants are constantly subjected by external conditions.   Helicoidal construction is also ideal for the spherical shells of spores, which may be subject to forces acting from any direction.   We note the parallel with the helicoidal walls of many animal eggshells.   The realization that the major secondary layers of wood fibres and tracheids may be connected by thin intervening regions of helicoid, may have implications for those involved in detailed theoretical studies on wood mechanics.

REFERENCES

Appelbaum, A. & Burg, S.P. (1971). Altered cell microfibrillar
    orientation in ethylene-treated Pisum sativum stems.
    Plant Physiology, 48, 648-652.
Bonfante-Fasolo, P. & Vian, B. (1984). Wall texture in the spore of a
    vesicular-arbuscular mycorrhizal fungus. Protoplasma, 120,
    51-60.
Bouligand, Y. (1965). Sur une disposition fibrillaire torsadée commune
    à plusieurs structures biologiques. Comptes Rendues de
    l'Acadamie des Sciences, Paris, 261, 4864-4867.
Chafe, S.C. (1974). Cell wall structure in the xylem parenchyma of
    Cryptomeria. Protoplasma, 81, 63-76.
Chafe, S.C. & Chauret, G. (1974). Cell wall structure in the xylem
    parenchyma of trembling aspen. Protoplasma, 80, 129-147.
Chafe, S.C. & Doohan, M.E. (1972). Observations on the ultrastructure
    of the thickened sieve cell wall in Pinus strobus L.
    Protoplasma, 75, 67-78.
Coulombre, J. & Coulombre, A. (1975). Corneal development. V. Treatment
    of five day old embryos of domestic fowl with 6-Diazo-5-oxo-
    L-norleucine (DON). Developmental Biology, 45, 291-303.
Doohan, M.E. & Newcomb, E.H. (1976). Leaf ultrastructure and δ^{13} values
    of three sea grasses from the Great Barrier Reef.
    Australian Journal of Plant Physiology, 3, 9-23.
Dunning, C.E. (1968). Cell wall morphology of long leaf pine latewood.
    Wood Science, 1, 65-76.
Emons, A.M.C. & Wolters-Arts, A.M.C. (1983). Cortical microtubules and
    microfibril deposition in the cell wall of root hairs of
    Equisetum hyemale. Protoplasma, 117, 68-81.
Gertel, E.T. & Green, P.B. (1977). Cell growth pattern and wall micro-
    fibrillar arrangement. Plant Physiology, 60, 247-254.
Green, P.B. (1954). The spiral growth pattern of the cell wall in
    Nitella axillaris. American Journal of Botany, 41, 403-409.
Grierson, J.P. & Neville, A.C. (1981). Helicoidal architecture of fish
    eggshell. Tissue & Cell, 13, 819-830.
Harada, H. (1965). Ultrastructure and organization of gymnosperm cell
    walls. In Cellular Ultrastructure of Woody Plants, ed.
    W.A. Côté, pp. 215-233. Syracuse: Syracuse University Press.
Hardham, A.R., Green, P.B. & Lang, J.M. (1980). Reorganization of
    cortical microtubules and cellulose deposition during leaf
    formation in Graptopetalum paraguayense. Planta, 149, 181-195.
Heath, I.B. (1974). A unified hypothesis for the role of membrane bound
    enzyme complexes and microtubules in plant cell wall
    synthesis. Journal of Theoretical Biology, 48, 445-449.
Hoffman, L.R. & Hofmann, C.S. (1975). Zoospore formation in Cylindrocapsa.
    Canadian Journal of Botany, 53, 439-451.
Kerr, T. & Bailey, I.W. (1934). The cambium and its derivative tissues.
    X. Structure, optical properties and chemical composition of
    the so-called middle lamella. Journal of the Arnold
    Arboretum, 15, 327-349.
Lang, J.M., Eisinger, W.R. & Green, P.B. (1982). Effects of ethylene on
    the orientation of microtubules and cellulose microfibrils
    of Pea epicotyl cells with polylamellate cell walls.
    Protoplasma, 110, 5-14.
Liese, W. (1965). The warty layer. In Cellular Ultrastructure of Woody
    Plants, ed. W.A. Côté, pp. 251-269. Syracuse: Syracuse
    University Press.

Lloyd, C.W. (1983). Toward a dynamic helical model for the influence of
    microtubules on wall patterns in plants. International Review
    of Cytology, 86, 1-51.
Mizuta, S. & Wada, S. (1982). Effects of light and inhibitors on
    polylamellation and shift of microfibril orientation in
    Boergesenia cell wall. Plant & Cell Physiology, 23, 257-264.
Mueller, S.C. & Brown, R.M. (1982a). The control of cellulose microfibril
    deposition in the cell wall of higher plants. I. Can
    directed membrane flow orient cellulose microfibrils?
    Indirect evidence from freeze-fractured plasma membranes
    of maize and pine seedlings. Planta, 154, 489-500.
Mueller, S.C. & Brown, R.M. (1982b). The control of microfibril
    deposition in the cell wall of higher plants. II. Freeze-
    fracture microfibril patterns in maize seedling tissues
    following experimental alteration with colchicine and
    ethylene. Planta, 154, 501-515.
Neville, A.C. (1975). Biology of the Arthropod Cuticle, pp. 1-448.
    Berlin: Springer-Verlag.
Neville, A.C. (1981). Cholesteric proteins. Molecular Crystals and
    Liquid Crystals, 76, 279-286.
Neville, A.C. & Caveney, S. (1969). Scarabaeid beetle exocuticle as an
    optical analogue of cholesteric liquid crystals.
    Biological Reviews, 44, 531-562.
Neville, A.C., Gubb, D.C. & Crawford, R.M. (1976). A new model for
    cellulose architecture in some plant cell walls.
    Protoplasma, 90, 307-317.
Neville, A.C. & Levy, S. (1984). Helicoidal orientation of cellulose
    microfibrils in Nitella opaca internode cells: ultrastructure
    and computed theoretical effects of strain reorientation
    during wall growth. Planta, 162, 370-384.
Neville, A.C. & Luke, B.M. (1969). Molecular architecture of adult
    locust cuticle at the electron microscope level. Tissue &
    Cell, 1, 355-366.
Neville, A.C. & Luke, B.M. (1971).A biological system producing a self-
    assembling cholesteric protein liquid crystal. Journal of
    Cell Science, 8, 93-109.
Palevitz, B.A. (1981). The structure and development of stomatal cells.
    In Stomatal Physiology, ed. P.G. Jarvis & T.A. Mansfield.
    Society for Experimental Biology Seminar Series, 8, 1-23.
    Cambridge: Cambridge University Press.
Parameswaran, N. & Liese, W. (1981). Occurrence and structure of
    polylamellate walls in some lignified cells. In Cell Walls
    '81, ed. D.G. Robinson & H. Quader. Proceedings of the
    Second Cell Wall Meeting Göttingen, pp. 171-188. Stuttgart:
    Wissenschaftliche Verlagsgesellschaft.
Parameswaran, N. & Liese, W. (1982). Ultrastructural localization of wall
    components in wood cells. Holz als Roh und Werkstoff, 40,
    145-155.
Pearlmutter, N.L. & Lembi, C.A. (1978). Localization of chitin in algal
    and fungal cell walls by light and electron microscopy.
    Journal of Histochemistry and Cytochemistry, 26, 782-791.
Pearlmutter, N.L. & Lembi, C.A. (1980). Structure and composition of
    Pithophora oedogonia (Chlorophyta) cell walls. Journal of
    Phycology, 16, 602-616.
Peng, H.B. & Jaffe, L.F. (1976). Cell wall formation in Pelvetia embryos.
    A freeze fracture study. Planta, 133, 57-71.

Preston, R.D. (1964). Structural plant polysaccharides. Endeavour, 23,
     153-159.
Preston, R.D. (1982). The case for multinet growth in growing walls of
     plant cells. Planta, 155, 356-363.
Probine, M.C. & Barber, N.F. (1966). The structure and plastic properties
     of the cell wall of Nitella in relation to extension growth.
     Australian Journal of Biological Sciences, 19, 439-457.
Probine, M.C. & Preston, R.D. (1961). Cell growth and the structure of
     mechanical properties of the cell in internodal cells of
     Nitella opaca. Journal of Experimental Botany, 12, 261-282.
Ray, P.M. (1967). Radioautographic study of cell wall deposition in
     growing plant cells. Journal of Cell Biology, 35, 659-674.
Rayle, D.L. (1973). Auxin-induced hydrogen ion secretion in Avena
     coleoptiles and its implications. Planta, 114, 63-73.
Reis, D., Mosiniak, M., Vian, B. & Roland, J.C. (1982). Cell walls and
     cell shape. Changes in texture correlated with an ethylene-
     induced swelling. Annales des Sciences Naturelles,
     Botanique et Biologie Vegetale, Paris, 4, 115-133.
Robinson, D.G. & Herzog, W. (1977). Structure, synthesis and orientation
     of microfibrils. III. A survey of the action of
     microtubule inhibitors on microtubules and microfibril
     orientation in Oocystis solitaria. Cytobiologie, 15, 463-474.
Roelofsen, P.A. & Houwink, A.L. (1953). Architecture and growth of the
     primary wall in some plant hairs and in the Phycomyces
     sporangiophore. Acta Botanica Neerlandica, 2, 218-225.
Roland, J.C. (1981). Comparison of arced patterns in growing and non-
     growing polylamellate cell walls of higher plants. In Cell
     Walls '81, eds. D.G. Robinson & H. Quader, pp. 162-170.
     Stuttgart: Wissenschaftliche Verglagsgesellschaft.
Roland, J.C. & Mosiniak, M. (1983). On the twisting pattern, texture and
     layering of the secondary cell walls of limewood. Proposal
     of an unifying model. International Association of Wood
     Anatomists Bulletin, 4, 15-26.
Roland, J.C., Reis, D., Mosiniak, M. & Vian, B. (1982). Cell wall texture
     along the growth gradient of the mung bean hypocotyl:
     ordered assembly and dissipative processes. Journal of Cell
     Science, 56, 303-318.
Roland, J.C., Vian, B. & Reis, D. (1977). Further observations on cell
     wall morphogenesis and polysaccharide arrangement during
     plant growth. Protoplasma, 91, 125-141.
Sargent, C. (1978). Differentiation of the cross-fibrillar outer
     epidermal wall during extension growth in Hordeum vulgare L.
     Protoplasma, 95, 309-320.
Sassen, M.M.A., Pluymaekers, H.J., Meekes, H. Th.H.M., De Jong-Emons,
     A.M.C. (1981). Cell wall texture in root hairs. In Cell Walls
     '81. Proceedings of the Second Cell Wall Meeting, Göttingen,
     eds. D.G. Robinson & H. Quader, pp. 189-197. Stuttgart:
     Wissenschaftliche Verlagsgesellschaft.
Satiat-Jeunemaitre, B. (1981). Texture et croissance des parois des deux
     épidermes du coléoptile de mäis. Annales des Sciences
     Naturelles, Botanique et Biologie Vegetale, Paris, 13, 163-
     176.
Schneppe, E., Stein, U., Deichgräber, G. (1978). Structure, function and
     development of the peristome of the moss, Rhacopilum
     tomentosum, with special reference to the problem of
     microfibril orientation by microtubules. Protoplasma, 97,
     221-240.

Shepherd, A.M., Clark, S.A. & Dart, P.J. (1972). Cuticle structure in the
       genus Heterodera. Nematologica, 18, 1-17.
Singer, S.J. & Nicolson, G.L. (1972). The fluid mosaic model of the
       structure of cell membranes. Science, 175, 720-730.
Smith, D.S., Telfer, W.H. & Neville, A.C. (1971). Fine structure of the
       chorion of a moth, Hyalophora cecropia. Tissue & Cell, 3,
       477-498.
Tang, R.C. (1973). The microfibrillar orientation in cell wall layers of
       Virginia pine tracheids. Wood Science, 5, 181-186.
Vian, B. (1982). Organized microfibril assembly in higher plant cells.
       In Cellulose and Other Natural Polymer Systems. Biogenesis,
       Structure and Degradation, ed. R.M. Brown, pp. 23-43.
       New York: Plenum.
Wardrop, A.B. & Harada, H. (1965). The formation and structure of the
       cell wall in fibres and tracheids. Journal of Experimental
       Botany, 16, 356-371.
Werbowyj, R.S. & Gray, D.G. (1976). Liquid crystalline structure in
       aqueous hydroxypropylcellulose solutions. Molecular
       Crystals and Liquid Crystals, 34, 97-103.

K. Roberts
John Innes Institute, Colney Lane, Norwich NR4 7UH, England

J. Phillips
John Innes Institute, Colney Lane, Norwich NR4 7UH, England

P. Shaw
John Innes Institute, Colney Lane, Norwich NR4 7UH, England

C. Grief
John Innes Institute, Colney Lane, Norwich NR4 7UH, England

E. Smith
John Innes Institute, Colney Lane, Norwich NR4 7UH, England
(Present address: School of Biological Sciences,
Macquarie University, N.S.W., Australia

Abstract. The structural basis of complexity in the plant
cell wall is discussed, and the importance of short cell-
wall oligosaccharides in a variety of functions emphasized.
The availability of probes for cell-wall components, in
particular oligosaccharides, is then examined, and it is
suggested that antibodies, and in particular monoclonal
antibodies, form ideal probes for both structural and
functional studies with plant cell walls.   Chlamydomonas
cell-wall glycoproteins are introduced as a well-
characterized model system, and the production of a
polyclonal antiserum against the major structural
glycopeptide is described.   Immunogold labelling studies
reveal the pathway of synthesis and transport from the
Golgi apparatus to the cell wall.   A family of monoclonal
antibodies has been raised against the cell-wall
glycoprotein 2BII, of which some appear to recognize
oligosaccharide antigens.   Two monoclonal antibodies, by
both LM and EM immunocytochemistry, are used to reveal
different developmental expression of their respective
antigens.

Key words: Immunology; cell wall; glycoproteins;
oligosaccharides; Chlamydomonas; monoclonal antibodies;
immunocytochemistry.

INTRODUCTION

In very broad terms there are two classes of question we can
ask about plant cell walls. We can demand, for example, to know in
detail about the nature of the polymers that go to make up the wall, how
they are made, how they are assembled and cross linked, how they turn
over and how they are modified during development. But alongside this
essential and detailed chemical approach, the instructive results of

which are seen in many of the chapters of this book, goes another class
of question which is harder to define.   We can ask about the degree and
significance of microheterogeneity within wall polymers, the extent to
which subtle modifications of sugars by, for example, acetylation,
methylation and sulphation, may modify the functional properties of the
polymers containing them.   We can ask about the spatial distribution of
particular polymers, oligosaccharides and glycoproteins within a wall
and within tissues, between primary walls and secondary walls, about the
sites of their synthesis and, perhaps most importantly, about the
biological function of the bewildering variety of chemical structures
found in the plant cell wall.

Compared with the increasing chemical literature in answer
to the first class of questions, answers to the second sort of problem
are relatively few.   With the major exception of the work of
Albersheim's group, who have consistently drawn our attention to the
important and wide ranging number of biological effects which can be
elicited by short cell-wall oligosaccharides, remarkably little
attention has been given to such functional considerations, and even less
to spatial questions.   There are several reasons for this imbalance in
the literature, but one significant one has been the lack of highly
specific probes for cell-wall components, such as short oligosaccharide
sequences, with which one can probe their function and distribution.   In
this chapter we outline our contention that antibodies, and in particular
monoclonal antibodies, against short sugar sequences hold out the
promise of providing us with the appropriate specificity and we
demonstrate that such probes can indeed be made, and can also provide us
with data on the location and function of wall oligosaccharide antigens.
But first let us look very briefly at the nature of the complexity we
find in the plant cell wall and some of the functions this complexity
might perform.

THE PLANT CELL WALL SHOWS MANIFEST COMPLEXITY

Plant cells live in remarkably complex little boxes.   This
extracellular matrix in the young growing cell is called the primary
cell wall, and needs to display properties of cohesiveness while at the
same time allowing for controlled expansion during growth.   As the cell
stops growing, the nature of the extracellular matrix changes as a
function of the mature cell type and becomes the essentially non-

extensible secondary cell wall. Most of the molecules that make up the
cell wall are high-molecular-weight polymers; mostly polysaccharides,
but also proteins and glycoproteins (Lamport & Catt 1981).   The
polysaccharides have traditionally been subdivided, both chemically and
enzymatically (Selvendran 1985), into 3 classes: cellulose,
hemicellulose and pectin.   The last two classes show enormous chemical
complexity, and detailed structures for only some of the constituents
have been described (see review by McNeil et al. 1984).   When one takes
into consideration the wealth of basic sugar building blocks used in
cell-wall construction, and bears in mind that they may exist as
pentoses and hexoses, have carbon 1 in the α or β configuration, and be
linked through a variety of other carbon atoms to each other and that,
unlike proteins, complex branched linkages can be made, it is not
surprising that it has been claimed "Recent research has shown that the
structures of many of the cell-wall polysaccharides are so complex that
even the most sophisticated technologies available today may not be
capable of completely delineating the primary structure of these polymers."
(McNeil et al. 1984).   Structural complexity in wall architecture,
however, does not end with sugar linkages.   Many sugars both in
polysaccharides and in glycoproteins are modified by other substituents.
Thus, sugar hydroxyl groups can be 0-acetylated, 0-methylated,
0-sulphated, or phosphorylated and carboxylic acids can be esterified.
Such modifications alter the structural properties and the net charge of
the polymers and, in turn, their biological properties.   For example,
the degree of methylesterification of pectin is developmentally
regulated (Northcote 1972) and this will affect the net charge and
cation binding properties of the wall (Aspinall 1980).   The degree of
pectin methylation has been suggested as a determining factor of
Agrobacterium adherence to its host cell wall (Rao et al. 1982).
Sugar sulphation in extracellular matrix glycoproteins was first found
in members of the Volvocales (Roberts et al. 1980).   One specific such
glycoprotein in Volvox has been found which appears to correlate with
the developmental program during embryogenesis (Wenzl & Sumper 1981).
This suggests a role for sulphated cell-surface glycoproteins as
functional signals during embryonic development.   In addition to
glycoproteins, sulphated polysaccharides are also found in the algae.
In Fucus, sulphated fucose-containing polysaccharides are located at a
specific point in the egg cell wall and are thought to be a major

determinant in the fixation of cell polarity during early development
(Quatrano et al. 1979).   In the Volvocales, but this time in
Chlamydomonas eugametos, it has been suggested that specific families of
sugars in the gamete cell-surface glycoconjugates are methylated
(mannose and glucose in the '-' mating strain and xylose and galactose
in the '+' mating strain, for example).   The evidence (Gerwig et al.
1984) needs to be extended, but the possibility that modified sugars may
provide the mechanism for ensuring specific cell-cell recognition is an
exciting one.

Further complexity within the cell wall is introduced when
the tertiary and quaternary structures of the wall polymers are taken
into account.   Besides the covalently bonded subunits, extensive
structure is imposed by non-covalent interactions, in particular
hydrogen bonding and salt bridges (Rees 1977).   Non-carbohydrate
linkages are also present, connecting certain categories of wall polymer.
Phenolic cross linkages are found both in pectin   (Fry 1982a) and cell-
wall glycoproteins (Fry 1982b) and the degree of cross linking
profoundly affects the cohesiveness of the wall.

SPECIFIC SUGAR SEQUENCES MAY MEDIATE BIOLOGICAL FUNCTIONS
The brief description above really only touches on the
enormously complex chemical and structural features of cell-wall
architecture, and does not even take into consideration the profound
differences between monocotyledonous and dicotyledonous plants, but the
real question that is apparent, even at this level of analysis, is:
what is all this complexity for?   Our first answer to this question
would probably relate the diversity of polymers and cross linkages to
the very varied structural requirements of the cell wall both in
different cell types and at different developmental stages, and
certainly this is one of the major reasons for the observed complexity.
Another answer, however, has been emerging over the last few years which
says that the specific information contained in a short oligosaccharide
sequence can be used by plants as a cell signalling mechanism.   This
idea has been experimentally tested by Albersheim and his group who have
identified a wide range of biological functions that can be elicitied
or mediated by short oligosaccharides released either chemically or
enzymically from the plant cell wall (Albersheim et al. 1983; Darvill &
Albersheim 1984).   These include different pectic fragments which can

elicit phytoalexin production in soybean, inhibit flowering in Lemna and
kill plant cells in culture, and a xyloglucan (York et al. 1984) which
inhibits 2,4D-induced growth in pea epicotyls (McNeil et al. 1984).
Other examples have been described in which short oligosaccharides, either
released from wall polysaccharides or as the side chains of cell-surface
glycoproteins, have informational potential.   Each example presumably
requires a specific receptor molecule, either another cell-wall
component or a plasma-membrane component, which can discriminate between
one sugar sequence and another that is closely related (Yoshikawa et al.
1983).   A cell-wall oligosaccharide, capable of inducing a powerful
proteinase inhibitor in tomato leaves, has been described by Bishop et al.
(1981) and the role of cell-wall constituents in the binding or
attachment of bacteria to their host cell wall has been reviewed by
Whatley and Sequeira (1981).

An important area in which specificity in a cell-cell
recognition event is of crucial importance is in sexual mating events.
In higher plants this has been mostly explored in relation to self-
incompatibility systems, and both in Prunus (Mau et al. 1982) and in
Brassica (Ferrari et al. 1981) a cell-surface glycoprotein has been
implicated in the important recognition event (reviewed by Harris et al.
1984).   In lower plants also it appears that glycoproteins and in
particular their specific oligosaccharide side chains, are involved in
mating specificity.   In both Chlamydomonas reinhardii (Adair et al.
1982; Cooper et al. 1983) and in Chlamydomonas eugametos (Musgrave et al.
1981; Lens et al. 1983) a minor extrinsic glycoprotein component of the
flagellar membrane has been identified as the sexual agglutination
factor.   Like all the extracellular glycoproteins of Chlamydomonas that
have been looked at (Roberts & Phillips, unpublished results), the
agglutinin is a hydroxyproline-rich glycoprotein (Cooper et al. 1983)
related in many ways to the structural glycoproteins of the cell wall
(Catt et al. 1976).   These cell-wall glycoproteins depend on their
specific oligosaccharide side chains for self assembly, into an
ordered wall structure, in itself a (self) recognition event (Hills et al.
1975).

We have seen then, that in a wide variety of examples, short
sugar sequences within the extracellular matrix of plants can perform
important biological functions and can exhibit a precise informational
content.   One of the problems that is holding up rapid progress in this

field is the difficulty of isolating, in pure form, many of these informational oligosaccharides.   One way round this problem would be if we had access to specific probes which could distinguish one oligosaccharide from another closely related one.

ANTIBODIES ARE PROBES THAT WILL RECOGNISE SPECIFIC SUGAR SEQUENCES

In principle, any molecule exhibiting specificity for sugars, either singly or as oligosaccharides, could be used as a probe for their location and function.   In practice, only antibodies have turned out to have sufficiently varied and specific binding properties to cope with complex oligosaccharides, although useful results have been obtained with a variety of other molecular probes such as enzymes and lectins.

Enzymes can theoretically be used as probes for the sugars that form their substrates and cell-wall degrading enzymes have indeed been used to locate such substrates.   By coupling a purified xylanase to colloidal gold particles it was possible to localize xylans to secondary cell walls in thin sections of young wood from lime trees (Vian et al. 1983).

Lectins are proteins or glycoproteins that show specific binding for sugars.   Although their specificity is relatively low, and they will bind single sugar residues, they have been used to some effect to explore the distribution and functions of such determinants (Knox & Clarke 1978).   Fluorescently labelled or gold-labelled lectins, for example, have been recruited as probes for localizing cell-surface architecture in a variety of algae as a function of the developmental status of the cell (Millikin & Weiss 1984; Roberts et al. 1982; Sengbusch et al. 1982).   Cell-surface glycoproteins were shown to be important in various developmental steps in Volvox by blocking experiments using the lectin Concanavalin A (Kurn 1981).

It is with antibodies, however, that really precise specificity can be exploited.   It was found long ago that bacterial cell-surface carbohydrates are intimately involved in our immune response to infection.   Immunospecificity has been found to depend not only on short sugar sequences such as a terminal glucuronic acid residue or a D-GlcUA; L-Rha disaccharide (Pazur et al. 1982) but also on the conformation of the carbohydrate antigen (Jennings et al. 1981). Specificity is very exact and antibodies can even distinguish between D

and L isomers of a sugar hapten.   Experiments with fractionated dextran
fragments suggest that the upper size limit for a carbohydrate antigen is
about 5 or 6 sugar residues (Kabat 1966) and other well-characterized
polysaccharide antigens are in agreement with this;   for example a mouse
lgA has been described which binds to a tetrasaccharide, 1,6 linked
β-D-galactotetraose (Glaudemans 1975).   In this respect antibodies
provide an extension of the specificity provided by lectins both in
terms of the size and range of oligosaccharides recognized (Feizi 1981).
Despite the advantages offered by an immunological approach to the
complexity of the plant cell wall, very few higher-plant cell
biologists have used this strategy.   In contrast, fungal cell walls
have long been recognised as very antigenic and considerable work has
gone into characterizing the mannans involved.   One of the immunodominant
determinants is a phosphorylated mannopentaose with an (α1-3)-linked
nonreducing terminal (Hamada et al. 1981).

There are two noteworthy examples of the use of cell-wall
antibodies in higher plants.   The first used antibodies to corn-seedling
cell-wall protein and found that the antiserum inhibited coleoptile
growth by 35%.   The authors (Huber & Nevins 1981) interpret the results
in terms of the importance of the mobility of cell-wall autolytic
enzymes in wall extension.   The second used an unusual and potentially
very powerful approach.   By raising antisera to a synthetic nonapeptide
(ser-pro-ser-pro$_4$-ser-pro) it was possible to detect the unhydroxylated
and unglycosylated cytoplasmic precursor of the major cell wall
glycoprotein in carrots (Smith 1981).   Antisera have also been raised
against non-structural components of the plant extracellular matrix.
Gleeson & Clarke (1980), using antibodies to the arabinogalactan-protein
of Gladiolus style mucilage, showed that the major antigenic determinant
was probably the short side-branches of 1-6-linked β galactose residues
bearing a terminal α-L-arabinose residue.   This approach to pollen-
stigma interaction has since been expanded (Harris et al. 1984).
This example illustrates the generalization that antibodies raised
against glycoproteins are increasingly being found to be directed
against the oligosaccharide side-chains and not the protein core.
Another good example of this is seen in the antisera raised against the
major cell-wall glycoprotein and the sex-specific flagellar glycoprotein
of Chlamydomonas eugametos both of which recognise short oligosaccharides on the glycoproteins (Lens et al. 1983; Musgrave et al. 1983).

If there have been few attempts to use the power of
antibodies to dissect the complexity of the plant cell wall, then there
have been even fewer attempts to use the remarkable specificity
provided by monoclonal antibodies.   It has been shown in animal systems
that monoclonal antibodies can be produced that will recognise a wide
variety of sugars and oligosaccharides;   becoming in effect a family of
custom-made 'super lectins'.   Thus, against the Lewis a human-blood-
group determinant, monoclonal antibodies have been produced that are
specific for a trisaccharide, a tetrasaccharide, α-fucose and β N-acetylglucosamine (Young et al. 1983).   There is a short report that
monoclonal antibodies against cell-wall carbohydrate antigens in Fucus
have been used to study their developmental regulation (Vreeland &
Laetsch 1984).   It has been suggested (Anderson et al. 1984) that the
production of monoclonal antibodies against crude cell-wall fractions
from plants may be complicated by the fact that terminal arabinose and
galactose seem to be very immunodominant, and that therefore a large
proportion of hybridomas produced will simply secrete antibodies
directed against these simple determinants.   We have not encountered
this problem with the family of monoclonal antibodies that we have
raised against the cell-wall glycoprotein of Chlamydomonas reinhardii
(Smith et al. 1984a).   Before describing these experiments on raising
antibodies to the Chlamydomonas cell wall, it is appropriate to describe
some of the aspects of the system which have made it an attractive model
for studying cell-wall assembly, the structure of hydroxyproline-rich
glycoproteins, and glycoprotein synthesis and secretion.

THE CHLAMYDOMONAS CELL WALL PROVIDES A USEFUL MODEL SYSTEM

We have been studying the cell wall of Chlamydomonas for
several years now and it is appropriate here to outline why we believe
that this research effort is warranted.   Chlamydomonas has a relatively
simple cell wall, composed of a thin fibrillar inner cell wall layer and
an outer double sandwich layer composed of two back-to-back monolayers
of glycoproteins in a two-dimensional crystalline array (Roberts et al.
1972, 1982; Hills et al. 1973; Roberts 1974).   Both these layers are
made simply of high-molecular-weight glycoproteins, rich in hydroxyproline,
arabinose, and galactose (Miller et al. 1972; Catt et al. 1976; Roberts
1979).   The major (94%) part of the cell wall by weight, the crystalline
layers, contains two glycoprotein components which can be easily

separated by gel chromatography in the presence of 1M sodium perchlorate.
These two glycoproteins have been designated 2BI and 2BII.   The latter
is the major structural component of the crystalline layer of the wall
and has been extensively characterized (Catt et al. 1976, 1978; Homer &
Roberts 1979; Roberts et al. 1980).   It has a M_r of approx. 300,000,
is sulphated, and runs on denaturing SDS gels as four major glycopeptides.
It has been visualized in the EM as a highly elongated molecule (Roberts
1981) which we assume in-vivo must fold up to form the complex shape
seen in the crystalline array.   The 2BII molecule alone can form this
array which we have demonstrated by in-vitro self-assembly experiments
(Catt et al. 1978).

   We have used this very convenient model system, which has
the twin advantages that the cell walls are easily obtained without cell
breakage, and that the structural glycoproteins are readily solubilized
intact from the wall, for three main lines of enquiry.   First, we have
used it as a model for biological self-assembly.   We have shown (Roberts
1974; Hills et al. 1975; Catt et al. 1978) that the cell wall can be
disassembled with chaotropic salts, and then reassembled again into a
morphologically and chemically similar wall, and that the factors
controlling this process can be unravelled.   Oligosaccharide side-chains
were shown (Catt et al. 1978) to be essential for the self-assembly
process.

   Second, we have used the system to investigate the structure
and chemistry of the glycoproteins, as a model for the hydroxyproline-
rich glycoproteins in higher-plant cell walls.   We have demonstrated
the asymmetric distribution of hydroxyproline within the polypeptide
backbone of 2BII and. the correlation of these domains with the presence
of the oligosaccharide side chains (Roberts 1979).   The presence of
polyproline II structure within 2BII has been demonstrated by circular
dichroism (Homer & Roberts 1979) and the presence of low levels of
sugar-0-sulphate esters has also been found (Roberts et al. 1980).
Methylation analysis (O'Neill & Roberts 1981) has revealed the presence
of several sugars in a wide variety of linkages.   The analysis suggests
that the average degree of polymerisation of the oligosaccharides is 6
and that many of them are branched.

   Third, we have used the fact that the glycoproteins in the
wall form a highly ordered 2-D crystalline lattice to develop high-
resolution computer-based methods of image analysis.   These methods

have enabled us to deduce a 2-D map of the Chlamydomonas reinhardii
cell wall to a resolution of 2nm (Roberts et al. 1982) and a 3-D model
of the wall of Lobomonas piriformis (Shaw & Hills 1982).

Many of the unique biological features of this cell wall can
be seen to depend on the presence of short oligosaccharide side chains
on the 2BII molecule.   It was in order to dissect the complexity of
such side chains that we decided to take an immunological approach to
the problem.   We were encouraged by the fact that we had already
successfully raised antibodies to intact cell walls during our
experiments on in-vitro self-assembly (Hills et al. 1975).   An added
incentive has been the realization that the mating recognition molecules
of Chlamydomonas, present on the flagellar membrane of gametes, are
probably very closely related molecules to 2BII (Musgrave et al. 1983;
Cooper et al. 1983), and that again their function appears to depend on
the carbohydrate portion of the molecule.

A POLYCLONAL ANTISERUM HAS BEEN RAISED AGAINST A SINGLE CELL-WALL GLYCOPEPTIDE

In order to investigate the relationships between the various
extracellular-matrix glycoproteins of Chlamydomonas that we have
described, to compare them with those of related algae, and to gain
information about the sites of their synthesis and secretion we have now
raised a rabbit antiserum to one of the glycopeptides of 2BII.

Cell walls at a concentration of 5 mg ml^{-1} in SDS sample
buffer, in the presence of mercaptoethanol, were run on a 4% PAGE slab
gel.   Side bands were cut off and rapidly stained with Coomassie blue
in order to locate the major glycopeptides.   The unstained portions of
gel, each containing one of the three main components, were cut out and
homogenized in 0.125M Tris, pH 6.8, + 0.1% SDS, and left shaking for
24 h for the glycopeptides to diffuse out of the gel.   The gel fragments
were then spun out and washed.   The supernatants were then dialysed and
rotary evaporated to half their original volume and samples rerun on
PAGE to check their purity (Fig. 1).   A sample of band 2 (Fig. 1), the
major glycopeptide, was used as the antigen and was injected, in the
presence of Freunds adjuvant, subcutaneously into the neck of a New
Zealand white rabbit.   After two booster injections serum was
collected from an ear vein and antibodies, precipitated by ammonium
sulphate, taken up in Tris-buffered saline at a concentration of

20 mg ml^{-1} and stored at -20°C in small aliquots.   Preimmune serum was treated similarly and used as a control in all experiments.

Fig. 1.   SDS-PAGE of cell wall of Chlamydomonas reinhardii. On the left is the cut-off edge of a 4% slab gel showing the main glycopeptides (bands 1-4).   On the right are the eluted glycopeptides 2-4 rerun on a second slab gel. Glycopeptide 2 was used as the antigen for raising a polyclonal antiserum.

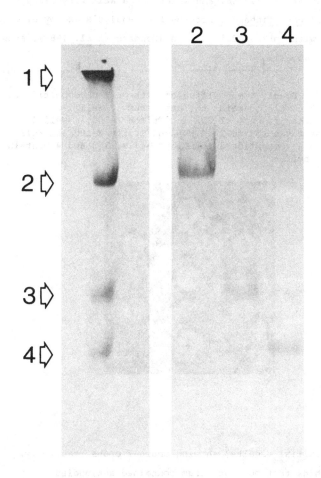

The antiserum was first tested against a variety of antigens in double immunodiffision tests (Chua & Blomberg 1979). With SDS-solubilized cell walls, 2BII, and the original band 2 antigen, two lines

of precipitation were found.   Precipitin lines are also found when the
antibody is tested against the original antigen, or 2BII, solubilized in
Triton X-100 (Fig. 2).   The antiserum was then tested against a variety
of antigens and was found to cross react with the minor structural
glycoprotein 2BI and with the cell wall of Lobomonas piriformis, a
related alga which has unrelated glycoproteins in its wall as revealed
by PAGE (Roberts et al. 1981).   Remarkably, a species of Chlamydomonas
which we isolated (CCAP No. 11/756) and which has a wall very closely
related to Chlamydomonas reinhardii both by PAGE analysis and by electron
microscopy coupled with optical diffraction (Roberts et al. 1982), shows
no cross reactivity at all.

Fig. 2.   Double immunodiffusion with anti-glycopeptide 2.
The centre well contains the antiserum.   Well 2 contains
2BII in the presence of SDS and Triton X-100.   Well 5
contains glycopeptide 2 in SDS and Triton X-100 and Well 6
contains glycopeptide 2 in SDS.   Wells, 1,3 and 4 contain
control buffer.

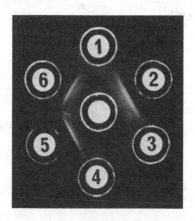

These initial results, showing unusual cross reactivity,
encouraged us to think that our antiserum contained antibodies
recognizing common determinants such as oligosaccharides, on the
variety of glycoproteins tested.   Preincubation of the antiserum with
a variety of monosaccharides resulted in some diminution in the intensity
of the precipitin line with arabinose and mannose, and particularly with
galactose.   The antiserum showed no cross reactivity either with potato

lectin, a hydroxyproline-containing glycoprotein with tetraarabinoside side-chains (a generous gift from A.K. Allen) or with the arabinogalactan-protein from Gladiolus style (Gleeson & Clark 1980).   Interestingly an antiserum against this style glycoprotein which recognises terminal arabinose and galactose residues (Gleeson & Clarke 1980) also appears not to cross react with our cell-wall material or with purified 2BII (Adrienne Clarke, personal communication).   This suggests that our cell-wall glycoproteins and the arabinogalactan-proteins of higher plants are antigenically quite different.

Our antiserum, however, showed some weak cross reactivity with the hydroxyproline-rich cell-wall glycoprotein which we isolated from carrot by the method of Chrispeels (1969), and produced strong precipitin lines with two of the soluble glycoproteins, unrelated to the cell-wall glycoproteins, that are secreted by Chlamydomonas into the culture medium in large amounts (Roberts & Phillips, unpublished results).

To analyse further the pattern of cross reactivity with the extracellular glycoproteins of Chlamydomonas we used 'Western' blot analysis.   PAGE gels of cell walls, or 2BII, were blotted onto nitrocellulose (Towbin et al. 1979), probed with the antiserum and the bound antibodies revealed with peroxidase-coupled goat anti-rabbit IgG (Miles) developed using 4-chloro-1-naphthol (Hawkes et al. 1982). All the bands on gels of cell-wall material that stain with periodic-acid-Schiff reagent or with periodic-acid-silver stain (Dubray & Bezard 1982) for carbohydrate, also stain with antiserum, strongly suggesting the presence of common sugar determinants on all the glycoproteins of the cell wall (Fig. 3).   Even with overloaded gels there are no cross-reacting antigens present in SDS extracts of whole soybean callus cells, hinting that the antigenic oligosaccharides present in 2BII may be unique to the lower plants.   The algae in the Volvocales have been classified into 4 classes on the basis of the glycoprotein structures present in their cell walls (Roberts et al. 1982).   Western blotting analysis showed that the single major glycoprotein band from Lobomonas cell walls (class IV) and from the cell wall of Chlamydomonas angulosa (class I) cross reacted with the antiserum, but that glycoproteins from the cell wall of C. moewusii (class II) and from C. asymmetrica (class III) did not cross react.

Western blotting also showed that antiserum (Ashford et al.

1982) against potato lectin cross reacts with the band II that was used
as our original antigen together with a few other minor bands.   This
would agree with the lectin result reported above.   Preincubation of
the blots with Concanavalin A did not alter the pattern of antibody
binding at all, even though ConA binds tightly to the cell wall (Roberts
et al. 1982), suggesting that terminal mannose groups, although present
(O'Neill & Roberts 1981), are not integral to the antigenic determinants.

Fig. 3   a) Western blot with anti-glycopeptide 2 of
cell walls of C. reinhardii after SDS/PAGE.   Numerous
glycopeptides cross react with the antiserum.
b) Western blot of purified flagella from C. reinhardii
with anti-glycopeptide 2.   Flagella are always
contaminated with cell-wall glycoproteins but major and
minor flagellar membrane glycoproteins (→) are clearly
seen to cross react with our antiserum.

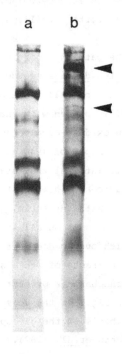

a   b

It is known that there are numerous glycoproteins associated
with the flagellar membrane in Chlamydomonas and that some at least of

these are hydroxyproline-rich glycoproteins (Cooper et al. 1983; Monk
et al. 1983; Lens et al. 1983; Musgrave et al. 1979).   We prepared
flagella from C. reinhardii by the method of Witman et al. (1972), and
showed by Western blot analysis of the membrane components (G. Rodwell,
personal communication) that the major high-molecular-weight flagellar-
membrane glycoprotein is also recognised by our antiserum (Fig. 3),
indicating antigenic affinities with the cell-wall glycoproteins.

ANTIBODIES CAN BE USED TO REVEAL THE LOCATION OF THE ANTIGENS AND THE PATHWAY OF SECRETION FOR THE EXTRACELLULAR GLYCOPROTEINS IN CHLAMYDOMONAS

Antibodies can be used to localize their respective antigens
at two levels, at the light microscope level and at the electron
microscope level.   We have used our antiserum in both an immunofluorescence study and in EM level immunocytochemistry.

Cells and walls for immunofluorescence studies were fixed in
freshly-prepared 2% paraformaldehyde, air-dried onto acetone-cleaned
coverslips, and immersed in methanol at -20°C.   They were first
incubated with the primary antiserum diluted 1:100 together with
2 mg ml^{-1} BSA and then in the FITC-coupled goat anti-rabbit lgG diluted
1:50, mounted in antifade mountant (City University, London) and
examined under a Zeiss epifluorescence microscope.

It was found that freshly prepared cell walls, when unfixed,
only bind antibody weakly, if at all, but that formalin-fixed walls
fluoresce brightly.   This agrees with a similar result found with cell-
wall antigens in C. eugametos (Musgrave et al. 1983).   However, intact
cells, when stained either directly or after fixation, show bright wall
fluorescence.   In addition to the wall fluorescence it was found that
both the flagellar collars (Roberts et al. 1975), and the flagella of
intact cells also fluoresce brightly (Fig. 4).   This latter result
would agree with the Western blotting data reported above.   Concanavalin
A, although binding to the cell wall, does not interfere with antibody
labelling.   In agreement with the immunoblotting data we found that the
cell wall around intact Lobomonas cells also fluoresces.   However,
unexpectedly, we also found that the cell wall of Chlamydomonas
asymmetrica fluoresced very brightly indeed with our antiserum, as did
the flagellar collars of Chlorogonium elongatum.

Closer examination of C. reinhardii cells at various stages

of their cell cycle revealed that the degree of fluorescence of the cell wall was clearly cell-cycle dependent.   Cast-off mother cell walls, walls surrounding single cells, and walls surrounding 4 daughter cells, all fluoresce much more weakly than the wall of young daughter cells within their mother cell wall (Fig. 5).   This apparent developmental regulation of antigen expression is at present mysterious.   We do not know, for example, whether the antigenic determinants are simply not there in the weakly fluorescent walls or whether they are present but masked in some way, as suggested for the so-called 'cryptic' antigens on the wall of C. eugametos (Musgrave et al. 1983).

Fig. 4.   Indirect immunofluorescence with anti-glycopeptide 2.   a,b: Cells of C. reinhardii; the cell wall fluoresces brightly and the flagella are also labelled.   c,d: Colony of Eudorina elegans seen in Nomarski optics and with immunofluorescence which shows labelling of the colony wall.

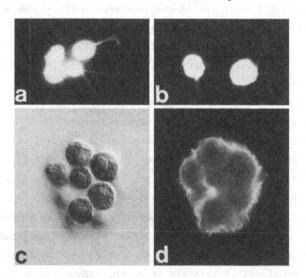

We have shown previously that all the multicellular members of the Volvocales that we have looked at, have cell walls that are crystalographically identical to that of C. reinhardii.   As their cell-wall glycoproteins are also electrophoretically similar we suggested that the multicellular forms may have evolved from a C. reinhardii-like

ancestor (Roberts et al. 1982).   We have now shown that, as might be
expected, the cell walls of Volvox aureus (LB 88/16), Eudorina elegans
(24/1a), Pandorina morum (60/1c) and Gonium pectorale (31/1a) all
fluoresce brightly with our antiserum (Fig. 4).

Fig. 5.   A gallery of cells of C. reinhardii at different
stages of the cell cycle, seen at the top using Nomarski
optics and beneath by immunofluorescence.

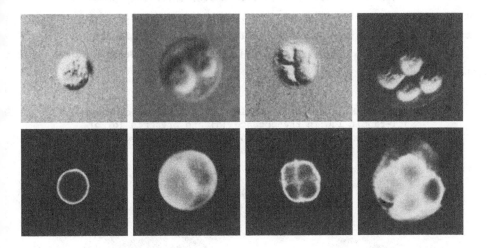

We went on to examine the nature of the antigenic sites in
C. reinhardii at the EM level.   Initially we fixed cells in glutaraldehyde,
and then, following extensive washing, incubated the cells with antiserum
(1:100) in the presence of 2 mg ml^{-1} BSA.   After washing they were then
incubated in ferritin-conjugated anti-rabbit IgG (Miles) at a dilution
of 1:40 in PBS for 30 min.   The cells were then washed, post-fixed in
osmic acid and then embedded for thin sectioning.   This immediately
confirmed that the antiserum bound to both faces of the cell wall, but
it also revealed that there were cross-reacting antigenic determinants
on the outer face of the plasma membrane in addition to the flagella
membrane staining (Figs. 6, 7).   In cells which had been permeabilized
after fixation with 0.1% Triton X-100, we also found ferritin labelling

in the contractile vacuole which is found adjacent to the basal bodies
of the two flagella.   To find glycoproteins in an organelle usually
simply thought of in terms of osmoregulation was relatively unexpected,
so we decided to use immunogold staining on thin sections to try to
localize the antigens more precisely within the cell.   Cells fixed
only in glutaraldehyde and embedded in either EPON or L.R. White resin,
were sectioned and collected on gold grids.   Sections were incubated in
our antiserum diluted 1:100 in TBS with 2 mg ml^{-1} BSA for 2 h.   They
were washed in TBS + 0.1% Triton X-100, incubated in freshly prepared
Protein A colloidal gold + 2 mg ml^{-1} BSA, washed in TBS + 0.1% Triton
X-100 and stained in uranyl acetate and lead citrate.   Colloidal gold
was made by the sodium citrate method and titrated with Protein A by the
method of Roth et al. (1978).

Fig. 6 a, b.   Flagella labelled with ferritin-conjugated
anti-rabbit lgG in the absence (a) and presence (b) of
anti-glycopeptide 2.   The antibody clearly labels the
'fuzzy-coat' outside the plasma membrane.

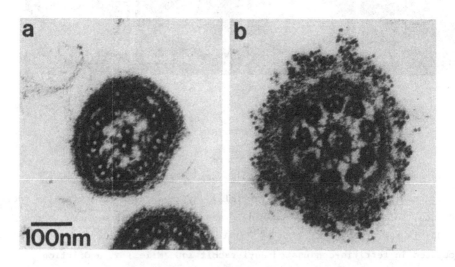

As with the ferritin labelling, colloidal gold particles
clearly localized antigens on the cell wall, plasma membrane and
flagella membrane (Fig. 8).   In addition they labelled the contactile
vacuole and the Golgi apparatus (of which there are usually two in

C. reinhardii in the small region of cytoplasm between the nucleus and
the pyrenoid in the chloroplast).   In addition there was label over the
numerous largish vesicles in the cytoplasm between the Golgi and the
anterior region of the cell where the contractile vacuole is found.
Statistical analysis (Roberts, unpublished results) showed a clear
gradation in the number of gold particles per unit area, increasing from
the Golgi apparatus to vesicles to the contractile vacuole to the cell wall.

Fig. 7 a, b.   Ferritin labelling shows that anti-
glycopeptide 2 labels not only the cell wall but the
plasma membrane (a) and also the contractile vacuole (b).
w = wall, pm = plasma membrane, cv = contractile vacuole.

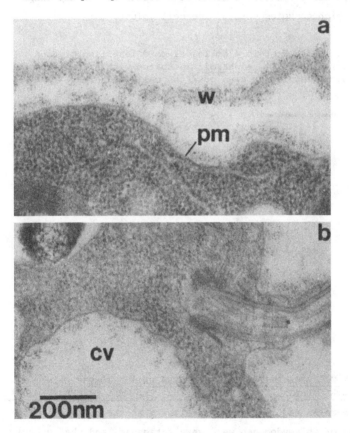

This has led us to the hypothesis that in Chlamydomonas the extracellular matrix glycoproteins are completed in the Golgi, are transferred to the largish cytoplasmic vesicles which transfer their contents to the contractile vacuole which excretes them along with water during the course of its regular contractions.   The glycoproteins of the growing wall would then have to diffuse around the cell, in the space between the wall and plasma membrane, before inserting themselves into their final location in the wall.   This novel method of exocytosis for cell wall components is not that surprising when it is remembered that most of the plasma membrane of the cell would be inaccessible to normal Golgi-vesicle exocytosis for steric reasons as a result of the cup-shaped chloroplast which occupies most of the cell volume.

Fig. 8.   Protein A/colloidal gold labelling of Epon sections of C. reinhardii incubated with anti-glycopeptide 2. Label is clearly seen over the cell-wall area.

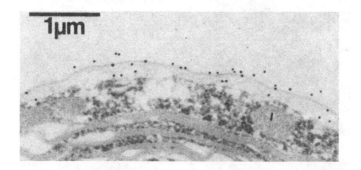

MONOCLONAL ANTIBODIES HAVE BEEN RAISED TO THE MAIN
STRUCTURAL GLYCOPROTEIN OF THE CELL WALL OF CHLAMYDOMONAS

When we had convinced ourselves that the cell-wall glycoproteins of Chlamydomonas were very antigenic, and that useful results could be obtained with our polyclonal antiserum, the decision was made to raise monoclonal antibodies against the purified 2BII glycoprotein, with the idea of providing antibodies specific to one or other of the oligosaccharide side-chains.   These could then be used to

probe in more detail their site of synthesis, their developmental
regulation, their location within the molecule and their chemical
structure.

Monoclonal antibodies (McAb) against 2BII were thus raised
by one of us in close collaboration with Drs. Galfe and Butcher at the
AFRC Monoclonal Antibody Center at Babraham, Cambs., U.K..   Their
production, screening and preliminary characterization is described
elsewhere (Smith et al. 1984a, 1984b).   Mild periodate oxidation of
2BII (40 mM sodium metaperiodate, pH 8.3, at 4°C for 4 h) resulted in
the loss of binding of almost all the chosen McAb.   None of the McAb
would recognise deglycosylated 2BII and a barium hydroxide digest of
2BII, which would be expected to contain free amino acids together with
the short oligosaccharides, 0-glycosidically linked to free hydroxyproline or serine, showed inhibition of binding of several of the McAb.
Taken together, these lines of evidence strongly suggest that some at
least of the 20 antibodies that were cloned recognise specific
oligosaccharide side-chains of 2BII (Smith et al. 1984a).

We have chosen two McAb to examine in more detail the
cellular location of their corresponding antigens.   On Western blots
MAC3 recognises band 2, 3 and 4 (see Fig. 9), while MAC6 recognises band
1 and (very weakly) band 2.

By immunofluoresence, two completely different images were
obtained using these two antisera.   MAC3 produced an even, bright
fluorescence of the entire cell wall and also lit up the flagella and
the flagellar collars.   MAC6 on the other hand produced a distinct
punctate fluorescence on the wall and did not label the flagella.
MAC3, like the polyclonal antiserum, revealed the cell-cycle-dependent
variation in intensity of wall fluorescence (Fig. 10), in which the walls
of young daughter cells, within the mother cell wall, fluoresce much more
brightly than at other stages.

Immunogold staining was then used to look at the ultrastructural basis of these very different patterns of staining with McAb
against the same antigen.   Cells were fixed in glutaraldehyde and
embedded in Lowicryl K4M resin, following the manufacturers schedule for
low-temperature embedding.   Thin sections were stained either by the
method of Robertson et al. (1984) or of Beezley et al. (1982), using
20nm-collodial-gold-labelled anti-rat lgG from Janssen Pharmaceutica.
MAC3 showed a staining pattern which closely resembles that found with

Fig. 9.   Western blots with monoclonal antibodies of
cell wall material from C. reinhardii.   The original
5% gel has been stained with periodic-acid-silver
staining for carbohydrate (a) and with glutaraldehyde-
silver staining for protein (b).   They both show
glycopeptides 1-4 together with various other minor
glycopeptides.   MAC3 stains bands 2-4, while MAC6
only stains band 1.

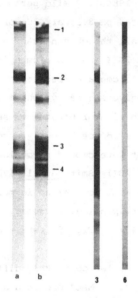

the polyclonal antiserum, a result to be expected as MAC3 was shown by
immunoblotting to recognise the band 2 antigen.   In addition to the
outer wall layers, MAC3 labelled the inner wall layer, the flagellar
collar, the plasma membrane and the flagellar membrane (Fig. 11).   As
before, clear labelling of the Golgi apparatus was found, strongly
suggesting that the mature glycoprotein is present in this compartment
before secretion via the contractile vacuole.   MAC6 on the other hand,
which predominantly binds to the very high molecular weight component in
band 1, shows discrete patches of labelling only on the outer surface of
the cell wall, a result in agreement with immunofluorescence data.   The
inner wall layer and the flagellar and plasma membranes were not labelled;
neither were the Golgi apparatus, vesicle, contractile vacuole system.

It would appear in this case that the antigenic determinant is only visible or accessible once the glycoprotein has been expressed on the outer face of the crystalline wall layer. This may be due to a variety of reasons such as post-secretory processing or the desulphation of sugar residues. Whatever the reasons for the developmental variation and the topological location of the antigens described here, it is clear that the McAb will help us to understand this complex system.

Fig. 10. Indirect immunofluorescence of _C. reinhardii_ using monoclonal antibodies. MAC6 clearly stains the cell wall but in a punctate manner (a) while MAC3 stains the wall evenly, the flagella collars and the flagella (b).

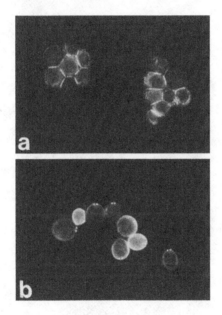

ANTIBODIES AS CELL-WALL PROBES - FUTURE PROSPECTS

Our results have shown that McAb can be raised against cell-wall glycoproteins, and that some of them recognise carbohydrate or carbohydrate-containing determinants. This approach, we feel, will be invaluable in understanding many of the biological and functional complexities of higher-plant cell-wall function, in particular in the

Fig. 11.   Immunogold staining of thin sections of
C. reinhardii using monoclonal antibodies.   MAC6
shows clustered labelling of the outer layer only of
the cell wall (a).   MAC3 in contrast ((b) and (c))
stains the inner and outer layers of the wall, the
plasma membrane and, the golgi apparatus (+), as well
as the flagellar collars and the flagellar membranes (c).

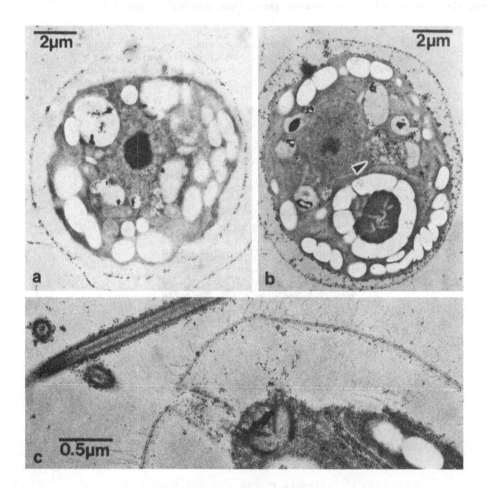

area of cell-cell recognition.   For example, the questions surrounding
the role of the cell wall in host-pathogen interaction would be amenable
to attack by such methods.   Questions of the cellular location of
proline hydroxylation and the sites of glycosylation of cell-wall
glycoproteins could also be answered.   The future also holds out the
possibility of raising McAb to chemically synthesized wall fragments,
such as short oligosaccharides, to map their synthesis, location and
function.   For cell-wall glycoproteins, antisera to defined portions of
the nascent protein core (Smith 1981) provides a way of obtaining the
appropriate mRNA and ultimately a way into the molecular biology of the
plant cell wall (Cooper et al. 1984).   And lastly, McAb against
specific regions (such as oligosaccharide side chains) of glycoproteins,
allow the direct mapping of these antigenic determinants on the isolated
molecules.   This process, called immunotopography by Adair, has been
used to locate epitopes on the sexual agglutinin of Chlamydomonas
(personal communication).

ACKNOWLEDGEMENTS
        We wish to thank G.J. Hills for help with several aspects of
this work, including electron microscopy, and many invaluable discussions,
G. Rodwell for work on the flagellar membrane antigens, and A. Hutchings,
G. Galfre, and G. Butcher (IAP, Babraham, Cambs.) for collaboration and
advice during the production of the monoclonal antibodies.

REFERENCES
Adair, W.S., Monk, B.C., Cohen, R., Hwang, C. & Goodenough, U.W. (1982).
        Sexual agglutinins from the Chlamydomonas flagellar membrane.
        Journal of Biological Chemistry, 257, 4593-4602.
Albersheim, P., Darvill, A.G., McNeil, M., Valent, B.S., Sharp, J.K.,
        Nothnagel, E.A., Davis, K.R., Yamazaki, N., Gollin, D.J.,
        York, W.S., Dudman, W.F., Darvill, J.E. & Dell, A. (1983).
        Oligosaccharins: naturally occurring carbohydrates with
        biological regulatory functions. In Structure and Function
        of Plant Genomes, eds. O. Ciferri & L. Dure. New York:
        Plenum.
Anderson, M.A., Sandrin, M.S. & Clarke, A.E. (1984). A high proportion
        of hybridomas raised to a plant extract secrete antibody to
        arabinose or galactose. Plant Physiology, 75, (in press).
Ashford, D., Allen, A.K. & Neuberger, A. (1982). The production and
        properties of an antiserum to potato (Solanum tuberosum)
        lectin. Biochemical Journal, 201, 641-645.
Aspinall, G.O. (1980). Chemistry of cell wall polysaccharides. In The
        Biochemistry of Plants, ed. J. Preiss, Vol. 3, pp. 473-500.
        New York: Academic Press.

Beezley, J.E., Orpin, A., & Adlam, C.A. (1982). A comparison of
    immunoferritin, immunoenzyme and gold labelled protein A
    methods for the localization of capsular antigen on frozen
    thin sections of the bacterium Pasteurella haemolytica.
    Histochemical Journal, 14, 803-810.

Bishop, P.D., Makins, D.J., Pearce, G. & Ryan, C.A. (1981). Proteinase
    inhibitor-inducing factor activity in tomato leaves residues
    in oligosaccharides enzymically released from cell walls.
    Proceedings of the National Academy of Sciences of the
    United States of America, 78, 3536-3540.

Catt, J.W., Hills, G.J. & Roberts, K. (1976). A structural glycoprotein,
    containing hydroxyproline, isolated from the cell wall of
    Chlamydomonas reinhardii. Planta, 131, 165-171.

Catt, J.W., Hills, G.J. & Roberts, K. (1978). Cell wall glycoproteins
    from Chlamydomonas reinhardii, and their self assembly.
    Planta, 138, 91-98.

Chrispeels, M.J. (1969). Synthesis and secretion of hydroxyproline
    containing macromolecules in carrot. Plant Physiology, 44,
    1187-1193.

Chua, N.H. & Blomberg, F. (1979). Immunochemical studies of thylakoid
    membrane polypeptides from spinach and Chlamydomonas
    reinhardii. Journal of Biological Chemistry, 254, 215-223.

Cooper, J.B., Adair, W.S., Mecham, R.P., Heuser, J.E. & Goodenough U.W.
    (1983). The Chlamydomonas agglutinin is a hydroxyproline-
    rich glycoprotein. Proceedings of the National Academy of
    Sciences of the United States of America, 80, 5898-5901.

Cooper, J.B., Chen, J.A. & Varner, J.E. (1984). The glycoprotein
    component of plant cell walls. In Plant Cell Walls:
    Structure Biosynthesis and Function, eds. M. Dugger &
    I. Ting. ASPP Monographs (in press).

Darvill, A.G. & Albersheim, P. (1984). Phytoalexins and their elicitors -
    a defense against microbial infection in plants. Annual
    Review of Plant Physiology, 35, 243-275.

Dubray, G. & Bezard, G. (1982). A highly sensitive periodic acid-silver
    stain for 1, 2-diol groups of glycoproteins and poly-
    saccharides in polyacrylamide gels. Analytical Biochemistry,
    119, 325-329.

Feizi, T. (1981). Carbohydrate differentiation antigens. Trends in
    Biochemical Sciences, 6, 333-335.

Ferrari, T.E., Bruns, D. & Wallace, D.H. (1981). Isolation of a plant
    glycoprotein involved with control of intercellular
    recognition. Plant Physiology, 67, 270-277.

Fry, S.C. (1982a). Phenolic components of the primary cell wall.
    Biochemical Journal, 203, 493-504.

Fry, S.C. (1982b). Isodityrosine, a new cross linking amino acid from
    plant cell-wall glycoprotein. Biochemical Journal, 204,
    449-455.

Gerwig, G.J., Kamerlin, J.P., Vleigenthart, J.F.G., Homan, W.L., Egmond,
    P.V. & van den Ende, H. (1984). Characteristic differences
    in monosaccharide composition of glycoconjugates from
    opposite mating types of Chlamydomonas eugametos.
    Carbohydrate Research, 127, 245-251.

Glaudemans, C.P.J. (1975). The interaction of homogeneous murine myeloma
    immunoglobulins with polysaccharide antigens. Advances in
    Carbohydrate Chemistry and Biochemistry, 31, 313-346.

Gleeson, P.A. & Clarke, A.E. (1980). Antigenic determinants of a plant proteoglycan, the Gladiolus style arabinogalactan-protein. Biochemical Journal, 191, 437-447.

Hamada, T., Nakajima, T. & Matsuda, K. (1981). Comparison of the mannan structure from the cell-wall mutant Candida sp. M-7002 and its wild type. European Journal of Biochemistry, 119, 373-379.

Harris, P.J., Anderson, M.A., Bacic, A. & Clarke, A.E. (1984). Cell-cell recognition in plants with special reference to the pollen stigma interaction. In Oxford Surveys of Plant Molecular and Cell Biology, ed. B.J. Miflin, Vol. 1, pp. 161-203. Oxford University Press.

Hawkes, R., Nidag, E. & Gordon, J. (1982). A dot immunobinding assay for monoclonal and other antibodies. Analytical Biochemistry, 119, 142-147.

Hills, G.J., Gurney-Smith, M. & Roberts, K. (1973). Structure, composition and morphogenesis of the cell wall of Chlamydomonas reinhardii. II. Electron microscopy and optical diffraction analysis. Journal of Ultrastructural Research, 43, 179-192.

Hills, G.J., Phillips, J.M., Gay, M.R. & Roberts, K. (1975). Self assembly of a plant cell wall in vitro. Journal of Molecular Biology, 96, 431-441.

Homer, R.B. & Roberts, K. (1979). Glycoprotein conformation in plant cell walls. Planta, 146, 217-222.

Huber, D.J. & Nevins, D.J. (1981). Wall protein antibodies as inhibitors of growth and of autolytic reactions of isolated cell wall. Physiologia Plantarum, 53, 533-539.

Jennings, H.J., Lugowski, C. & Kasper, D.L. (1981). Conformational aspects critical to the immunospecificity of the type III Group B Streptococcal polysaccharide. Biochemistry, 20, 4511-4518.

Kabat, E.A. (1966). The nature of an antigenic determinant. Journal of Immunology, 97, 1-11.

Knox, R.B. & Clarke, A.E. (1978). Localization of proteins and glycoproteins by binding to labelled antibodies and lectins. In Electron Microscopy and Cytochemistry of Plant Cells, ed. J.L. Hall, pp. 149-185. Amsterdam: Elsevier/North Holland.

Kurn, N. (1981). Altered development by the multicellular alga Volvox carteri caused by lectin binding. Cell Biology International Reports, 5, 867-875.

Lamport, D.T.A. & Catt, J.W. (1981). Glycoproteins and enzymes of the cell wall. In Encyclopedia of Plant Physiology, eds. W. Tanner & F.A. Loewus, Vol. 3, pp. 133-164. Springer-Verlag.

Lens, P.L., Olofsen, F., van Egmond, P., Musgrave, A. & van den Ende, H. (1983). Isolation of an antigenic determinant from flagellar glycoproteins of Chlamydomonas eugametos. Archives of Microbiology, 135, 311-314.

Mau, S.L., Raff, J. & Clarke, A.E. (1982). Isolation and partial characterization of components of Prunus avium L. styles including an antigenic glycoprotein associated with a self-incompatibility genotype. Planta, 156, 505-516.

McNeil, M., Darvill, A.G., Fry, S.C. & Albersheim, P. (1984). Structure and function of the primary cell walls of plants. Annual Review of Biochemistry, 53, 625-663.

Miller, D.H., Lamport, D.T.A. & Miller, M. (1972). Hydroxyproline hetero-oligosaccharides in Chlamydomonas. Science, 176, 918-920.

Milliken, B.E. & Weiss, R.L. (1984). Distribution of ConA binding
     carbohydrates during mating in Chlamydomonas.
     Journal of Cell Science, 66, 223-239.
Monk, B.C., Adair, W.S., Cohen, R.A. & Goodenough, U.W. (1983).
     Topography of Chlamydomonas: fine structure and polypeptide
     components of the gametic flagellar membrane surface and the
     cell wall. Planta, 158, 517-533.
Musgrave, A., Homan, W.L., van den Briel, W., Lelie, N., Schol, D.,
     Evo, L. & van den Ende, H. (1979). Membrane glycoproteins of
     Chlamydomonas eugametos flagella. Planta, 145, 417-425.
Musgrave, A., van Eijke, E., te Welscher, R., Broekman, R., Lens, P.,
     Homan, W. & van den Ende, H. (1981). Sexual agglutination
     factor from Chlamydomonas eugametos. Planta, 153, 362-369.
Musgrave, A., de Wildt, P., Broekman, R. & van den Ende, H. (1983).
     The cell wall of Chlamydomonas eugametos. Immunological
     aspects. Planta, 158, 82-89.
Northcote, D.H. (1972). Chemistry of the Plant Cell Wall. Annual Review
     of Plant Physiology, 23, 113-132.
O'Neill, M.A. & Roberts, K. (1981). Methylation analysis of cell wall
     glycoproteins and glycopeptides from Chlamydomonas
     reinhardii. Phytochemistry, 20, 25-28.
Pazur, J.H., Dreher, K.L., Kubrick, R.L. & Erikson, M.S. (1982). The use
     of affinity chromatography for the isolation of two sets of
     anti-carbohydrate antibodies with specificity for a
     Streptococcal glycan. Analytical Biochemistry, 126, 285-294.
Quatrano, R.S., Brawley, S.H. & Hogsett, W.E. (1979). The control of the
     polar deposition of a sulphated polysaccharide in Fucus
     zygotes. In Symposium of the Society for Developmental
     Biology, eds. S. Subtelny & I.R. Konigsberg, pp. 77-96.
Rao, S.S., Lippincott, B.B. & Lippincott, J.A. (1982). Agrobacterium
     adherence involves the pectic portion of the host cell wall
     and is sensitive to the degree of pectin methylation.
     Physiologia Plantarum, 56, 374-380.
Rees, D.A. (1977). Polysaccharide Shapes. London: Chapman & Hall.
Roberts, K. (1974). Crystalline glycoprotein cell walls of algae: their
     structure, composition and assembly. Philosophical
     Transactions of the Royal Society of London, Series B, 268,
     129-146.
Roberts, K. (1979). Hydroxyproline: its asymmetric distribution in a
     cell wall glycoprotein. Planta, 146, 275-279.
Roberts, K. (1981). Visualizing an insoluble glycoprotein. Micron, 12,
     185-186.
Roberts, K., Gurney-Smith, M. & Hills, G.J. (1972). Structure,
     composition and morphogenesis of the cell wall of
     Chlamydomonas reinhardii. I. Ultrastructure and preliminary
     chemical analysis. Journal of Ultrastructural Research, 40,
     599-613.
Roberts, K., Phillips, J.M. & Hills, G.J. (1975). Structure, composition
     and morphogenesis of the cell wall of Chlamydomonas
     reinhardii. VI. The flagellar collar. Micron, 5, 341-357.
Roberts, K., Gay, M.R. & Hills, G.J. (1980). Cell wall glycoproteins
     from Chlamydomonas reinhardii are sulphated. Physiologia
     Plantarum, 49, 421-42.
Roberts, K., Shaw, P.J. & Hills, G.J. (1981). High resolution electron
     microscopy of glycoproteins: the crystalline cell wall of
     Lobomonas. Journal of Cell Science, 51, 295-321.

Roberts, K., Hills, G.J. & Shaw, P.J. (1982). Structure of algal cell walls. In Electron Microscopy of Proteins, ed. J.R. Harris, Vol. 3, pp. 1-40. New York, London: Academic Press.

Robertson, J.G., Wells, B., Bisseling, T., Farnden, K.J.F. & Johnston, A.W.B. (1984). Immunogold localization of leghaemaglobin in the plant cytoplasm in nitrogen-fixing root nodules of pea. Nature, 311, 254-256.

Roth, J., Bensayan, M. & Orci, L. (1978). Ultrastructural localization of intracellular antigens by the use of protein A-gold complexes. Journal of Histochemistry and Cytochemistry, 26, 1074-1081.

Sengbusch, P.V., Mix, M., Wachholz, I. & Manshard, E. (1982). FITC-labelled lectins and calcafluor white ST as probes for the investigation of the molecular architecture of cell surfaces. Studies on Conjugatophycean species. Protoplasma, 111, 38-52.

Shaw, P.J. & Hills, G.J. (1982). Three dimensional structure of a cell wall glycoprotein. Journal of Molecular Biology, 162, 459-471.

Smith, E., Roberts, K., Hutchings, A. & Galfre, G. (1984a). Monoclonal antibodies to the major structural glycoprotein of the Chlamydomonas cell wall. Planta, 161, 330-338.

Smith, E., Roberts, K., Butcher, G.W. & Galfre, G. (1984b). Monoclonal antibody screening: two methods using antigens immobilized on cellulose. Analytical Biochemistry, 138, 119-124.

Smith, M.A. (1981). Characterization of carrot cell wall protein. II. Immunological study of cell wall protein. Plant Physiology, 68, 964-968.

Towbin, H., Stachelin, T. & Gordon, J. (1979). Electrophoretic transfer of proteins from polyacrylamide gels to nitrocellulose sheets: procedures and some applications. Proceedings of the National Academy of Sciences of the United States of America, 76, 4350-4354.

Vian, B., Briollouet, J.M. & Satiat-Jeunemaitre, B. (1983). Ultra-structural visualization of xylans in cell walls of hardwood by means of xylanase-gold complex. Biologie Cellulaire, 49, 179-182.

Vreeland, V. & Laetsch, W.M. (1984). Intracellular synthesis of Fucus cell wall antigens. Plant Physiology, 75, 62a.

Wenzl, S. & Sumper, M. (1981). Sulfation of a cell surface glycoprotein correlates with the developmental program during embryogenesis of Volvox carteri. Proceedings of the National Academy of Sciences of the United States of America, 78, 3716-3720.

Whatley, M.H. & Sequeira, L. (1981). Bacterial attachment to plant cell walls. Recent Advances in Phytochemistry, 15, 213-240.

Witman, G.B., Carlson, K., Berliner, J. & Rosenbaum, J.L. (1972). Chlamydomonas flagella. I. Isolation and electrophoretic analysis of microtubules, matrix, membranes and mastigonemes. Journal of Cell Biology, 54, 507-539.

York, W.S., Darvill, A.G. & Albersheim, P. (1984). Inhibition of 2,4-Dichlorophenoxyacetic acid-stimulated elongation of pea stem segments by a xyloglucan oligosaccharide. Plant Physiology, 75, 295-297.

Yoshikawa, M., Keen, N.T. & Wang, M.C. (1983). A receptor on soybean membranes for a fungal elicitor of phytoalexin accumulation. Plant Physiology, 73, 497-506.

Young, W.W., Johnson, H.S., Tamura, Y., Karlsson, K-A., Larson, G., Parker, J.M.R., Khare, D., Spohr, U., Baker, D.A., Hindsgaul, O. & Lemieux, R.U. (1983). Characterization of monoclonal antibodies specific for the Lewis a human blood group determinant. Journal of Biological Chemistry, 258, 4890-4894.

6 HYDROXYPROLINE-RICH GLYCOPROTEIN BIOSYNTHESIS: A COMPARISON WITH THAT OF COLLAGEN

D.G. Robinson
Abteilung Cytologie, Pflanzenphysiologisches Institut,
Universität Göttingen, Untere Karspüle 2, D-3400 Göttingen,
FRG

M. Andreae
Abteilung Cytologie, Pflanzenphysiologisches Institut,
Universität Göttingen, Untere Karspüle 2, D-3400 Göttingen,
FRG

A. Sauer
Abteilung Cytologie, Pflanzenphysiologisches Institut,
Universität Göttingen, Untere Karspüle 2, D-3400 Göttingen,
FRG

Abstract. The presence of significant amounts of hydroxyproline is characteristic of collagen and some extracellular
plant glycoproteins. This imino acid is nontranslatable
and is formed instead by the hydroxylation of peptidyl
proline. This reaction is thus a posttranslational
modification of the polypeptide being synthesized. Another
important modification is the glycosylation of the polypeptide, which is in each case an O-glycosylation, but in
collagen involves hydroxylysine rather than hydroxyproline
residues. This chapter attempts to draw attention to
differences and similarities between the two types of
glycoprotein in terms of their structure, the enzymology of
their hydroxylation and glycosylation and the intracellular
location of these reactions. We conclude that the
similarities are only superficial.

Key words: Arabinogalactan proteins, arabinosyl transferase,
collagen, ER, extensin, GA, prolyl hydroxylase.

INTRODUCTION

Hydroxyproline is an imino acid characteristic of some
extracellular glycoproteins in plant cells. Because collagen, the major
extracellular polymer of animal cells, is also a glycoprotein rich in
hydroxyproline it is logical to ask whether parallels may also exist in
terms of their synthesis. Although much is known about the structure
and biosynthesis of collagen, a comparison of this sort is only now
possible as a result of recent research done with plants.

A. A COMPARISON OF STRUCTURES

Both collagen and the extracellular hydroxyproline-rich

glycoproteins (HRGPs) of plant cells represent families of related
polymers.   In both cases one of the essential differences between the
members of each family is the degree and type of glycosylation.   There
are seven different types of collagen polypeptide chains (named α1(I-V)
and α2(I and V)), each containing about 1000 amino acids.   From these,
five different trimer helical configurations are formed, three of which
(types I-III) exist morphologically as fibrils (Bornstein & Sage, 1980).

        In contrast to the HRGPs, both lysine and proline residues
are hydroxylated in collagen (see Table 1) and it is the former which
becomes glycosylated (see Table 2).   Thus, as a rule, collagen types
with high amounts of carbohydrate usually show a greater degree of lysyl
hydroxylation.   On the other hand, the hydroxylation of proline appears
to be relatively constant in the various collagen types, with about half
(approx. 100) of the residues being involved.   This is not surprising
since the interchain hydrogen bonding which occurs as a result leads to
a stabilization of the triple helix without which the molecule cannot be
secreted (see Grant & Jackson, 1976, for details).   Collagen glycosylation
is also different from that of the HRGPs, with either a single galactose
or a galactose-glucose dimer being attached (Kivirikko & Myllylä, 1979).

Table 1:   Peptidyl hydroxylation of collagen and HRGPs.

FEATURE	COLLAGEN	HRGPs
1. AAs hydroxylated	4-Proline	4-Proline
	(3-Proline)	
	Lysine	
2. % total AAs as Pro	10 %[1]	variable (up to 60 %)[2]
3. % total AAs as Hyp	0.5-4 %[1]	variable (up to 60 %)[2]
4. Degree of proline hydroxylation	45-60 %[3]	variable (5-50 %)[2]

References

1 - Grant & Jackson (1976)   2 - Selvendran & O'Neill (1981)

3 - Davidson & Berg (1981)

As a group the HRGPs are more heterogeneous than the collagens; indeed, the collective term HRGP may be too artificial for some workers.   We can, however, differentiate two major types: structural glycoproteins of the "extensin" type and the so-called "arabinogalactan proteins" (AGPs).   Probably the only thing they have in common is that both types are relatively rich in hydroxyproline; otherwise they differ appreciably, both in amino acid composition and in the nature and degree of glycosylation.

The extensin type of HRGP includes the structural cell-wall glycoproteins of higher plants (Lamport, 1980) and the Chlamydomonads (Catt et al. 1976) as well as soluble lectins of the Solanaceae (Leach et al. 1982).   In each case the polypeptide takes on a left-handed polyproline II helix conformation (Allen et al. 1978; Homer & Roberts 1979), the stability of which is maintained by the carbohydrate side chains (van Holst & Varner 1984).   In the cell wall, individual chains of this type may become cross-linked with one another via isodityrosine residues (Fry 1982; Epstein & Lamport 1984), making them relatively insoluble.   Unlike collagen, where the 0-glycosylation is restricted

Table 2: Glycosylation of collagen and HRGPs

FEATURE	COLLAGEN	HRGPs
1. % Carbohydrate in molecule	0.2-10 %[1]	Extensin - 50 %[2] Chlamydomonas - 65 % AGPs - 80 %
2. 0-Glycosides	Hyl-Gal-Glu[3] Hyl-Gal	Hyp-$(Ara)_{1-4}$[4] Ser-Gal[4] Hyp-$(Gal)_2$-Ara_n[5]
3. % Glycosylation of hydroxy AAs	20-90 %[6]	90 % (dicots)[7] 30 % (monocots)[7]
4. N-Glycosylation? (Asn-GlcNAcMan)	Yes (restricted to carboxylpropeptide)	Not known

References

1 - Schofield et al. (1971)   2 - Selvendran & O'Neil (1981)
3 - Spiro (1976)   4 - Lamport (1980)
5 - Fincher et al. (1983)   6 - Davidson & Berg (1981)
7 - Lamport & Miller (1971)

to one amino acid, in the extensin type of HRGPs two are involved:
hydroxyproline which bears up to 4 arabinose residues and serine which
has a single galactose attached to it (see Fig. 1A and Table 2).   It is
of note that the terminal arabinose in the tetraarabinoside is, in
contrast to the others, in the α-configuration and the linkage is 1→3
rather than 1→2 (Akiyama et al. 1980).

The polypeptide chains of AGPs also have a portion in the
polyproline II configuration (van Holst & Fincher 1984).   Circumstantial
evidence exists for the presence of four such chains per AGP molecule
(see Fincher et al. 1983).   The carbohydrate side chains attached to
hydroxyproline in the AGPs are much more extensive than in extensin.   The
major sugar residues are galactose and arabinose but other sugars including
uronic acids may also be present.   In one of the best studied examples,
the AGP from ryegrass endosperm (Clarke et al. 1979), the side chains are
a branched galactan with terminal arabinose residues (see Fig. 1B).

B.   POSTTRANSLATIONAL MODIFICATIONS

Collagen is synthesized as a molecule termed preprocollagen
(see Fig. 2).   As might be expected for a secretory glycoprotein, this
polypeptide has a signal peptide at the amino terminal.   Interestingly
this is about 2-3 times larger than most signal sequences, and its
removal through the action of endopeptidase(s) may occur in more than
one step (Palmiter et al. 1979).   This step represents the first in a
series of events which are collectively termed posttranslational
modifications.   We can conveniently subdivide them into two groups:
those occurring within the cell and those which take place extracellularly
(see Davidson & Berg 1981, for a review).

After the cleavage of the signal sequence, which occurs upon
entry into the endoplasmic reticulum (ER) and before the translation of
the procollagen molecule has been completed, the hydroxylation of
proline and lysine residues is begun.   At the same time the N-
glycosylation of asparagine residues with a high mannose type of oligosaccharide and involving a dolichol phosphate intermediate occurs at the
carboxy terminal end of the molecule (Clark 1979).   These three events,
together with the O-glycosylation of hydroxylysine which follows them,
require that the procollagen be in a nonhelical conformation.   The
procollagen chains at this stage are loosely associated with one another.
The formation of the triple helix is initiated by the creation, mainly

Fig. 1A.   A possible structure of the extensin type of
            HRGP as based on studies by Akiyama et al. (1980)
            and Lamport et al. (1983).
   1B.   A possible structure for the AGP from ryegrass
          endosperm as given by Clarke et al. (1979).

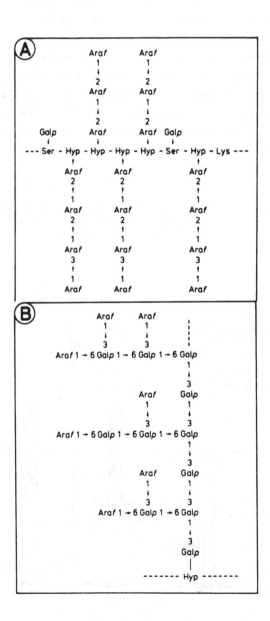

between neighbouring carboxyl propeptides, of interchain disulfide
bridges.   All of the enzymes involved in these reactions, including a
disulfide isomerase (Freedman et al. 1978), are present in the
endomembrane system.   The site of triple helix formation however may
vary, probably in accordance with the degree of hydroxylation and
glycosylation of the particular collagen type under investigation
(Bornstein & Sage 1980).

    After exocytosis the amino acid and carboxyl propeptides are
removed by independently acting peptidases (Davidson et al. 1979).   These
events can occur immediately upon release from the cell or may be
delayed for up to 24 h (Fessler & Fessler 1978).   Subsequent to this,
and constituting the last posttranslational modification, is the
polymerization of the collagen molecules into a collagen fibril (see
Fig. 2).

Fig. 2. Changes occurring during the biosynthesis of
collagen.   Structural domains in preprocollagen are
indicated at the left of the figure.

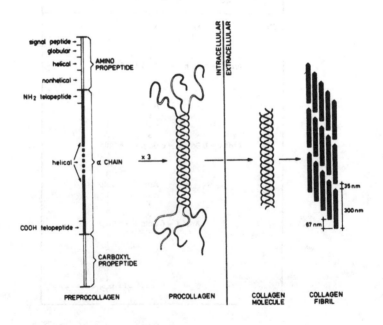

The biosynthesis of HRGPs in plant cells is also associated
with the endomembrane system, soluble hydroxyproline containing
macromolecules having been detected in both ER and Golgi apparatus (GA)
fractions (Wienecke et al. 1982; Samson et al. 1983).   There is also
indirect evidence available for the existence of a signal peptide.
In-vitro translation experiments in the absence of microsomal vesicles
(Smith 1981) have resulted in the production of a proline-rich polypeptide somewhat larger than a proline-rich polypeptide which can be
isolated from the walls of cells incubated in vivo with α,α'dipyridyl
to prevent prolyl hydroxylation.

Evidence that the hydroxylation of proline is begun on
nascent polypeptide chains does not exist for HRGPs; indeed, the early
experiments of Sadava & Chrispeels (1971a) indicate that this event is
restricted to chains which are no longer associated with polysomes.
Likewise there is no evidence for the N-glycosylation of asparagine
residues in HRGPs.   Although the pertinent experiments have not yet
been undertaken it is a provoking thought that this event might occur at
a terminal portion of the polypeptide, a portion which will later be
excised before release from the cell in analogy to the propeptides of
collagen.   Apart from the O-glycosylation reactions we know of only one
further posttranslational modification, namely, in the case of extensin,
the cross-linking of monomers via an isodityrosine bridge (Epstein &
Lamport 1984; Fry 1982).   This occurs extracellularly and is probably
mediated via a cell-wall-located peroxidase.

C. PROLYL HYDROXYLASE

The enzyme prolyl hydroxylase is one of the most well known
and studied of enzymes from animal cells.   Its properties are known
exactly (Cardinal & Udenfriend 1974);   it has been isolated and
purified many times (e.g. Berg & Prockop 1972) and antibodies against it
were first obtained almost 15 years ago (McGee et al. 1971).   Prolyl
hydroxylase from plants has also been studied, but by no means as
frequently.   The first attempt at its characterization was carried out
by Sadava & Chrispeels (1971b), and almost a decade passed by until the
next report became available (Tanaka et al. 1980).   Since then two
further reports have appeared (Tanaka et al. 1981; Cohen et al. 1983)
and we, ourselves, have just completed a paper on this enzyme (Sauer &
Robinson 1984).

Prolyl hydroxylase (and lysyl hydroxylase for that matter) is a dioxygenase, or more precisely a multifunctional oxygenase. It requires α-ketoglutarate as cosubstrate and has ascorbate and Fe^{2+} as cofactors (see Fig. 3). There are a number of ways by which one can measure its activity:

a) by offering an underhydroxylated substrate prepared in vivo through the action of the iron chelator α,α'dipyridyl. When this substrate is labelled with 3,4-^{3}H-proline, the release of ^3H_2O as a result of the hydroxylation can be used as a measure of enzyme activity in vitro.

b) an interesting variation on a) has been introduced by Kawasaki (1980) and employs an endogenous underhydroxylated substrate (also prepared by the action of α,α'dipyridyl) which is then hydroxylated in vitro through the addition of Fe^{2+}.

Instead of using a radioactively labelled substrate, assays a) and b) can be performed colorimetrically after acid hydrolysis of the hydroxylated substrate.

c) Measuring the release of $^{14}CO_2$ from the Cl position of αKGA. This is a very sensitive assay for prolyl hydroxylase, providing the unspecific degradation of the substrate is not too high.

The properties of prolyl hydroxylase are presented in Table 3. A fundamental difference between the enzyme from animal and plant sources lies in the nature of the primary substrate. Whereas the animal enzyme recognizes proline in any polymeric form of the amino acid triplet -X-Pro-Gly-, the plant enzyme prefers a repeating series of at least 4 proline residues. Indeed, Tanaka et al. (1981) have shown that the latter enzyme actually recognizes the secondary structure, the polyproline II helix, of the substrate, which is a most unusual feature. Interestingly polyproline is an inhibitor for animal prolyl hydroxylase (see for example Rhoads & Udenfriend 1968). However, animal substrates such as $(Pro-Pro-Gly)_{5 \text{ or } 10}$ can be hydroxylated by the plant enzyme at rates of about 1/5-1/4 of that when polyproline is used as substrate (Tanaka et al. 1980; Cohen et al. 1983).

There are conflicting statements as to the relative degree of enzyme association with the membranes of plant cells. Whereas Tanaka et al. (1980) were able to solubilize appreciable amounts of their prolyl hydroxylase by sonifying Triton X-100-treated membrane preparations from Vinca rosea, Cohen et al. (1983) using ryegrass cells

Fig. 3. The hydroxylation of poly-L-proline by prolyl
hydroxylase: a mixed functional oxygenase (prolyl peptide,
2-oxoglutarate: oxygen oxidoreductase EC 1.14.11.2).

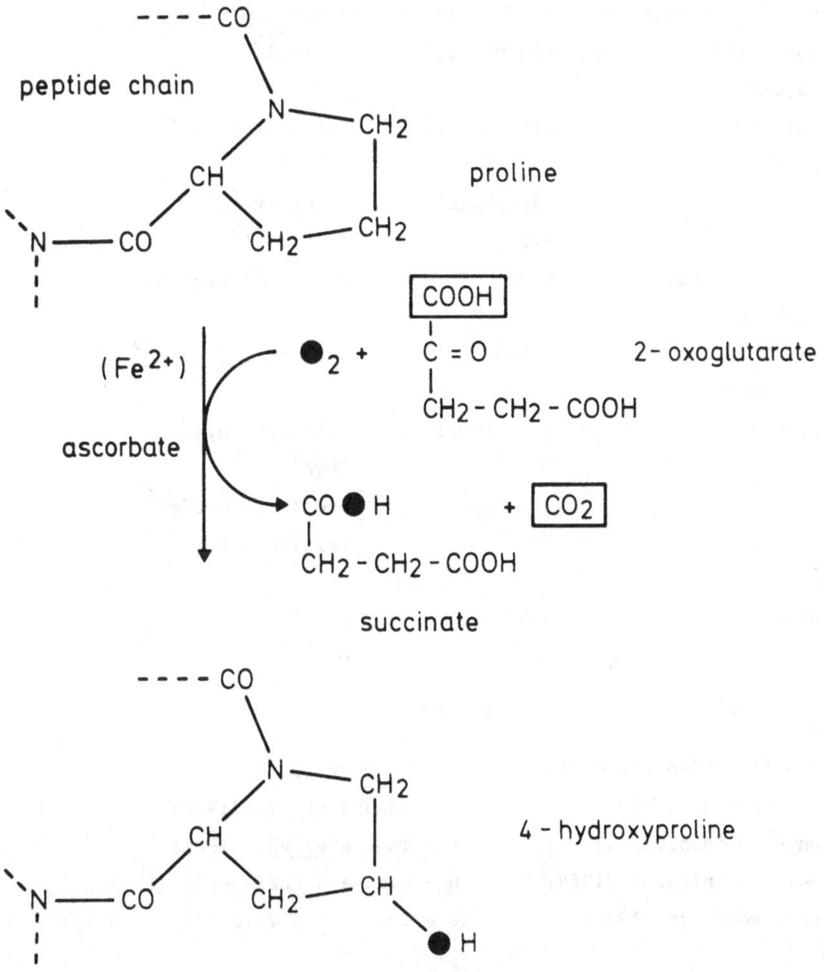

Table 3: Characteristics of prolyl hydroxylase from plant and animal sources.

PROPERTY	ANIMAL (COLLAGEN)	PLANT (ANGIOSPERM)
1. Substrates		
a) Natural hydroxylation sequon(s)	-x-(Pro)-Gly-[1]	-Ser-(Pro)-(Pro)-(Pro)-(Pro)[2]-
b) Polymeric alternatives	$(Pro-Pro-Gly)_n$[3]	Poly L-Proline[4]
c) Oligomeric alternatives	Bradykinin	$(Pro)_8$; Gly- Pro_8[4]
d) Configuration for prolyl hydroxylation	random coil[5]	helical[6]
2. Cosubstrate	ketoglutarate[7]	ketoglutarate[4,9]
3. Cofactors	Fe^{2+}[1]	Fe^{2+}[8,9]
	Ascorbate[1]	Ascorbate[8,9]
4. pH Optimum	7.2-7.4[1]	6.8-7.0[4,9]
5. Km - with KGA as substrate	0.01 mM[1]	0.09 mM[4]; 0.25 mM[9]
- with Prolyl substrate	0.1 mM[1]	0.23 mM[4]; 0.04 mM[9,10]
6. Inhibitors	α,α'dipyridyl[1]	α,α'dipyridyl[8]
	EDTA[1]	EDTA[7]
		Phosphate buffer[4]
		Tricine buffer[9]
		NaCl[4,7]
7. Stimulators	BSA[1]	BSA[4,9]
	DTT[1]	DTT[4,9]

References

1 - Cardinale & Udenfriend (1974)   2 - Lamport (1980)
3 - Berg & Prockop (1973)   4 - Tanaka et al. (1980)
5 - Murphy & Rosenbloom (1973)   6 - Tanaka et al. (1981)
7 - Rhoads & Udenfriend (1968)   8 - Sadava & Chrispeels (1971b)
9 - Sauer & Robinson (1984)   10 - Cohen et al. (1983)

were unable to solubilize any activity by this treatment.   We have
recently examined the situation with membrane preparations from maize
roots (Sauer & Robinson 1984) and Chlamydomonas (Blankenstein &
Robinson, unpublished observations), and have found that by sonifying
tritonated material, 20 % and 60 % respectively of the activity remains
membrane-bound.   Cohen et al. (1983)claim that "sonication of membranes
in the presence of Triton X-100 is essential for activity".   This is an
absolute requirement which is not met with in the measurement of prolyl
hydroxylase in animal systems (see for example Harwood et al. 1974),
nor have we been able to confirm it with our plant systems.   We have,
however, observed that sonification gives rise to at least a doubling of
activity in any particular membrane preparation.

        Perhaps one of the reasons for the differences between our
findings and those of Cohen et al. (1983) is that the latter authors
used a colorimetric assay and we have employed principally the $^{14}CO_2$
release method.   We have also used a colorimetric assay but find that
it is not sensitive enough when small amounts of membrane protein (e.g.
in single-gradient fractions) are present.   Furthermore, polyproline,
at concentrations greater than 0.5 mg per assay, itself contributes to
the development of a complex with Ehrlich's reagent absorbing at the
same wavelength as that of the hydroxyproline-Ehrlich's reagent complex
(Sauer & Robinson 1984).   Finally endogenous amounts of hydroxyproline-
containing macromolecules have to be separately determined and subtracted.
These points, together with the drawback that the colorimetric assay
takes about 48 h to complete, make its use for the analysis of gradients
fractions somewhat prohibitive.

        We have already published the results of our attempts at
localizing the prolyl hydroxylase activity in the membrane systems of
suspension-cultured sugar-cane cells and maize roots using the
colorimetric assay (Robinson et al. 1984).   Although sonification was
not used at this stage in our investigations our results were in
agreement with those of Cohen et al. (1983) in allocating the majority
of the activity to fractions rich in IDPase activity (i.e. to GA-rich
fractions).   These results contrast strongly with findings from animal
systems (Harwood et al. 1974;   Olsen et al. 1975) whereby the rough ER
is clearly indicated as the site of prolyl hydroxylase activity.   Our
most recent results, obtained with the $^{14}CO_2$ release method, change the
picture somewhat.   As can be seen in Fig. 4, prolyl hydroxylase is

Fig. 4. The distribution of prolyl hydroxylase activity in
sucrose density gradients of sugar-cane homogenates as
compared with that of ER (cytochrome c reductase, CCR), GA
(inosine diphosphatase, IDPase), PM (steroyl glucosyl
transferase, SGT), and mitochondrial (cytochrome c oxidase,
CCO) marker enzymes. Profiles are given for homogenates and
gradients prepared with 0.1 mM and 4 mM Mg^{2+}.

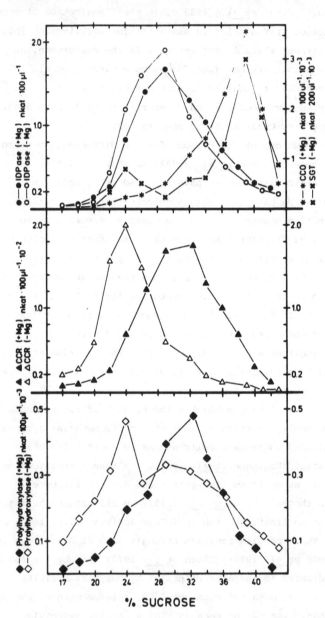

present in ER fractions, as warranted by the activity profiles in low
and high Mg^{2+} sucrose gradients.   Taking into account the relative
amounts of protein in the fractions and employing a cross-contamination
analysis (see Table 4), we can show that there is not only relatively
more enzyme in the ER portion of the gradient but the specific activity
is even higher than for the GA region.   However, there is no question
that appreciable amounts are associated with GA membranes;   much more
than can be explained by cross-contamination with ER membranes.

The significance of this bimodal enzyme distribution is not
at the moment clear.   It does not necessarily mean that the GA-based
enzyme actually hydroxylates peptidyl proline in situ, since product and
enzyme may be spatially separated from one another.   The presence of
some ER marker enzymes in GA membranes of liver cells has been
demonstrated unequivocally (Howell et al. 1978) and has been interpreted
by Rothman (1981) as indicating a crude export of both product and
enzyme, from the ER, followed by the retrieval of the enzyme from the GA
(see Fig. 5).   Indirect support of this may be forthcoming from
immunocytochemical studies, which should allow us to determine whether
or not the enzyme in the GA is localized primarily in the cis as
opposed to the trans cisternae.

D.   O-GLYCOSYLATION

The collagen hydroxylsyl glycosyltransferases have been the
subject of numerous investigations.   They have been successfully
characterized and also isolated and purified (see Kivirikko & Myllylä
1979, for a review).   Since the sugars involved are different from those
in the glycosylation of HRGPs, we need here only to devote our attention
to their intracellular location.

It is generally agreed that the glycosylation of pro-collagen
can continue up to the time of triple-helix formation.   When this event
is prevented glycosylation continues (Uitto et al. 1978).   On the other
hand the glycosylation of hydroxylysine has been shown to be initiated
on nascent polypeptide chains (Brownell & Veis 1975).   Both in-vivo
labelling experiments with ^{14}C glucose (Oohira et al. 1979) and the
determination of enzyme activities in vitro (Harwood et al. 1975) suggest
that hydroxylysine glycosylation occurs in both ER and GA fractions.   The
latter authors have shown the galactosyl transferase to predominate in
rough ER fractions but that the glucosyl transferase activity is more or

Table 4: Distribution of prolyl hydroxylase amongst membrane fractions of sugar cane cells.

Fraction[1]	Relative activities of marker enzymes[2] (Total = 100 %)				Relative activity of Prolyl hydroxylase (Total = 100 %)		Protein (mg)
	CCR	IDPase	SGT	CCO	Measured	Corrected[3]	
ER	62.5	20.2	16.0	5.7	45.3	57	0.58
GA	37.5	67.9	27.4	43.6	49.1	43	1.03
PM	-	11.9	56.6	50.7	5.6	-	1.7

1 - in sucrose density gradients ER=16-25 %; GA=25-36 %; PM=36-44 %

2 - Values taken from -Mg^{2+} gradients depicted in Fig. 4. For abbreviations, see legend to Fig. 4.

3 - because prolyl hydroxylase was absent from PM fractions a cross-contamination correction for 2 parameters adapted from Wirtz et al. (1980) was used. Accordingly

$$\text{prolyl hydroxylase ER}_{corr} = \frac{\text{prohy ER}_{meas} - (b \cdot \text{prohy GA}_{meas})}{1 - a}$$

$$\text{prolyl hydroxylase GA}_{corr} = \frac{\text{prohy GA}_{meas} - (a \cdot \text{prohy ER}_{meas})}{1 - b}$$

$$\text{whereby } a = \frac{\text{CCR}_{GA}}{\text{CCR}_{total}} \ ; \quad b = \frac{\text{IDPase}_{ER}}{\text{IDPase}_{total}}$$

less equally distributed between rough and smooth microsomal sub-
fractions.   This distribution is of course compatible with a terminal
glycosylation using glucose.

Until we began our investigations, all of the previously
published data pointed to the GA as being the site of hydroxyproline
arabinosylation.   Kawasaki (1981) incubated suspension-cultured
tobacco cells in vivo with ^{14}C-arabinose and found that the label was
incorporated, with little conversion into other sugars, into polysaccharides and hydroxyproline-containing glycoproteins held with the GA.

Fig. 5.   A possible interpretation of the fact that in
plant cells prolyl hydroxylase is associated with both ER
and GA membranes.   This drawing is based on an idea of
Rothman (1981) and suggests a return flow, via vesicles,
of enzyme (stippled) back to the ER allowing the product,
a HRGP (rods), to continue through the GA.

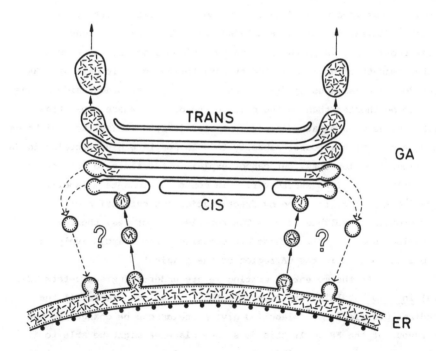

Unfortunately, the GA fractions appeared to be heavily contaminated with
ER membranes.   Attempts at localizing hydroxyproline arabinosyl
transferase activity in vitro have also confirmed the important
glycosylating nature of the GA in plant cells (Gardiner & Chrispeels
1975; Owens & Northcote 1981).   We were therefore most surprised when
we looked at the results of our first attempts at localizing this
enzyme on sucrose gradients of sugar-cane cell homogenates.   It was
quite clear that the activity profile is not identical with that of the
GA marker IDPase.   In the absence of data from high Mg^{2+} gradients we
initially interpreted our results as indicating the secretory vesicle
location of this enzyme (Robinson et al. 1984).   We now have the
necessary information (see Fig. 6) and are compelled to change our
interpretation.   The arabinosyl transferase activity shows a very
convincing Mg^{2+}-shifting, corresponding to that of ER membranes.

        The analysis of the products of our in-vitro transferase
assays also gave us somewhat unexpected results, in that over 80 % of
the label was incorporated into a Hyp-monoarabinoside (results not shown
here).   Similar results were obtained with membrane fractions from
maize roots.   It is known that monocotyledonous species have much
higher amounts of unglycosylated hydroxyproline in their cell wall HRGP
than do dicotyledonous species.   Moreover, their arabinose side chains
tend to be shorter than is the case in the dicots, where hyp-tetra-
arabinosides predominate (Lamport & Miller 1971).   These facts might be
in agreement with a poorer capacity to arabinosylate glycoproteins in GA
fractions from monocots, but cannot explain our failure to demonstrate
any activity in vitro whatsoever.   In contrast, long-term labelling
with ^{14}C-arabinose followed by fractionation and extraction of
glycoproteins insoluble in 20 % TCA reveals more or less the reverse
situation (see Fig. 7).   These are contained almost exclusively in
membranes located in the GA region of the gradient.

        At the moment of writing we are undertaking short-term (2-8
min) in-vivo labelling experiments with ^{14}C-arabinose to see if the
gradient profile for ^{14}C-labelled glycoproteins can be changed to one
in favour of the ER.   If this does take place we might be able to
interpret our results as indicating a sort of "core glycosylation" in
the ER followed by a more extensive "terminal" arabinosylation in the GA.
One notes that there are different linkage types in the arabinoside side
chains of extensin (see Fig. 1A) which might suggest different arabinosyl

transferases and also their different intracellular locations.

Fig. 6.   The distribution of hydroxyproline-arabinosyl transferase activity in sucrose density gradients of sugar-cane homogenates as compared with that of an ER marker enzyme (cytochrome c reductase, CCR).   Profiles are given for gradients prepared with 0.1 mM and 3 mM Mg^{2+}).

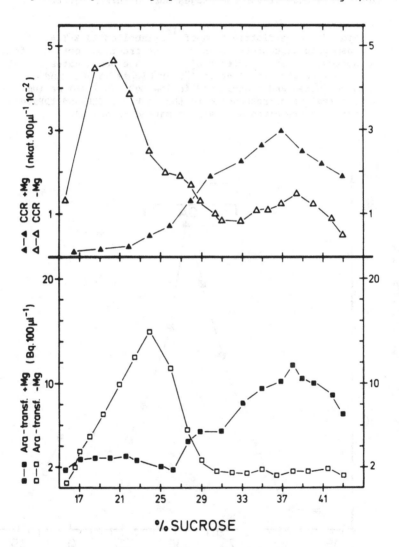

ACKNOWLEDGEMENTS

This work was supported by funds from the Deutsche
Forschungsgemeinschaft and the Stiftung Volkswagenwerk.   We appreciate
greatly the help of Dr. W. Lang (Univ. Kaiserslautern) in the analysis
of our in-vitro transferase products.   Ms. H. Freundt is thanked for
her technical assistance at various stages during this project.

Fig. 7.   The distribution of ^{14}C-labelled 20 % TCA-
insoluble glycoproteins in membrane fractions obtained from
sucrose-density gradients of sugar-cane homogenates.   Cells
were incubated for 1 hr in ^{14}C-arabinose (5 µCi) and
homogenized and centrifuged in low-Mg^{2+} (0.1 mM) medium.
The regions corresponding to the peaks of CCR and IDPase
activity are given as brackets marked ER and GA.

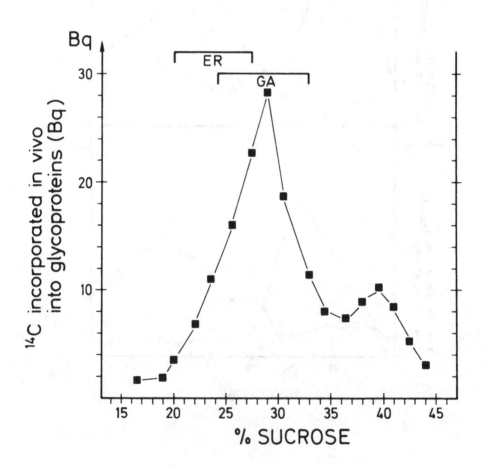

REFERENCES

Akiyama, Y., Mori, M. & Kato, K. (1980). ^{13}C-NMR analysis of
hydroxyproline arabinoside from Nicotiana tabacum.
Agricultural and Biological Chemistry, 44, 2487-2489.

Allen, A.K., Desai, N.N., Neuberger, A. & Creetz, J.M. (1978).
Properties of potato lectin and the nature of its
glycoprotein linkages. Biochemical Journal, 171, 665-674.

Berg, R.A. & Prockop, D.J. (1972). Affinity column purification of
protocollagen proline hydroxylase from chick embryos and
further characterization of the enzyme. Journal of
Biological Chemistry, 248, 1175-1182.

Bornstein, P. & Sage, H. (1980). Structurally distinct collagen types.
Annual Review of Biochemistry, 49, 957-1003.

Brownell, A.G. & Veis, A. (1975). The intracellular location of the
glycosylation of hydroxylysine of collagen. Biochemical and
Biophysical Research Communications, 63, 371-377.

Cardinale, G.J. & Udenfriend, S. (1974). Prolyl hydroxylase. In
Advances in Enzymology, ed. A. Meister, pp. 243-301.
New York: John Wiley & Sons.

Catt, J.W., Hills, G.J. & Roberts, K. (1976). A structural glycoprotein,
containing hydroxyproline isolated from the cell wall of
Chlamydomonas reinhardii. Planta, 131, 165-171.

Clark, C.C. (1979). The distribution and initial characterization of
oligosaccharide units on the COOH-terminal propeptide
extensions of the pro- 1 and pro- 2 chains of type I
procollagen. Journal of Biological Chemistry, 254, 10798-
10802.

Clarke, A.E., Anderson, R.L. & Stone, B.A. (1979). Form and function of
arabinogalactans and arabinogalactan-proteins. Phytochemistry,
18, 521-540.

Cohen, P.B., Schibeci, A. & Fincher, G.B. (1983). Biosynthesis of
arabinogalactan-proteinin Lolium multiflorum (Ryegrass)
endosperm cells.   III. Subcellular distribution of prolyl
hydroxylase. Plant Physiology, 72, 754-758.

Davidson, J.M., McEnear, L.S.C. & Bornstein, P. (1979). Procollagen
processing.   Limited proteolysis of COOH-terminal extension
peptides by a cathepsin-like protease secreted by tendon
fibroblasts. European Journal of Biochemistry, 100, 551-558.

Davidson, J.M. & Berg, R.A. (1981). Posttranslational events in collagen
biosynthesis. In Methods in Cell Biology, Vol. 23, eds.
A.R. Hand & C. Oliver, pp. 119-136. New York: Academic Press.

Epstein, L. & Lamport, D.T.A. (1984). An intramolecular linkage
involving isodityrosine in extensin. Phytochemistry, 23,
1241-1246.

Fessler, J.H. & Fessler, L.I. (1978). Biosynthesis of procollagen.
Annual Review of Biochemistry, 47, 129-162.

Fincher, G.B., Stone, B.A. & Clarke, A.E. (1983). Arabinogalactan-
proteins: structure, biosynthesis and function. Annual Review
of Plant Physiology, 34, 47-70.

Freedman, R.B., Newell, A. & Walklin, C.M. (1978). Paradoxical detergent
effects on microsomal protein disulfide isomerase. FEBS
Letters, 88, 49-52.

Fry, S.C. (1982). Isodityrosine, a new cross-linking amino acid from
plant cell wall glycoprotein. Biochemical Journal, 204,
449-455.

Gardiner, M. & Chrispeels, M.J. (1975). Involvement of the Golgi
    apparatus in the synthesis and secretion of hydroxyproline-
    rich cell wall glycoproteins. Plant Physiology, 55, 536-541.
Grant, M.E. & Jackson, D.S. (1976). The biosynthesis of procollagen.
    Essays in Biochemistry, 12, 77-113.
Harwood, R., Grant, M.E. & Jackson, D.S. (1974). Collagen biosynthesis.
    Characterization of subcellular fractions from embryonic
    chick fibroblasts and the intracellular localization of
    protocollagen prolyl and protocollagen lysyl hydroxylases.
    Biochemical Journal, 144, 123-130.
Harwood, R., Grant, M.E. & Jackson, D.S. (1975). Studies on the
    glycosylation of hydroxylysine residues during collagen
    biosynthesis and the subcellular localization of collagen
    galactosyltransferase and collagen glucosyltransferase
    in tendon and cartilage cells. Biochemical Journal, 152,
    291-302.
Holst, van G.-J. & Fincher, G.B. (1984). Polyproline II conformation in
    the protein component of arabinogalactan-protein from Lolium
    multiflorum. Plant Physiology, 75, 1163-1164.
Holst, van G.-J. & Varner, J.E. (1984). Reinforced polyproline II
    conformation in a hydroxyproline-rich cell wall glycoprotein
    carrot root. Plant Physiology, 74, 247-251.
Homer, R.B. & Roberts, K. (1979). Glycoprotein conformation in plant
    cell walls. Circular dichroism reveals a polyproline II
    structure. Planta, 146, 217-222.
Howell, K.A., Ito, A. & Palade, G.E. (1978). Endoplasmic reticulum
    marker enzymes in Golgi fractions - what does this mean?
    Journal of Cell Biology, 79, 581-589.
Kawasaki, S. (1980). Studies on the Golgi apparatus of suspension-
    cultured tobacco cells. Ph.D. Thesis. University of Tokyo,
    Japan.
Kawasaki, S. (1981). Synthesis of arabinose-containing cell wall
    precursors in suspension-cultured tobacco cells. I.
    Intracellular site of synthesis and transport. Plant & Cell
    Physiology, 22, 431-442.
Kivirikko, K.I. & Myllylä, R. (1979). Collagen glycosyltransferases.
    International Review of Connective Tissue Research, 8,
    23-72.
Lamport, D.T.A. (1980). Structure and function of plant glycoproteins.
    In The Biochemistry of Plants 3, ed. J. Preiss, pp. 501-541.
    New York: Academic Press.
Lamport, D.T.A. & Miller, D.H. (1971). Hydroxyproline arabinosides in
    the plant kingdom. Plant Physiology, 48, 454-456.
Lamport, D.T.A., Katona, L. & Roerig, S. (1973). Galactosylserine in
    extensin. Biochemical Journal, 133, 125-131.
Leach, J.E., Cantrell, M.A. & Sequeira, I. (1982). A hydroxyproline-rich
    bacterial agglutinin from potato: extraction, purification,
    and characterization. Plant Physiology, 70, 1353-1358.
McGee, O'D.J., Langness, U. & Udenfriend, S. (1971). Immunological
    evidence for an inactive precursor of collagen proline
    hydroxylase in cultured fibroblasts. Proceedings of the
    National Academy of Sciences of the United States of America,
    68, 1585-1589.
Murphy, L. & Rosenbloom, J. (1973). Evidence that chick tendon procollagen
    must be denatured to serve as substrate for proline hydroxylase.
    Biochemical Journal, 135, 249-251.

Olsen, B.R., Berg, R.A., Kishida, Y. & Prockop, D.J. (1975). Further characterization of embryonic tendon fibroblasts and the use of immunoferritin techniques to study collagen biosynthesis. Journal of Cell Biology, 64, 340-355.

Oohira, A., Nogami, H., Kusakabe, A., Kimita, K. & Suziku, S. (1979). Structural differences among procollagens associated with rough and smooth microsomes from chick embryo cartilage. Journal of Biological Chemistry, 254, 3576-3583.

Owens, R.J. & Northcote, D.H. (1981). The location of arabinosyl: hydroxyproline transferase in the membrane system of potato tissue culture cells. Biochemical Journal, 195, 661-667.

Palmiter, R.D., Davidson, J.M., Gagnon, J., Rowe, D.W. & Bornstein, P. (1979). NH_2-terminal sequence of the chick proα1(I) chain synthesized in the reticulocyte lysate system. Journal of Biological Chemistry, 254, 1433-1436.

Rhoads, R.E. & Udenfriend, S. (1968). Decarboxylation of α-ketoglutarate coupled to collagen proline hydroxylase. Proceedings of the National Academy of Sciences of the United States of America, 60, 1473-1478.

Robinson, D.G., Andreae, M., Glas, R. & Sauer, A. (1984). Intracellular localization of hyp-rich glycoprotein biosynthesis. In Proceedings 7th Annual Winter Symposium in Botany, University of California, Riverside, eds. S. Bartnicki-Garcia & W.M. Dugger, pp. 254-267.

Rothman, J.E. (1981). The Golgi apparatus: two organelles in tandem. Science, 213, 1212-1219.

Sadava, D. & Chrispeels, W.J. (1971a). Intracellular sites of proline hydroxylation in plant cells. Biochemistry, 10, 4290-4294.

Sadava, D. & Chrispeels, W.J. (1971b). Hydroxyproline biosynthesis in plant cells. Peptidyl proline hydroxylase from carrot root discs. Biochimica et Biophysica Acta, 227, 278-287.

Samson, M., Klis, F.M., Sigon, C.A.M. & Stegwee, D. (1983). Localization of arabinogalactan proteins in the membrane system of etiolated hypocotyls of Phaseolus vulgaris L. Planta, 159, 322-328.

Sauer, A. & Robinson, D.G. (1984). Intracellular localization of posttranslational modifications in the synthesis of hydroxyproline-rich glycoproteins. I. Peptidyl proline hydroxylation in maize roots. Planta (in press).

Schofield, J.D., Freeman, I.L. & Jackson, D.S. (1971). The isolation and amino acid and carbohydrate composition of polymeric collagens prepared from various human tissues. Biochemical Journal, 124, 467-473.

Selvendran, R.R. & O'Neil, M.A. (1982). Plant Glycoproteins. In Plant Carbohydrates I. Encyclopedia of Plant Physiology, 13A, eds. F. Loewus & W. Tanner, pp. 515-583. Berlin, FRG: Springer Verlag.

Smith, M.A. (1981). Characterization of carrot cell wall protein. I. Effect of α,α'-dipyridyl on cell wall protein synthesis and secretion in incubated carrot discs. Plant Physiology, 68, 956-963.

Spiro, R.G. (1973). Glycoproteins. Advances in Protein Chemistry, 27, 349-467.

Tanaka, M., Shibata, H. & Uchida, T. (1980). A new prolyl hydroxylase acting on poly-L-proline from suspension cultured cells of Vinca rosea. Biochimica et Biophysica Acta, 616, 188-198.

Tanaka, M., Sato, K. & Uchida, T. (1981). Plant prolyl hydroxylase
    recognizes poly(L-proline)II helix. Journal of Biological
    Chemistry, 256-11397-11400.
Uitto, V.J., Uitto, J., Kao, W.W.-Y. & Prockop, J. (1978). Procollagen
    polypeptides containing cis-4-hydroxy-L-proline are over-
    glycoslyated and secreted as nonhelical pro-α-chains.
    Archives of Biochemistry and Biophysics, 188, 214-221.
Wienecke, K., Glas, R. & Robinson, D.G. (1982). Organelles involved in
    the synthesis and transport of hydroxyproline-containing
    glycoproteins in carrot root discs. Planta, 155, 58-63.
Wirtz, W., Stitt, M. & Heldt, H.-W. (1980). Enzymatic determination of
    metabolites in the subcellular compartments of spinach
    protoplasts. Plant Physiology, 66, 187-193.

D.H. Northcote
Department of Biochemistry, University of Cambridge,
Tennis Court Road, Cambridge, CB2 1QW, England

Abstract. Control of cell wall formation depends upon
several aspects of the processes for the secretion of
macromolecules from the cell.   The polysaccharides apart
from cellulose are synthesised within the lumen of the
endomembrane system, principally in the Golgi apparatus.
The syntheses both quantitatively and qualitatively are
primarily governed by the levels of synthase activities
in the appropriate part of the membrane system and by the
transport into the lumen of the endomembranes of the
soluble nucleoside diphosphate sugar precursors.   The
formation of these precursors is controlled to some extent
by biochemical feedback modulation.   Not much alteration
in the epimerase activities which interconvert the glucose
and galactose series of precursors takes place, although
there is some regulation by alteration of the level of the
enzyme activities bringing about the conversion of
UDPGlc to UDPXyl.

Once formed, the polysaccharides are packaged in vesicles,
probably in discrete compartments of the Golgi cisternae,
and moved under controlled conditions to the plasmamembrane
where a membrane-fusion process dependent on Ca^{2+} takes
place.   The vesicles are partly directed by microtubules
at the plasmamembrane, and the number of vesicles present
exceeds the number fusing, so that the rate of fusion is a
control process for the net secretion of material into the
wall.

Cellulose synthesis is controlled by the availability of
UDPGlc at the plasmamembrane surface in conjunction with
an organised relationship between the synthetic system and
acceptor glucosyl radical, so that an intact membrane is
necessary for $\beta 1 \rightarrow 4$ bonds to be formed rather than $\beta 1 \rightarrow 3$
linkages.   The orientation of the cellulose microfibrils
is partly directed by microtubules.

The lignin precursors, the hydroxycinnamyl alcohols, are
formed in association with the membranes of the endoplasmic
reticulum, and they are probably packaged in vesicles for
transport into the wall.   The regulation of the synthesis
of the hydroxycinnamyl alcohols is highly controlled by the
levels of several enzymes for phenylpropanoid synthesis,
including phenylalanine ammonia lyase.   Polymerization

occurs within the wall by reaction of free radicals formed
by a peroxidase/ascorbate oxidase system, which may be
regulatory.   The formation of some cross linkages between
tyrosine residues of polypeptides (isodityrosine) and by
ferulic diester bonds between polysaccharides is also
brought about by the oxidation system in the wall.

Key words: Cell differentiation; xylem formation;
cellulose synthesis; pectin and hemicellulose synthases;
PAL; lignin synthesis; vesicle fusion; secretion;
glycoproteins.

INTRODUCTION

The secretion of the plant cell wall at the external surface
of the plant cell depends partly on a coordinated and partly on a
sequential synthesis of various materials, mainly polysaccharide and
lignin.   These substances vary in amounts at different times of the
growth and development of the cells.   The secretion also depends on the
distribution and movement of membrane-bounded vesicles and their fusion
at specific areas of the cell surface at particular times.   The phenomenon
is therefore very similar to the secretion of glycoproteins in animal
cells (Northcote 1983).

In considering control of the development of the cell wall,
four different processes must be discussed: (i) a modulation of the
amounts of material that is continually being formed;   (ii) the switching
on and off of particular material at various stages of the development;
(iii) the mechanisms of the processes within the cell which lead to the
movement of precursors to the synthesising systems;   (iv) the packaging
of material and its deposition at specific sites.   The first two
processes are biochemical and the third and fourth processes are
cytological.   I propose to deal with these two aspects separately even
though to a certain extent they overlap since the distinction is for the
most part clear and the separation emphasises the importance of each
aspect.

BIOCHEMICAL ASPECTS OF CONTROL

1) Polysaccharide formation

The initial precursors for polysaccharide formation are the
nucleoside diphosphate sugars.   These are either formed from the sugar
phosphates and the nucleoside triphosphates by pyrophosphorylases or they
are produced by interconversions from the nucleoside diphosphate sugars.

The sugars arise initially from glucose, sucrose or myoinositol as the main sources (Fig. 1).   The nucleoside diphosphate sugars involved in pectin synthesis are formed by epimerases from the corresponding compound derived directly from UDPGlc (Fig. 1) (Northcote 1969).   The activities of the epimerases hardly change during the differentiation of the xylem cells, even at the stages of secondary wall formation when pectin is no longer being synthesised for deposition into the wall (Table 1) (Dalessandro & Northcote 1977c).   Xylan synthesis arises from UDPXyl and UDPGlcA and during secondary thickening, when the amount of xylem deposited in the wall is considerably increased, the activities of the enzymes which produce these two precursors (UDPGlc dehydrogenase and UDPGlcA decarboxylase) are increased (Table 1).   In addition, the activity of UDPGlc dehydrogenase is monitored by a biochemical feed-back inhibition by the level of UDPXyl (Dalessandro & Northcote 1977).

However, the main biochemical control of the amount of pectin and hemicellulose formed seems to be the activities of the synthases which are membrane-bound and which occur at the lumen of the endomembrane system.   These activities increase or decrease in relation to the amount of the polysaccharide that is formed at any one stage of the development of the cell (Table 2) (Dalessandro & Northcote 1981b; Bolwell & Northcote 1981).   For the xylan and arabinan synthases it can be shown that these variations in activity are controlled at the levels of transcription and translation since the expected rise in the activities either at secondary thickening for the former or at cell division and growth in wall area for the latter can be inhibited by inhibitors of transcription (actinomycin D) and translation (D-2-(4-methyl-2,6-dinitroanilino)-N methylpropionamide) (Bolwell & Northcote 1983).   The arabinan synthase in bean tissue has been shown to be inhibited by a monoclonal antibody raised against a mixed cell-membrane preparation. This antibody indicated, by a binding assay to the membranes of the cell, that the amount of enzyme present at the membrane increased during the time at which the increased activity was apparent (Bolwell & Northcote 1984).

As well as this prime mechanism for the regulation of these enzyme activities there may be further biochemical controls imposed by the energy status of the cell, since xylan synthase is modulated by the levels of nucleoside mono- and di-phosphates (Dalessandro & Northcote, 1981b).

The xylans and arabinans of the cell wall are heteropolymers.

Fig. 1. Interconversions of nucleoside diphosphate sugars and control points for polysaccharide and glycoprotein synthesis and transport during wall formation.

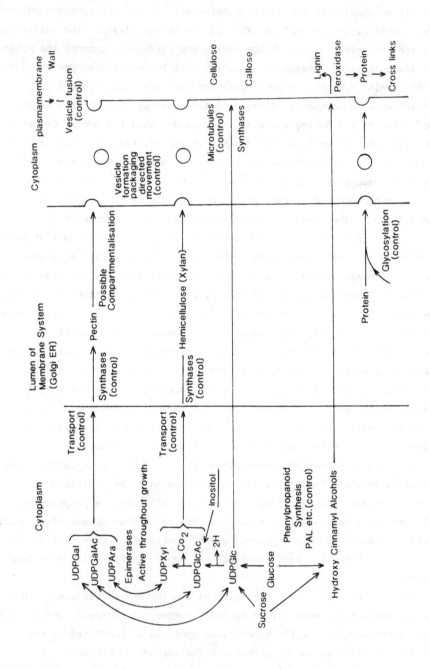

Table 1   Changes in enzyme activities of nucleoside diphosphate sugar
interconversions during differentiation of cambium to xylem
in sycamore (Dalessandro & Northcote 1977).

Enzyme	Activities of enzymes (nmol/min/mg protein)		
	Cambium	Differentiating Xylem	Differentiated Xylem
UDP-D-galactose-4-epimerase	16.3	19.9	12.1
UDP-D-xylose-4-epimerase	1.0	1.3	0.9
UDP-D-glucose dehydrogenase	6.4	10.0	18.0
UDP-D-glucuronate decarboxylase	132.8	313.5	303.7

Table 2   Synthase activities in differentiating cells   (Bolwell & Northcote 1981; 1983; 1984; Dalessandro & Northcote 1981 and unpublished results of Bolwell, Dalessandro & Northcote).

Source of enzyme	Xylan synthase activity (nmol/min/mg protein)	Arabinan synthase activity (nmol/min/mg protein)	Polygalacturonan synthase activity (nmol/min/mg protein)
Sycamore trees			
Cambium cells	0.7	-	7.8
Differentiating xylem cells	1.6	-	3.9
Differentiated xylem cells	4.2	-	1.7
Hypocotyls of bean			
6 days after imbibition	0.07	0.15	-
6-7 days after imbibition Rapid extension growth	0.17	0.60	-
9-10 days after imbibition Stele formation	0.45	0.20	-
Bean callus tissue			
Subcultured on maintenance medium			
At subculture	0.03	0.02	-
5-8 days after subculture Period of cell division	0.10	0.52	-
Subcultured on induction medium			
At subculture	0.02	0.02	-
5-8 days after subculture	0.12	0.20	-
9-12 days after subculture Period of xylem formation	0.40	0.10	-

They are branched polysaccharides which contain monosaccharides other
than xylose or arabinose.   During the step-wise addition of the xylose
or arabinose to the polymer, stages must be reached which include the
branch point and/or the different sugar.   Continued growth of the
polymer might therefore be controlled by the insertion of the particular
sugar or the formation of the branch-linkage at stages in the assembly
during the extension of the main or the branch chains of the molecule.
This will be true whether the polymer is synthesised directly from the
nucleoside diphosphate sugars or whether oligosaccharides are assembled
on intermediates such as dolichyl diphosphate oligosaccharides for
subsequent transfer to the polymer (see below).

2)   Cellulose and callose synthesis
The formation of these polymers takes place at the plasmamembrane.   The intact membrane is needed for microfibrillar synthesis.
If the membrane is damaged callose synthesis predominates.   The rate
of cellulose synthesis in vitro, with isolated membranes, is usually
much lower than that in vivo.   Callose synthesis is nearly always much
greater with in vitro preparations than it is in vivo (Delmer 1983).
Callose is used in plants to plug wounds and it can be rapidly removed,
probably by a hydrolytic or phosphorylytic system.

We have shown that with the same membrane preparation, using
UDPGlc as a donor, the synthesis of $\beta1\rightarrow3$ linkages can be increased by
freezing and thawing the membranes, i.e. damage and disorientation of the
membranes (Jacob & Northcote 1984).   It is possible therefore that the
same enzyme system carried on the membrane can synthesise either $\beta1\rightarrow3$ or
$\beta1\rightarrow4$ links (a transglucosylase system specific to the secondary alcoholic
groups of the acceptor) and that the particular assembly which occurs
depends on the orientation of pre-existing polysaccharide, oligosaccharide
or monosaccharide held on the membrane in relation to the synthetic or
transglucosylase system organised on the plasmamembrane surface.   In
this way, hydroxyl on position C-3 or C-4 of the glucose becomes
available for glycosylation because of the way in which the acceptor
glucose molecule is held on the membrane.   The acceptor molecule could
probably attach to the membrane via the hydroxyl on C-2 and C-4 would be
available to the directed donor molecule when held in the correct
orientation.   When the membrane is damaged the orientation is disturbed
and more $\beta1\rightarrow3$ linkages are formed.   (It is of interest to note that with

the enzymic system of <u>Acetobacter</u> <u>xylinium</u> some 1→2 glucan can be
formed using an <u>in</u> <u>vitro</u> membrane preparation.)   If it is further
postulated that the breakdown of callose is continually taking place and
that this reaction is inhibited by tissue damage, then it is possible to
reconcile experiments which show callose to be a possible precursor of
cellulose (Meier <u>et al</u>. 1981) and also to indicate why there could be
small amounts of callose at the immediate surface of the wall next to the
plasmamembrane (Fig. 2) (Waterkeyn 1981).

Fig. 2.   Possible interrelation of cellulose and callose
formation mediated by an intact plasmamembrane (see text
for discussion).

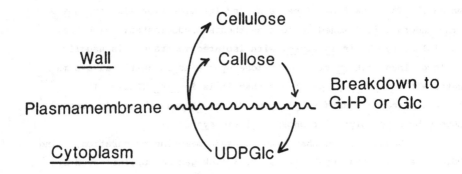

3)   <u>Lignin formation</u>
        Lignin formation occurs during secondary thickening of the
wall and it can be induced in tissue cultures of bean by the application
of naphthalene acetic acid (NAA) and kinetin in the ratio of 5:1.   These
tissue cultures can therefore be used to monitor the formation of the
lignin and the extent of the differentiation (Haddon & Northcote 1975).
        The starting point for the formation of the hydroxycinnamyl
alcohols which act as the immediate building units for lignin is the
phenylalanine ammonia lyase (PAL) reaction.   During the differentiation
of plant tissue cultures, when nodules of xylem and phloem are formed,

the activity of PAL increases to a maximum during the onset of the
differentiation and then falls off to a value that is usually higher
than the level at the beginning of the differentiation, and this level
of activity is maintained over the initial period of the differentiation
(Haddon & Northcote 1975; Fukuda & Komamine 1982; Kuboi & Yamada 1978;
Bevan & Northcote 1979).

The careful application of inhibitors of transcription
(actinomycin D) and translation (D-2-(4-methyl 2,6-dinitroanilino)-N
methyl-propionamide) have shown that for the increase in activity of PAL
both transcription and translation are necessary (Jones & Northcote 1981).
Other enzymes associated with lignin synthesis (S-adenosylmethionine :
caffeic acid 3-0 methyl transferase and cinnamic acid-4 hydroxylase)
show an increase in activity during differentiation, and it appears as if
there might be a coordinated induction of enzymes necessary for the
synthesis of the lignin precursors at the time of lignin deposition
(Haddon & Northcote 1976; Amrhein & Zenk 1970; Hahlbrock & Wellmann
1973).   The type of lignin deposited may also be controlled to some
extent by the transmethylases, since the enzyme present in gymnosperms
is relatively inactive on 5-hydroxyferulate and thus the syringyl type
of lignin is not formed (Grisebach 1981).   Before reduction of the
hydroxycinnamic acids to the corresponding alcohols the acids are
activated by a CoA ligase.   This enzyme may also control the type of
lignin that is eventually produced, since isoenzymes which occur in
different proportions in different plants and in different tissues of the
same plant have different affinities and activities for p-coumaric,
ferulic and sinapic acids.   The distribution of the specific ferulic
acid 5-hydroxylase may also control the production of sinapic acid and
hence its incorporation into lignin (Grand 1984).

The PAL activity which is necessary for lignin formation
occurs in the cytoplasm or bound to the cytoplasmic surface of the
endoplasmic reticulum membranes (Wagner & Hrazdina 1984).   The cinnamic
acid that is produced is probably carried in the lipid of the membrane,
and it is sequentially hydroxylated by membrane-bound hydroxylases
(Hanson & Havir 1981; Butt & Lamb 1981) so that a channelling route
from phenylalanine to p-coumaric acid is established.   The lignin
precursors may be packaged in vesicles and these are transported to the
wall by fusion with the plasmamembrane.   The polymerisation takes place
within the wall after oxidation of the hydroxycinnamyl alcohols by

peroxidase to mesomeric phenoxy radicals.   The half-life of these free
radicals is very short before they react together to give lignin and to
form linkages between lignin and the polysaccharides of the wall
(Freudenberg 1965; Freudenberg 1968).   The activities of the isoenzymes
of peroxidase found in the wall are therefore possibly control points for
the lignification, and in some cases these have been found to increase
in activity during lignification (Fukuda & Komamine 1982).

    4)   The action of the growth factors
        Since the change of the ratio of the growth factors (NAA and
kinetin) in the growth medium of the tissue cultures induces the
differentiation (Haddon & Northcote 1975; Bevan & Northcote 1979), it is
possible that these substances control to some extent the transcription
and translation of the various enzymes involved in the synthesis of the
polysaccharides and the lignin precursors.
        The effects of NAA and kinetin on the induction of PAL
activity during the time course of differentiation can be separated in
time of application of the two growth substances (Bevan & Northcote 1979).
It is possible therefore that they may act at two separate loci.   It
can also be shown by gel electrophoresis of proteins that the
application of NAA to the tissue culture after subculture, brings
about the induction of new proteins so that a new pattern of protein
synthesis occurs within the cells.   Subculture is essential for the
change (Bevan & Northcote 1981a).   Kinetin application, even after
subculture, has no effect on the pattern of protein synthesis (Bevan &
Northcote 1981b).
        In many different tissues, the composition of the wall pectin
changes during development and differentiation (Asamizu et al. 1984).
These changes are related to alterations in the physical properties of
the wall that occur as the functions of the cell alter during its
development (Northcote 1972).   The pectin properties are considerably
modified by the presence of blocks of neutral arabinogalactans joined
to the polygalacturonorhamnan backbone (Stoddart & Northcote 1967).
The type of pectin synthesised can be modified by the application of
growth factors such as 2,4 dichlorophenoxyacetic acid to a tissue
culture, and the pattern of the synthesis of the arabinan is influenced
(Rubery & Northcote 1970).   The pectin synthesised by a plant cell can
be influenced during normal growth by a variety of experimental

procedures which includes application of plant growth factors
(Asamizu et al. 1983), plasmolysis (Boffey & Northcote 1975) and cell
wall removal (protoplast formation) (Hanke & Northcote 1974).   The
enzymes necessary for synthesis of the individual polysaccharides of the
pectin and the transglycosylases by which they become linked, especially
the possibility of transferring arabinogalactan portions to the main
uronic chain, are therefore of considerable importance.   These enzymes
not only control the texture of the wall but are targets for the
controlling factors of its development.

5)   Protein in the wall

Most of the proteins found in the cell wall are glycoproteins.   These range from enzymes such as isoenzymes of peroxidase,
phosphatase and amylase to the hydroxyproline-rich glycoproteins
(Gould & Northcote 1984).   The hydroxyproline-rich glycoproteins may
be classified on the basis of the size of their sugar prosthetic groups.
The soluble lectins and agglutinins and the insoluble wall glyco-
proteins have small oligosaccharides of arabinose (α-L-Araf(1$\rightarrow$3)-0-β-L-
Araf(1$\rightarrow$2)-0-β-L-Araf(1$\rightarrow$2)-0-β-L-Araf-1$\rightarrow$Hyp) linked to the hydroxyproline
and also single galactose units attached to serine (Lamport 1980; Allen
et al. 1978; Muray & Northcote 1978), while the arabinogalactan proteins
are mainly large molecular weight polysaccharides attached to protein,
the resultant molecule being about 80-90% carbohydrate (Fincher et al.
1983).   Hydroxylation of peptidyl proline occurs as a post-translational
process in the endoplasmic reticulum (Wienecke et al. 1982), and the
addition of the small arabinosyl oligosaccharides probably occurs within
the Golgi apparatus without the necessity for assembly on a lipid
intermediate (Owens & Northcote 1981).   However, with the large
molecular weight arabinogalactan protein a lipid carrier might be
involved, especially as repeating subunits within the arabinogalactan
portion of the molecule have been detected.   Oligosaccharide linked to
polyisoprenyl-pyrophosphate contained arabinose (Hayashi & Maclachlan
1984a) and galactose (Hayashi & Maclachlan 1984b);   the galactose was
linked by 1$\rightarrow$6, 1$\rightarrow$4 and 1$\rightarrow$3 bonds and the arabinose was 1$\rightarrow$5 linked.
These isoprenyl diphosphate oligosaccharides were formed by membranes of
pea cells when they were incubated with the appropriate radioactive UDP
sugar compounds, and they could serve as precursors of the
arabinogalactan-glycoproteins.

The possible inclusion of lipid-linked intermediates in the transglycosylations involved in glycoprotein and polysaccharide formation provides a further step at which control of the synthase system can be exercised.   However, although lipid intermediates are well established for the formation of N-linked glycoproteins and, in some instances, for polysaccharides where glycoproteins can function as intermediates during the formation of the polysaccharides (Green & Northcote 1978; Green & Northcote 1979a,b), there is, at present, no evidence for a direct transfer from lipid-oligosaccharide onto polysaccharide.   One of the important consequences of the participation of lipid-oligosaccharide intermediates is that the sequence of sugars formed on the lipid can be successively transferred so that a repeating ordered sequence of the mixed sugars in the oligosaccharide could occur in the polymer (Ielpi et al. 1981a,b; Dixon & Northcote 1982; Couso et al. 1982).

6)   Peroxidase and the cross linkages between polymers

In the rigid secondary wall of woody tissue, lignin replaces water present in the growing wall and the matrix becomes hydrophobic and rigid.   Many hydrogen bonds occur between the polysaccharides at the microfibrillar-matrix interface and between the components of the matrix. The wall becomes a composite in which the polysaccharide polymers are enclosed in a cross-linked lignin cage (Northcote 1972).   The lignin in the matrix is also covalently linked to polysaccharide by the reaction of the mesomeric phenoxy radicals which form the lignin and which are produced as a result of the oxidation of the hydroxycinnamyl alcohols by peroxidase.

The tyrosine residues of proteins of the wall may also be oxidised by peroxidase to give cross linkages of isodityrosine (Fry 1982; Cooper & Varner 1983).   These linkages are known to occur intra-molecularly (Smith et al. 1984; Epstein & Lamport 1984) and they may occur intermolecularly to form covalent links between polypeptides. There is some evidence to suggest that soluble hydroxyproline-rich polypeptides such as the lectins and agglutinins may be precursors of the insoluble cell wall glycoproteins, which become insoluble due to the formation of intermolecular isodityrosine linkages (Cooper & Varner 1983; Kawasaki 1982), although these intermolecular bonds have not yet been identified.   The degree of coupling may be controlled by the

peroxidase/ascorbate oxidase system localised in the wall.   The
formation of a cross-linked cage of glycoprotein is believed by some
workers to limit plant cell wall extensibility.

The existence of ferulic acid in the cell wall of some plants
allows the possibility of oxidation to give diferulic acid ester
linkages joining polysaccharide chains (Fry 1983; Tanner & Morrison 1983),
and this also gives rise to further speculation on the mechanisms which
can limit cell wall extensibility.   However, regardless of the presence
of these intermolecular covalent bonds, in the growing wall (unlignified)
the main cohesion between the constituents, especially the polysaccharides, must be the hydrogen bonding that allows a flexibility to
the structure of the wall which depends on its most variable feature,
the water content.

CYTOLOGICAL ASPECTS OF CONTROL

1)   Polysaccharide formation

The nucleoside diphosphate sugars which are the precursors
of the polysaccharides are formed and interconverted by soluble enzymes
in the cytoplasm outside the endomembrane system (Dalessandro &
Northcote 1977).   Some epimerase activity and some of the other
enzymes for the interconversion of the sugar nucleotides are found
associated with membranes (Dalessandro & Northcote 1981a).   It has been
shown that many soluble enzymes found experimentally within the cytoplasm
are either loosely bound to membranes in an organised manner or they
form organised aggregates (Masters 1981).   In this way, sequences of
directed reactions can be achieved, and these can be controlled and
modulated by direct biochemical feed-back mechanisms.

The bulk of the transglycosylation reactions for hemicellulose and pectin polysaccharides of the cell wall occur in the Golgi
apparatus, but it is possible that initial glycosylations may occur in
the endoplasmic reticulum if protein and polyisoprenoid carriers are
involved (Northcote 1982).   The transport processes for the nucleoside
diphosphate sugars are not known, but as in animal tissues there may be
at least two systems.   The mechanism of transport used would depend
upon the site, either endoplasmic reticulum or Golgi apparatus, and also
whether polyisoprenoids were involved in a transglycosylation-transport
process or whether protein transporters of the nucleotide-sugars were
present (Northcote 1984).   Whatever the mechanism, the transport may be

an important control point for polysaccharide production (Fig. 1), since
although the enzymes for polymerisation are developed within the lumen
of the membranes, the synthases will not function unless the substrates
are available to them.   Their full potential and activity can be
restricted in any part of the endomembrane system, both quantitatively
and qualitatively, by the transport of nucleoside diphosphate sugars.
These considerations may account for the ambiguities encountered between
experiments with in vitro and in vivo systems.   It is possible, for
instance, to detect polysaccharide synthase systems in characterised
parts of the endomembranes isolated from broken cells, and it can be
suggested from a knowledge of this enzyme complement, that the membranes
in vivo will synthesise particular polysaccharides.   However when intact
cells are used and fed radioactive monosaccharides, these particular
polysaccharides sometimes cannot be found at the membrane site where the
synthase activities are known to occur (Bolwell & Northcote 1983a).
Thus, although a particular part of the endomembrane system has the
potential for the polysaccharide synthesis, this does not always occur
in vivo.   One of the reasons for this difference may be the control of
the transport of the precursors.   This will be true not only for parts
of the system such as the endoplasmic reticulum and Golgi apparatus but
also for the plasmamembrane, which can receive enzymes via the vesicles
which fuse with it from the Golgi apparatus.

2)   Cellulose microfibrils

It is generally assumed that the cellulose microfibrils,
which are made up of $\beta1\rightarrow4$ glucan chains, are spun out at the surface
of the plasmamembrane by a moving enzymic system.   The glucan chains
are arranged in a definite order with respect to one another within
the microfibril, and their conformation is stabilized by inter- and
intra-molecular hydrogen bonds.   An organised array of particles, seen
both at the E and P faces of freeze-fractured plasmamembrane as
complementary clusters, is thought to traverse the membrane, and since
they occur at the ends of the microfibrils (terminal complex) (Mueller &
Brown 1980; Giddings et al. 1980) they could constitute the synthetic
system for the microfibril.   However, this complex, although it is
probably protein, has not yet been shown to be enzymic.   The microfibrils are oriented in definite directions at various stages of cell
growth.   The orientation of the microfibrils as they are deposited is

normally related to that of the underlying microtubules in the cytoplasm just at the plasmamembrane (Wooding & Northcote 1964; Pickett-Heaps & Northcote 1966).   It is assumed that the microtubules serve to direct the movement of the terminal complex and that the movement is generated either in conjunction with the cytoskeleton or by the alignment, almost crystallization, of the β1→4 glucan chains into organised microfibrils at the cell surface (Herth 1980; Heath & Seagull 1982).

3) Vesicle fusion

Studies on wall regeneration on protoplasts and in plasmolysed cells and also on the new cell walls formed at the cell plate during cytokinesis have indicated the sequence of wall formation.   Usually the matrix is deposited first by vesicle fusion, and into this matrix the microfibrils are woven (Northcote 1977; Northcote 1979b).   The early matrix is rich in pectin and its composition varies during these initial stages.

All the polysaccharide deposited into the wall depends on vesicle fusion either for the direct transfer of the polymers formed within the endomembrane system, or for the incorporation of enzyme into the plasmamembrane or into the wall.   The vesicles arise from the Golgi apparatus and are transported to particular areas of the cell surface. The secreted material from the Golgi apparatus is almost certainly sequestered to particular cisternae and perhaps even to particular parts of a cisterna (Quinn et al. 1983; Tartakoff 1982; Griffiths et al. 1983; Rindler et al. 1984).   By analogy with model systems suggested for animal cells, the sequestering of material within the Golgi cisternae and the fusion of the vesicle with pre-existing membrane at the cell surface is dependent upon receptors (Sly & Fischer 1982; Pastan & Willingham 1983; Farquhar 1983).   These are capable of recognition of ligands either on or contained within the membranes.

In some cells it can be seen that vesicle fusion is dependent upon a characteristic distribution of particles (proteins) within the membrane (Satir et al. 1973; Burwen & Satir 1977; Da Silva & Nogueira 1977).   The vesicles produced from the Golgi apparatus may also be directed to particular sites at the cell surface by microtubules (Northcote 1971).   This is clearly illustrated during the development of xylem elements when a spiral or reticulate secondary wall is laid down and during its formation the microtubules are distributed just under

the plasmamembrane at the sites of the thickening (Wooding &
Northcote 1964; Pickett-Heaps & Northcote 1966).

Part of the control of vesicle fusion at the surface is
mediated by the ionic atmosphere at the membrane, and Ca^{2+} is
necessary for the fusion to occur (Loister & Layter 1973; Dahl et al. 1978;
Baydoun & Northcote 1980a,b,1981).   Membrane fractions of the cells
of maize root tips can be obtained which allow an in vitro study of
membrane fusion.   It is possible using this system to identify a
characteristic protein, which can be partially removed by trypsin,
which influences Ca^{2+} mediated fusion.   The rate of vesicle fusion can
be a limiting process for the rate of cell wall formation, since at any
one time the number of vesicles ready for fusion exceeds the number that
are fusing and depositing material into the wall (Morris & Northcote
1977).   In this way the composition and amount of wall material
deposited may respond very quickly to a stimulus at the cell surface
which allows the rate of vesicle fusion to vary.   A new steady state
would then be achieved that produced the requisite number of vesicles
necessary to maintain the new rate of fusion.   The turnover of the
Golgi apparatus can be very fast in plant cells, and times of 5-40
minutes have been calculated (Robinson & Kristen 1982).

REFERENCES

Allen, A.K., Dsai, N.N., Neuberger, A. & Creeth, J.M. (1978).
        Properties of potato lectin and the nature of its
        glycoprotein linkages. Biochemical Journal, 171, 665-674.
Amrhein, N. & Zenk, M.H. (1970). Concomitant induction of phenylalanine
        ammonia-lyase and cinnamic acid 4-hydroxylase during
        illumination of excised buckwheat hypocotyls.
        Naturwissenschaften, 57, 312.
Asamizu, T., Nakano., N., Nishi, A. (1983). Changes in non-cellulosic
        cell-wall polysaccharides during the growth of carrot cells
        in suspension cultures. Planta, 158, 166-174.
Asamizu, T., Nakano., N., Nishi, A. (1984). Pectic polysaccharides in
        carrot cells growing in suspension culture. Planta, 160,
        469-473.
Baydoun,E.A.H. & Northcote, D.H. (1980a). Isolation and characterization
        of membranes from the cells of maize root tips. Journal of
        Cell Science, 45, 147-167.
Baydoun,E.A.H. & Northcote, D.H. (1980b). Measurement and characteristics
        of fusion of isolated membrane fractions from maize root tips.
        Journal of Cell Science, 45, 169-186.
Baydoun,E.A.H. & Northcote, D.H. (1981). The extraction from maize
        (Zea mays) root cells of membrane-bound protein with Ca^{2+}-
        dependent ATPase activity and its possible role in membrane
        fusion in vitro. Biochemical Journal, 193, 781-792.

Bevan, M. & Northcote, D.H. (1979). The interaction of auxin and
    cytokinin in the induction of phenylalanine ammonia-lyase
    in Phaseolus vulgaris. Planta, 147, 77-81.
Bevan, M. & Northcote, D.H. (1981a). Some rapid effects of synthetic
    auxins on mRNA levels in cultured plant cells. Planta, 152,
    32-35.
Bevan, M. & Northcote, D.H. (1981b). Subculture-induced protein
    synthesis in tissue cultures of Glycine max and Phaseolus
    vulgaris. Planta, 152, 24-31.
Boffey, S.A. & Northcote, D.H. (1975). Pectin synthesis during wall
    regeneration of plasmolysed tobacco leaf cells. Biochemical
    Journal, 150, 433-440.
Bolwell, G.P. & Northcote, D.H. (1981). Control of hemicellulose and
    pectin synthesis during differentiation of vascular tissue
    in bean (Phaseolus vulgaris) callus and in bean hypocotyl.
    Planta, 152, 225-233.
Bolwell, G.P. & Northcote, D.H. (1983a). Arabinan synthase and xylan
    synthase activities of Phaseolus vulgaris. Subcellular
    localization and possible mechanism of action. Biochemical
    Journal, 210, 497-507.
Bolwell, G.P. & Northcote, D.H. (1983). Induction by growth factors of
    polysaccharide synthases in bean cell suspension cultures.
    Biochemical Journal, 210, 509-515.
Bolwell, G.P. & Northcote, D.H. (1984). Demonstration of a common
    antigenic site on endomembrane proteins of Phaseolus
    vulgaris by a rat monoclonal antibody. Planta, in press.
Burwen, S.J. & Satir, B.H. (1977). A freeze-fracture study of early
    membrane events during mast cell secretion. Journal of
    Cell Biology, 73, 660-671.
Butt, V.S. & Lamb, C.J. (1981). Oxygenases and the metabolism of plant
    products. In The Biochemistry of Plants. Secondary Plant
    Products, ed. E.E. Conn, 7, 627-665. London: Academic Press.
Cooper, J.B. & Varner, J.E. (1983). Insolubulization of hydroxyproline-
    rich cell wall glycoprotein in aerated carrot root slices.
    Biochemical and Biophysical Research Communications, 112,
    161-167.
Couso, R.O., Ielpi, L., Garcia, R.C., Dankert, M.A. (1982).
    Biosynthesis of polysaccharides in Acetobacter xylinium -
    sequential synthesis of a heptasaccharide diphosphate
    prenol. European Journal of Biochemistry, 123, 617-627.
Dahl, G., Schudt, C., Gratzl, M. (1978). Fusion of isolated myoplast
    plasma membranes. Biochimica et Biophysica Acta, 514, 105-116.
Dalessandro, G. & Northcote, D.H. (1977). Changes in enzymic activities
    of nucleoside diphosphate sugar interconversions during
    differentiation of cambium to xylem in sycamore and poplar.
    Biochemical Journal, 162, 267-279.
Dalessandro, G. & Northcote, D.H. (1981a). Xylan synthetase activity
    in differentiated xylem cells of sycamore trees (Acer
    pseudoplatanus). Planta, 151, 53-60.
Dalessandro, G. & Northcote, D.H. (1981b). Increase of xylan synthetase
    activity during xylem differentiation of the vascular
    cambium of sycamore and poplar trees. Planta, 151, 61-67.
Da Silva, P.P. & Nogueira, L. (1977). Membrane fusion during secretion.
    Journal of Cell Biology, 73, 161-181.
Delmer, D.P. (1983). Biosynthesis of cellulose. Advances in
    Carbohydrate Chemistry, 41, 105-153.

Dixon, W.T. & Northcote, D.H. (1982). Glycolipids and glycoproteins
    during maize-root slime synthesis. In preparation.
Epstein, L. & Lamport, D.T.A. (1984). An intramolecular linkage
    involving isodityrosine in extensin. Phytochemistry, 23,
    1241-1246.
Fincher, G.B., Stone, B.A., Clarke, A.E. (1983). Arabinogalactan-
    proteins: structure, biosynthesis and function. Annual
    Review of Plant Physiology, 34, 47-70.
Freudenberg, K. (1965). Lignin: its constitution and formation from
    p-hydroxycinnamyl alcohols. Science, 148, 595-600.
Freudenberg, K. (1968). The constitution and biosynthesis of lignin.
    In Constitution and Biosynthesis of Lignin, ed.
    A. Kleinzeller. Molecular Biology, Biochemistry and
    Biophysics, 2, 45-122. Berlin: Springer-Verlag.
Fry, S.C. (1982). Isodityrosine, a new cross linking amino acid from
    plant cell wall glycoprotein. Biochemical Journal, 204,
    449-455.
Fry, S.C. (1983). Feruloylated pectins from the primary cell wall:
    their structure and possible functions. Planta, 157, 111-123.
Fukuda, H. & Komamine, A. (1982). Lignin synthesis and its related
    enzymes as markers of tracheary-element differentiation in
    single cells isolated from the mesophyll of Zinnia elegans.
    Planta, 155, 423-430.
Giddings, T.H., Brower, D.L., Staehelin, L.D. (1980). Visualization of
    particle complexes in the plasma membrane of Micrasterias
    denticulata associated with the formation of cellulose fibrils
    in primary and secondary cell walls. Journal of Cell Biology,
    84, 327-339.
Gould, J. & Northcote, D.H. (1984). Characteristics of plant surfaces.
    In Bacterial Adhesion: Mechanisms and Physiological
    Significance, in press, eds. M.M. Fletcher & D.C. Savage.
    New York: Plenum Press.
Grand, C. (1984). Ferulic acid 5-hydroxylase: a new cytochrome P-450-
    dependent enzyme from higher plant microsomes involved in
    lignin synthesis. FEBS Letters, 169, 7-11.
Green, J.R. & Northcote, D.H. (1978). The structure and function of
    glycoproteins synthesised during slime-polysaccharide
    production by membranes of the root-cap cells of maize
    Zea mays. Biochemical Journal, 170, 599-608.
Green, J.R. & Northcote, D.H. (1979a). Polyphenyl phosphate sugars
    synthesised during slime-polysaccharide production by
    membranes of the root-cap cells of maize Zea mays.
    Biochemical Journal, 178, 661-671.
Green, J.R. & Northcote, D.H. (1979b). Location of fucosyl transferases
    in the membrane system of maize root cells. Journal of Cell
    Science, 40, 235-244.
Griffiths, G., Quinn, P., Warren, G. (1983). Dissection of the Golgi
    complex. 1. Monensin inhibits the transport of viral
    membrane proteins from medial to trans Golgi cisternae in
    baby hamster kidney cells infected with Semliki Forest
    virus. Journal of Cell Biology, 96, 835-850.
Grisebach, H. (1981). Lignins. In The Biochemistry of Plants. Vol. 7,
    Secondary Plant Products, ed. E.E. Conn, pp. 457-478.
    London: Academic Press.
Haddon, L.E. & Northcote, D.H. (1975). Quantitative measurement of the
    course of bean callus differentiation. Journal of Cell
    Science, 17, 11-26.

Haddon, L. & Northcote, D.H. (1976). Correlation of the induction of
    various enzymes concerned with phenylpropanoid and lignin
    synthesis during differentiation of bean callus Phaseolus
    vulgaris L.. Planta, 128, 255-262.
Hahlbrock, K. & Wellmann, E. (1973). Light-independent induction of
    enzymes related to phenylpropanoid metabolism in cell
    suspension cultures from parsley. Biochimica et Biophysica
    Acta, 304, 702-706.
Hanke, D.E. & Northcote, D.H. (1984). Cell wall formation by soybean
    callus protoplasts. Journal of Cell Science, 14, 29-50.
Hanson, K.R. & Havier, E.A. (1981). Phenylalanine ammonia lyase. In
    The Biochemistry of Plants. Vol. 7, Secondary Plant
    Products, ed. E.E. Conn, pp. 577-625. London: Academic Press.
Hayashi, T. & Maclachlan, G. (1984a). Biosynthesis of pentosyl lipids by
    pea membranes. Biochemical Journal, 217, 791-803.
Hayashi, T. & Maclachlan, G. (1984b). Glycolipids and glycoproteins
    formed from UDP-Galactose by pea membranes. Phytochemistry,
    23, 487-492.
Heath, I.B. & Seagull, R.W. (1982). Oriented cellulose fibrils and the
    cytoskeleton: a critical comparison of models. In The
    Cytoskeleton in Plant Growth and Development, ed. C.W. Lloyd,
    pp. 163-182. London: Academic Press.
Herth, W. (1980). Calcofluor white and congo red inhibit chitin
    microfibril assembly of Poterisochromonas. Evidence for a
    gap between polymerization and microfibril formation.
    Journal of Cell Biology, 87, 442-450.
Ielpi, L., Couso, R., Dankert, R. (1981a). Lipid-linked intermediates
    in the biosynthesis of xanthan gum. FEBS Letters, 130, 253-
    256.
Ielpi, L., Couso, R., Dankert, R. (1981b). Xanthan gum biosynthesis.
    Biochemical and Biophysical Research Communications, 102,
    1400-1408.
Jacob, S. & Northcote, D.H. (1984). Unpublished work.
Jones, D.H. & Northcote, D.H. (1981). Induction by hormones of
    phenylalanine ammonia-lyase in bean-cell suspension cultures.
    Inhibition and superinduction by actinomycin D. European
    Journal of Biochemistry, 116, 117-125.
Kawasaki, S. (1982). Synthesis of arabinose-containing cell wall
    precursor in suspension-cultured tobacco cells. II. Partial
    purification and some physical characterization of an
    intracellular precursor of cell wall glycoprotein. Plant
    and Cell Physiology, 23, 1443-1452.
Kuboi, T. & Yamada, Y. (1978). Regulation of the enzyme activities
    related to lignin synthesis in cell aggregates of tobacco
    cell culture. Biochimica et Biophysica Acta, 542, 181-190.
Lamport, D.T.A. (1980). Structure and function of plant glycoproteins.
    In The Biochemistry of Plants. Vol. 3, Carbohydrates:
    Structure & Functions, ed. J. Preiss, pp. 501-541. London:
    Academic Press.
Loister, Z. & Layter, A. (1973). The mechanism of cell fusion. Journal
    of Biological Chemistry, 248, 422-432.
Masters, C.J. (1981). Interactions between soluble enzymes and
    subcellular structure. Critical Reviews in Biochemistry, 11,
    105-143.
Meier, H., Buchs, L., Buchala, A.J., Homewood, T. (1981). (1→3)-β-D-
    glucan (callose) is a probable intermediate in biosynthesis
    of cellulose of cotton fibres. Nature, 289, 821-822.

Morris, M.R. & Northcote, D.H. (1977). Influence of cations at the
     plasma membrane in controlling polysaccharide secretion
     from sycamore suspension cells. Biochemical Journal, 166,
     603-618.
Mueller, S.C. & Brown, R.M. (1980). Evidence for an intramembrane
     component associated with a cellulose microfibril-
     synthesising complex in higher plants. Journal of Cell
     Biology, 84, 315-326.
Muray, R.H.A. & Northcote, D.H. (1978). Oligoarabinosides of
     hydroxyproline isolated from potato lectin. Phytochemistry,
     17, 623-629.
Northcote, D.H. (1969). The synthesis and metabolic control of
     polysaccharides and lignin during the differentiation of
     plant cells. Essays in Biochemistry, 5, 90-137.
Northcote, D.H. (1971). Organisation of structure, synthesis and
     transport within the plant during cell division and growth.
     Symposia of the Society for Experimental Biology, 25, 51-69.
Northcote, D.H. (1972). Chemistry of the plant cell wall. Annual Review
     of Plant Physiology, 23, 113-132.
Northcote, D.H. (1977). The synthesis and assembly of plant cell walls;
     possible control mechanisms. In Cell Surface Reviews, eds.
     G. Poste & G.L. Nicholson, Vol. 4, pp. 717-739. Pub.
     North-Holland, Amsterdam.
Northcote, D.H. (1979b). Polysaccharides of the plant cell during its
     growth. In Polysaccharides in Food, eds. J.M.V. Blanshard
     & J.R. Mitchell, pp. 3-13. Pub. Butterworth, London.
Northcote, D.H. (1982). Macromolecular aspects of cell wall
     differentiation. In Encyclopaedia of Plant Physiology.
     Vol. 14A, Nucleic-Acids and Proteins in Plants I, eds.
     D. Boulter & B. Parthier, pp. 637-655. Pub. Springer-Verlag,
     Berlin.
Northcote, D.H. (1983). Cell membranes and glycoprotein synthesis.
     Pub. The Royal Society, London, ed. D.H. Northcote.
     Philosophical Transactions of the Royal Society of London,
     Series B, 300, 107-235.
Northcote, D.H. (1984). Cell organelles and their function in
     biosynthesis of cell wall components: control of cell wall
     assembly during differentiation. In Biosynthesis and
     Biodegradation of Wood Components, in press, ed. T. Higuchi.
     Pub. Academic Press, San Diego.
Owens, R.J. & Northcote, D.H. (1981). The location of arabinosyl:
     hydroxyproline transferase in the membrane system of potato
     tissue culture cells. Biochemical Journal, 195, 661-667.
Pickett-Heaps, J.D. & Northcote, D.H. (1966). Relationship of cellular
     organelles to the formation and development of the plant cell
     wall. Journal of Experimental Botany, 17, 20-26.
Quinn, P., Griffiths, G., Warren, G. (1983). Dissection of the Golgi
     complex. II. Density separation of specific Golgi functions
     in virally infected cells treated with monensin. Journal
     of Cell Biology, 96, 851-856.
Rindler, M.J., Ivanov, I.E., Plesken, H., Rodriguez-Boulan, E.,
     Sabatini, D. (1984). Viral glycoproteins destined for
     apical or basolateral plasma membrane domains traverse the
     same Golgi apparatus during their intracellular transport
     in doubly infected Madin-Darby canine kidney cells.
     Journal of Cell Biology, 98, 1304-1319.

Robinson, D.G. & Kristen, U. (1982). Membrane flow via the Golgi
    apparatus of higher plant cells. International Review of
    Cytology, 7, 89-127.
Rubery, P.H. & Northcote, D.H. (1970). The effect of auxin (2,4
    dichlorophenoxyacetic acid) on the synthesis of cell wall
    polysaccharides in cultured sycamore cells. Biochimica et
    Biophysica Acta, 222, 95-108.
Satir, B., Schooley, C., Satir, P. (1973). Membrane fusion in a model
    system. Journal of Cell Biology, 56, 153-176.
Smith, J.J. (1984). Isolation of extensin precursors by direct elution
    of intact tomato cell suspension cultures. Phytochemistry,
    23, 1233-1239.
Stoddart, R.W. & Northcote, D.H. (1967). Metabolic relationships of
    the isolated fractions of the pectic substances of
    actively growing sycamore cells. Biochemical Journal, 102,
    194-204.
Tanner, G.R. & Morrison, I.M. (1983). Phenolic-carbohydrate complexes
    in the cell walls of Lolium perenne. Phytochemistry, 22,
    1433-1439.
Tartakoff, A.M. (1982). The role of subcompartments of the Golgi complex
    in protein intracellular transport. Philosophical
    Transactions of the Royal Society of London, Series B, 300,
    173-184.
Wagner, G.J. & Hrazdina, G. (1984). Endoplasmic reticulum as a site of
    phenylpropoid and flavonoid metabolism in Hippeastrum.
    Plant Physiology, 74, 901-906.
Waterkeyn, L. (1981). Cyto-chemical localisation and function of the
    3-linked glucan callose in the developing cotton fibre
    cell wall. Protoplasma, 106, 49-67.
Wienecke, K., Glas, R., Robinson, D.G. Organelles involved in the
    synthesis and transport of hydroxyproline-containing
    glycoproteins in carrot root discs. Planta, 155, 58-63.
Wooding, F.B.P. & Northcote, D.H. (1964). The development of the
    secondary wall of xylem in Acer pseudoplatanus.
    Journal of Cell Biology, 23, 327-337.

8 ARE LIPID-LINKED GLYCOSIDES REQUIRED FOR PLANT POLYSACCHARIDE
BIOSYNTHESIS?

G. Maclachlan
Department of Biology, McGill University, 1205 Dr. Penfield
Avenue, Montreal, Quebec, Canada   H3A 1B1

Abstract. Plant membranes are capable of incorporating
saccharides from sugar nucleotides into glycolipids as well
as into glycoproteins and polysaccharides. The question
raised here is whether any of these glycolipids are
intermediates (precursors) in the pathways for synthesis
of non-lipid products. The known saccharides in weakly-
charged lipid-monophosphate-monosaccharides in plants are
reviewed, as well as those in highly charged lipid-
pyrophosphate-oligosaccharides. It is concluded that the
evidence supports many of these charged glycolipids as
intermediates for plant glycoprotein biosynthesis but that
very few have been clearly identified to date as required
for cell-wall polysaccharide biosynthesis.

Key words: Cell wall, dolichol, ER, GDP-mannose,
glycolipid, lipid-linked saccharides, pea membranes,
plasmamembrane, polyisoprenol, polysaccharide, UDP-arabinose,
UDP-galactose, UDP-glucose, UDP-N-acetyl glucosamine,
UDP-xylose.

INTRODUCTION

The notion that glycolipids may act as precursors for the

biosynthesis of polysaccharides was introduced well over 20 years ago by

Colvin and co-workers (Colvin 1959; Khan & Colvin 1961 a,b) in

conjunction with studies on cellulose biosynthesis by crude enzyme

preparations of Acetobacter xylinum. When enzyme preparations were

incubated with ethanolic extracts of bacteria or higher plants,

fibrillar materials resembling cellulose microfibrils were observed to

develop in the mixtures. In addition, Kjosbakken & Colvin (1973) and

Garcia et al. (1974), using membrane preparations from A. xylinum,

demonstrated the formation from UDP-glucose of charged glucolipids,

including a component tentatively identified as cellobiosyl-lipid.
Thus, a potential pathway was proposed for biosynthesis of cellulose from
UDP-glucose via a membrane-bound glucolipid intermediate (reviewed by
Colvin, 1980).

Since that time, a large body of evidence has accumulated
(Waechter & Lennarz 1976; Tonn & Gander 1979; Elbein 1979) to show
beyond doubt that bacteria form cell-wall peptidoglycans and capsular
polysaccharides from sugar nucleotides via lipid intermediates, where
the lipid moiety is an unsaturated polyprenyl phosphate or pyrophosphate
with approx. 11 prenyl groups (bactoprenyl, undecaprenyl).   In addition,
lipid-saccharides in which the lipid component is esterified poly-
glycerophosphate are known to be intermediates in the synthesis of
bacterial teichoic acids (Tonn & Gander 1979).   Animal and plant
membranes can also assemble lipid-linked oligosaccharides which are used
as substrates for en-bloc transfer of the oligosaccharide to N-linked
glycoproteins (Parodi & Leloir 1979).   The lipid moiety is a polyprenyl
phosphate or pyrophosphate which is usually α-saturated and contains
16-20 prenyl groups (dolichol (dol) type).   In some instances (Alam &
Hemming 1971; 1973), the lipid may be a totally unsaturated polyprenol
with a shorter chain-length (ficaprenol, betulaprenol type).   Thus,
Colvin's original suggestion that glycolipids act as intermediates for
polymer synthesis is certainly sustained for the biosynthesis of many
glycoproteins.   The evidence is much less complete or convincing, however,
that such intermediates are also necessary for the biosynthesis of cell-
wall polysaccharides.

Colvin's proposal remains attractive if only because the
presumed initial substrates (sugar phosphates or nucleotides) must be
generated intracellularly while the final products accumulate extracellularly, implying vectorial transfer of the glycone moiety through a
lipid membrane.   One mechanism to bring this about could be glycosyl
transfer via an intermediate through the membrane, be this the plasmamembrane, ER, dictyosome, or secretory vesicle.   Such a transfer could
be envisaged as taking place if the hydrophilic glycone were temporarily
rendered effectually hydrophobic by covalent attachment to and assembly
on the cytoplasmic surface of a membrane-bound lipid, followed by
transmembrane movement of the sugar residues in the lipid-linked form
and transglycosylation to growing acceptors on the luminal or outer
plasmamembrane surface (see Snider & Rogers 1984).   Alternatively, the

enzyme that is responsible for the transfer of glycone from the lipid to
final acceptor could itself effect a transmembrane translocation of
glycone (Snider et al. 1982).   These and other models for transmembrane
movement of activated lipid-linked saccharides are a matter for current
debate (e.g., see Hanover & Lennarz 1982), and it may be that different
membrane types employ different mechanisms to effect vectorial sugar
transfers.   Mechanisms may also exist which would not necessarily
require glycolipid intermediates, e.g., one such process involves a
system of alternating vesiculation and vesicle fusion, which can transfer
precursors from one side of a membrane to another (Paiement et al. 1982).

      In the case of cellulose biosynthesis by bacteria, the
linkages and nature of glucolipids which may be formed from UDP-glucose
by bacterial membranes have not yet been characterized with certainty,
nor have such components been observed to turn over during cellulose
formation.   Moreover, UDP-glucose:1,4-β-glucan 4-β-glucosyltransferase
has recently been solubilized with detergent from Acetobacter membranes
(Aloni et al. 1983), and no evidence could be found for any lipid
intermediates in this highly active system.   Thus, many researchers
with Acetobacter at this point of time conclude that the balance of
evidence indicates that lipid intermediates are not required for
bacterial cellulose biosynthesis (Delmer 1982; Delmer et al. 1983;
Aloni et al. 1983).

      With respect to the biosynthesis of higher-plant poly-
saccharides, the formation of intracellular storage polysaccharides,
such as starch and fructosan, are clearly catalysed by soluble-enzyme
preparations which use sugar phosphates and/or sugar nucleotides
directly without the mediation of lipid intermediates.   It is in the
biosynthesis of extracellular polysaccharides, or intraorganelle
polysaccharides which are destined to become extracellular, by
transglycosylases that are membrane-bound, that the question of a lipid
intermediate has most pertinence.   It remains theoretically possible
that, as discussed above for N-linked glycoprotein synthesis, mono- or
oligosaccharides could be assembled first as intermediate glycolipids on
the cytoplasmic face of an organelle or plasmamembrane and used as
glycosyl donors to a growing polysaccharide on the other face.   This
possibility has been most dramatically illustrated in an artificial
model system (Haselbeck & Tanner 1981; 1982), wherein water-soluble
GDP-(^{14}C)-mannose was supplied to lecithin liposomes containing dolichyl-

phosphate and purified yeast mannosyltransferase.   The labelled mannose
was not only transferred from the medium to the lipid phase by
transmannosylation to dolichyl-phosphate mannose, but it could also
be transferred via the lipid intermediate to internal GDP to reform GDP-
mannose if the acceptor were also present within the liposome.

Those cell-wall polysaccharides containing repeating
oligomers or side chains are particularly suited to be candidates for
biosynthesis via en-bloc transfer from a preformed lipid-linked oligosaccharide.   Thus, for example, homopolysaccharides like cellulose or
mixed-linkage glucans (Nevins et al. 1978; Woodward et al. 1983;
Staudte et al. 1983), are composed of repeating simple oligosaccharide
units, i.e., cellobiose or cellotriosyl-1,3-β-glucose and cellotetrasyl-
1,3-β-glucose, respectively.   In the case of the extracellular
polyglucan, pullulan, which is found in basidiomycetes, charged lipid-
linked short-chain oligosaccharides containing α-1,4 and α-1,6 linked
glucose units have been isolated (Taguchi et al. 1973; Catley & McDowell
1982) and proposed as precursors of identical sequences in this homopolymer.
Many heteropolysaccharides also contain one or more repeating
oligosaccharide subunit(s) which are randomly distributed within a given
chain.   Examples include xyloglucan which, in sycamore, soybean and
pea, is almost entirely composed of equal amounts of a heptasaccharide
and a nonasaccharide (Albersheim 1975; Hayashi et al. 1980; Hayashi &
Maclachlan 1984a).   The structures of these repeating units are shown
in Figure 1.   Similarly, arabinoxylans (Banberji & Rao 1963; McNeil
et al. 1975), galactomannans (McCleary 1979) and rhamnogalacturonans
(McNeil et al. 1973) all appear to contain repeating subunits.   In
addition, many pectic materials, as well as hemicelluloses, e.g.,
arabinogalactans (Clarke et al. 1979; Fincher et al. 1979), are highly
branched and it is possible that repeated branch-oligosaccharides in
particular are derived by block-transfer from lipid-linked precursors.
One type of common branching unit which is repeated in certain
arabinogalactan-proteins is illustrated in Figure 1.

The possibility should also be considered that the initial
acceptor (primer) for cell-wall polysaccharide synthesis is a protein,
i.e., that glycoproteins or proteoglycans are polysaccharide precursors.
Indeed, this has been proposed (Whelan 1976) as a general mechanism for
synthesis of all polysaccharides.   The evidence that this mechanism
operates in the biosynthesis of β-glucans by plants has been reviewed

(Maclachlan 1982), and led to the conclusion that the question of the
nature of the initial acceptor must remain open at the present state of
our understanding.   Whether arabinogalactan proteins are precursors for
free arabinogalactans is also not clear as yet (Clarke et al. 1979;
Fincher et al. 1983).   Whatever the initial primer may be, however, it
is the question of whether lipid intermediates are involved in
polysaccharide-chain elongation that is considered here.

Figure 1. Typical repeating units in xyloglucans and branch-
units in certain arabinogalactans.   The nona- and
heptasaccharides are found in xyloglucans from pea, soybean
and sycamore, and this particular branched oligosaccharide
is found in timothy pollen arabinogalactan protein.   The
xylose branches are attached by α-1,6 linkages to a glucan
backbone containing β-1,4 linkages.   Fucose and galactose
units are linked α-1,2 and β-1,2, resp.   Treatment of the
xyloglucan with endo-1,4-β-glucanase (E.C. 3.2.1.4) generates
equal amounts of a nonasaccharide and heptasaccharide, plus
small amounts of hexadecasaccharide.   Further treatment
with mixed glycosidases produces fucose, galactose, glucose
and the disaccharide, isoprimeverose (Hayashi & Maclachlan
1984d).   The galactose backbone is 1,6 linked and the
arabinose side chains are all furanosyl with some α-1,3
linkages and blocks of up to 5 α-1,5 linkages (Haavik et al.
1982).

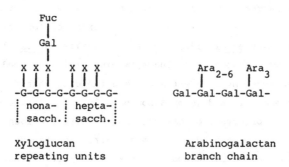

Xyloglucan   Arabinogalactan
repeating units   branch chain

NEUTRAL LIPID-LINKED GLYCOSIDES IN PLANTS

There are a great many uncharged glycolipids formed in plants
(Elbein 1980), many of which are considered to be structural components
of particular membranes, e.g. mono- and digalactosyl diglycerides of
chloroplasts, or soluble detoxification products, e.g. cardiac
glycosides.   The majority do not yet have clear functions attributed to
them, and few, if any, appear to possess properties which could qualify
them as potential substrates for transglycosylation reactions in

polysaccharide biosynthesis.

To consider only those neutral glycolipids that contain glucose, the steryl glucosides and acylated steryl glucosides are probably universally distributed in plant membranes where, however, they may function only as stabilizing integral components of the membrane (Eichenberger 1977; Heftmann 1977; Grunwald 1978).   Attempts to use steryl glucosides as substrates for glycosyl transferases to glucans have not succeeded (e.g. (Kauss 1968).   Other lipid-linked mono-glucosides, e.g. flavonoid glucosides, phenolic glucosides, etc., are also very widely distributed in plants, but these are generally water soluble and non membrane-bound.   They are regarded as storage products and there is no evidence that the glucoside bond has sufficient free energy of hydrolysis to act as substrate in transglucosylation reactions (Hosel 1981).   Glucosides of indole-3-acetic acid (Michalczuk & Bandurski 1982), herbicides (Schmitt & Sandermann 1982), cyclitols (Kemp & Loughman 1974) and other acids or alcohols are likewise soluble products and are not known to act as precursors for any other product. They may function in detoxification or solubilization and transport reactions, but not in further biosynthesis.

There are a few reports of neutral lipid-linked glucose-containing oligosaccharides, whose functions probably merit further investigation.   For example, there appear to exist cellobiosides of sesquiterpene (Nakano et al. 1983) and sterol (Palter et al. 1972) whose functions are unknown.   Diglucosylmyoinositol is formed in low yield from UDP-glucose by bean extracts, and the disaccharide appears to be non-reducing and β-linked (Kemp & Loughman 1974).   It was suggested that this may be a precursor of callose, but no direct evidence for this possibility has been presented.   The most intriguing reports concern the existence in peas of large amounts of kaemferol-triglucoside, where the glucose linkages are 1,2-β (Furuya & Galston 1965; Singh & Maclachlan 1983).   This is the only known oligosaccharide linkage in plants where the free energy of hydrolysis is as high as the linkage in sucrose and sugar nucleotides (Sutter & Grisebach 1975).   Kaemferol is the major flavonoid in peas and its triglucoside is formed by repeated transglucosylation from UDP-glucose by soluble pea enzymes (Jourdan & Mansell 1982).   It is one of the major soluble products generated from glucose, fuctose or sucrose by isolated pea stems (Maclachlan & Singh 1983).   We have conducted repeated tests using isolated pea membranes and purified highly radioactive pea kaemferol-triglucoside as substrate, but without the slightest indication that the trisaccharide moiety is used in transglucosylation reactions, even under reaction conditions where UDP-glucose is readily metabolized further.

GLYCOSYLATED MONOPHOSPHATE LIPIDS

There are numerous reports of the synthesis in plants of monoglycosylated monophosphoryl lipids which bear a partial negative charge and therefore dissolve best in lipid solvents that contain some water, e.g. the classical solvent of Bligh & Dyer (1959) (chloroform: methanol: water :: 10:10:3). Standard methods for the isolation and characterization of these lipids have been reviewed by Brett (1980). Such glycolipids bind to columns of positively charged supports, e.g. DEAE-cellulose or DEAE-sepharose, from which they can be eluted (after washing to remove neutral lipid) by extraction with lipid solvent containing dilute salt, e.g. 5mM ammonium formate. On investigation, such products often possess properties that resemble authentic dolichyl monophosphoryl saccharide.   These include characteristic mass and NMR spectra (Evans & Hemming 1973; Mankowski et al. 1976; Delmer et al. 1978) and chromatographic mobilities.   There are TLC methods which distinguish mono- and di-phosphorylated polyprenols from sugar phosphates, sugar nucleotides and neutral lipids (Herscovics et al. 1977).   Other TLC and HPLC solvents have been developed (Wells et al. 1981; Eggens et al. 1983) to separate free polyisoprenols and their derivatives according to chain length.   The polyisoprenyl phosphoryl saccharides are very susceptible to acid hydrolysis (e.g. 0.01N trifluoroacetic acid in 50% propanol at 100°C releases at least 90% of the free sugar within 10 min).   Mild alkaline degradation (0.1N NaOH, 65°C) generally results in formation of anhydro sugars, indicating a β-glycosidic linkage (Pont-Lezica et al. 1975; Herscovics et al. 1977).   Stability in 50% phenol (1 h, 70°C) indicates that the α-isoprenyl is saturated, i.e. a dolichyl type of isoprenoid (Garcia et al. 1974; Delmer et al. 1978).

It should be cautioned here that not all moderately charged glycolipids in plants are necessarily polyisoprenyl monophosphoryl derivatives.   The class of glycolipids referred to by Elbein (1980) as phytoglycolipids may also show some of the same characteristics, though generally these products, e.g. the glycophosphosphingolipids, are not as readily hydrolysed by dilute acid and have very different TLC

chromatographic mobilities (Kaul & Lester 1975).   They also commonly
contain complex oligosaccharides rather than a single saccharide as the
glycone, and a variety of sugar components including N-acetyl
glucosamine, glucuronic acid, inositol and other sugars such as
galactose, arabinose and fucose (Carter et al. 1964).   The functions of
such glycolipids in plants are unknown.

Table 1 summarizes data from a number of experiments
conducted with pea-stem membranes in this laboratory whereby the monosaccharide moieties of sugar nucleotides were found to be incorporated
into weakly charged lipid fractions where the product has properties
resembling dolichyl monophosphoryl monosaccharide.   Of the sugar
nucleotides tested, substantial yields of dol-P-glucose, mannose and
xylose have all been recorded.   There are many other reports of dol-P-
glucose and dol-P-mannose formed by membranes from peas (e.g. Daleo &
Pont-Lezica 1977; Brett & Leloir 1977) and other plant sources (see
reviews by Elbein, 1979; 1980).   Three sugar nucleotides, UDP-arabinose,
UDP-galactose and UDP-N-acetylglucosamine failed to be utilized by pea
membranes to form detectable amounts of the corresponding dolichyl-
phosphate sugar.   There are no reports in the literature of dol-P-
arabinose, dol-P-galactose or dol-P-N-acetylglucosamine in plants,
although dol-P-galactose appears to be synthesized in Acetobacter
(Romero et al. 1977) and it may be a component of galactolipid in plants
(Mellor & Lord 1979a).

In the case of the monophosphorylated glycolipids formed by
pea membranes from UDP-glucose and GDP-mannose, the addition of authentic
dolichyl phosphate or ficaprenyl phosphate to the reaction mixture results
in a great increase in the yield of monoglycosylated product (Pont-Lezica
et al. 1976;   Durr et al. 1979).   Similar results have been reported
with membranes from other tissues, e.g. mung bean (Alam & Hemming 1971;
Lehle et al. 1978) and cotton (Ericson et al. 1978).   The site of
biosynthesis of these glycolipids is clearly the endoplasmic reticulum in
pea stems (Hopp et al. 1979; Durr et al. 1979) as it is in most other
plant tissues (Nagahashi & Beevers 1978; Lehle et al. 1978; Mellor &
Lord 1979b).   There are also reports (Bowles et al. 1977; Chadwick &
Northcote 1980) of synthesis by membrane fractions rich in Golgi or
plasmamembrane.   With respect to pea-stem membranes, using a new
technique to isolate and purify sheets of plasmamembrane from pea
protoplasts (Polonenko & Maclachlan 1984), we have found (Gordon, R.,

unpublished) that such membranes show no capacity whatever to synthesize
dol-P-glucose from UDP-glucose as does ER, even though they are active
in forming β-glucan.

Table 1.   Incorporation of the Glycone of Sugar Nucleotides into
           Various Fractions of Total Lipid Following Incubation with
           Pea-Stem Membranes.

| Substrate | % Distribution of Glycone* | | | Reference |
	Neutral lipid	Weakly charged	Highly charged	
UDP-Ara	40	0	60	Hayashi & Maclachlan 1984b
UDP-Gal	65	0	35	Hayashi & Maclachlan 1984a
UDP-Glc	90	10	?	Durr et al. 1979
UDP-GlcNAc	0	0	100	Durr et al. 1979
GDP-Man	0	60	40	Durr et al. 1979; Bailey et al. 1979
UDP-Xyl	75	25	0	Hayashi & Maclachlan 1984b

*Sugar nucleotides (10-20μM) were incubated for 10-30 min with total
membranes (0.5 mg protein) from growing region of pea epicotyl at
pH 7-7.5, room temp., plus Mg^{++} (10mM).   Values refer to % distribution
of the particular glycone supplied and do not include incorporation
after epimerization to other sugars.   The products soluble in
chloroform: methanol: water :: 10:10:3 were fractionated on columns of
DEAE-cellulose or DEAE-sepharose into fractions that eluted with this
solvent (neutral) or solvent containing 5 mM ammonium formate (weakly
charged) or 200 mM formate (highly charged).

        The free energy of hydrolysis of the glycoside linkage
between mannose and dolichyl phosphate is comparable to that in sugar
nucleotides, since its synthesis from GDP-mannose is freely reversible
(Kauss 1969; Forsee & Elbein 1973; Elbein 1980).   This indicates that
such lipid-linked saccharides preserve the sugar-transfer potential of
the sugar nucleotides and renders them good candidates for donors in
further transglycosylation reactions.   Both dol-P-mannose and dol-P-
glucose are commercially available with the glycone radioactive, and
experiments are now possible with these as direct substrates.   Precautions

should be taken, however, to ensure that these preparations are indeed
dolichol derivatives, for tests in this laboratory have shown that
different sources vary in purity and degree of degradation.

The monophosphoryl mannose and glucose derivatives of
naturally-occurring polyisoprenoids in animal systems have been isolated
radioactive and in quantity and shown to act as effective precursors for
complex glycolipids and glycoproteins in some plant systems (Ericson &
Delmer 1977; Hopp et al. 1979; Pont-Lezica et al. 1978).   One potential
difficulty in such tests is to be certain that endogenous dinucleotides
in crude membrane preparations are not acting as preferential acceptors
of saccharides supplied as dolichyl phosphate saccharides, with the
former then acting as direct substrates in subsequent reactions.   The
only definitive way to be sure that dol-P-saccharides are direct donors
to other products is to purify and solubilize the relevant glycosyltransferases and assay for substrate affinities.   This has been done,
for example, with solubilized glycoprotein glycosyltransferase from yeast
membranes (Sharma et al. 1981) and with chitin synthase (Ruiz-Herrera &
Bartnicki-Garcia 1974; Duran & Cabib 1978).   The latter system
utilized UDP-N-acetylglucosamine as substrate and showed no evidence for
a lipid intermediate.   Similar results were reported (Aloni et al. 1983)
for the UDP-glucose-requiring 1,4-β-glucan synthase of Acetobacter.   To
date, such tests have not been reported with solubilized higher plant
systems.

Using membrane preparations from pea stem, and UDP-(^{14}C)-
glucose as substrate, we have been able to confirm the observations of
Pont-Lezica et al. (1976) that one of the glycolipids that is synthesized
has properties identical to those of dol-P-glucose.   However, in our
tests, this product showed no kinetic evidence for turnover in β-glucan
synthase preparations and, when collected in quantity, purified and
supplied back to fresh pea membranes, it was not utilized as a
substrate for any detectable further reactions.   Likewise, commercially
available (New England Nuclear) dolichol-phosphoryl (6-^{3}H)-glucose was
not metabolised by these pea preparations.   Similar negative results
were reported by Helsper (1979) with a membrane fraction from pollen
tubes.   We have concluded (Camirand et al., unpublished) that pea
membrane systems that synthesize β-glucans from UDP-glucose do not
require polyprenol-monophosphoryl saccharides as obligatory
intermediates.

GLYCOSYLATED PYROPHOSPHATE LIPIDS

Polyisoprenyl pyrophosphoryl saccharides are formed by an
initial transfer of sugar-1-phosphate from sugar nucleotides to
polyisoprenyl-monophosphate.   The glycoside in sugar nucleotides is
α-linked and transferred intact to the polyisoprenyl phosphate,
probably on the outer surface of the ER vesicles (Snider & Rogers 1984)
which are readily visible in crude plant membrane preparations (Shore &
Maclachlan 1975; Polonenko & Maclachlan 1984).   The glycoside linkage
has a free energy of hydrolysis comparable to sugar nucleotides and
accordingly is a good candidate for participation in further transglycosylation reactions.   These products are therefore particularly
interesting as potential precursors for β-linked polysaccharide.

Results in Table 1 indicate that many sugar nucleotides gave
rise to such highly charged glycolipids when they were incubated with
pea membranes in different experiments conducted in this laboratory.
Analyses of the glycone components of such lipids in this and many other
laboratories have repeatedly confirmed that plant membranes contain all
of the transglycosylases required to assemble complex lipid-linked
oligosaccharides.   UDP-N-acetylglucosamine and GDP-mannose are
particularly effective substrates, which give rise to the typical lipid-
linked core oligosaccharide that is found in N-linked glycoproteins
(Elbein 1979; Parodi & Leloir 1979).   Plant membranes, including those
from pea stems, readily assemble and accumulate polyisoprenyl
pyrophosphoryl chitobiose which can then be mannosylated with 1, 3, 5 or
more mannose units (see Table 2).   In the pea system, the size of the
lipid-linked oligosaccharide assembled from UDP-N-acetylglucosamine and
GDP-mannose did not exceed $GlcNAc_2$-Man_{5-6} (Bailey et al. 1979).   This
may mean that this product was formed entirely on the cytoplasmic surface
of ER vesicles by direct mannosyl transfer from GDP-mannose, since
translocation to the luminal surface appears to be necessary for further
mannosylation, with dol-P-mannose as substrate (Snider & Rogers 1984).

The assembled lipid-linked mannose-rich oligosaccharide may
be "capped" in plant membranes by the addition of a limited number of
glucose units at the non-reducing end derived from dol-P-glucose (Bailey
et al. 1979; Staneloni et al. 1980; Lehle 1981; Hori et al. 1982), a
step which appears to be necessary for the subsequent transfer of this
oligosaccharide to selected asparagine residues in nascent glycoproteins
(Staneloni et al. 1981).   Actual transfer, using mannose-labelled plant

lipid-linked oligosaccharide as substrate, has been demonstrated with membrane-bound enzymes from soya roots or pea cotyledons plus endogenous polypeptide acceptors (Staneloni et al. 1981).   Transfer has also been effected by plant enzymes to exogenously supplied peptides or derivatives (Mellor et al. 1979).   Generally, the nascent glycoprotein is then "processed" further to its final structure by removal of the terminal glucosides and addition of other sugars (e.g. see Lehle 1980).   Thus, the terminal glucose units are transitory and few, if any, glycoproteins survive their passage through the ER/Golgi endomembrane system with glucose still part of the glycone.   Indeed, deglucosylation may be a required step for efficient maturation of secreted N-linked glycoproteins (Lodish & King 1984).

Table 2.   Incorporation of the Glycone of Sugar Nucleotides into Highly Charge Lipid-linked Oligosaccharides by Pea-Stem Membranes.

Substrate	Lipid-PP-product(s)	Reference
UDP-Ara	- (unknown)$_{7-10}$-(α-1,5-ara)$_7$	Hayashi & Maclachlan 1984b
UDP-Gal	- (some 1,6-gal + unknown)	Hayashi & Maclachlan 1984a
UDP-Glc	- (glcNAc$_2$-man$_{5-6}$-glc$_{2-3}$)	Bailey et al. 1979
UDP-GlcNAc	- (glcNAc)$_{1-2}$	Bailey et al. 1979
GDP-Man	- (glcNAc$_2$-man$_{1,3,5+}$) as above	Bailey et al. 1979; Durr et al. 1979

Reaction conditions are described in Table 1.   Sizes of oligosaccharides were determined by gel-permeation chromatography on Biogel P$_2$ columns (Biorad. Inc.) or Sepharose CL-6B, (Pharmacia), of dilute acid-hydrolysates of lipid-linked oligosaccharides eluted from DEAE cellulose columns.

With respect to glucose-containing highly-charged lipid-linked saccharides which could be candidates for polysaccharide biosynthesis, there have been three reports of such products formed from UDP-glucose, all using crude pea membranes as source of enzyme.   Brett & Northcote (1975) reported synthesis by pea-root membrane of a lipid-pyrophosphoryl oligosaccharide containing both 1,4 and 1,3 linkages.   Romero &

Pont-Lezica (1976) reported the formation by pea-stem membrane of
polyisoprenyl-pyrophosphoryl glucose and later, the same laboratory
(Pont-Lezica et al. 1978) provided evidence that this reaction mixture
also contained a lipid-linked oligosaccharide with a glycone chain
length equivalent to 7-8 glucose units.   The former was proposed as
a precursor for cellulose, and the latter for pea lectin   (a glucoprotein).
Both UDP-glucose and liver dol-P-glucose could act as precursors for
these dol-PP-saccharides.

        Several researchers in this laboratory have made repeated
efforts to duplicate these results using pea-stem membranes and either
UDP-glucose or dol-P-glucose (New England Nuclear) as substrate.   These
efforts have not led to the isolation of any trace of dol-PP-glucose or
dol-PP-gluco-oligosaccharides, even under conditions where other sugar
nucleotides gave rise to major yields of highly charged lipid-linked
oligosaccharides (Tables 1 and 2).   To be sure, as reported by
Pont-Lezica et al. (1976), small amounts of radioactivity are usually
recovered in the fractions of CMW extracts that bind strongly to DEAE-
cellulose columns (Durr et al. 1979).   This material does not co-
chromatograph on TLC with dol-PP-saccharide in solvents (e.g. chloroform :
methanol : water :   10:10:3 v/v) that separate such products from UDP-
glucose.   It is highly susceptible to acid hydrolysis, but the only
saccharide that we have been able to detect in the products is free
glucose.   It seems likely, therefore, that highly-charged radioactive
material which may be found in CMW extracts of experiments with labelled
sugar nucleotides, could be contaminated with unused substrate.

        The addition of UDP-xylose to soybean or pea-stem membrane
preparations that are actively synthesizing dol-P-glucose and
polysaccharides from UDP-(^{14}C)-glucose results in the synthesis of
xyloglucan, provided Mn^{++} ions are present in the system (Hayashi &
Matsuda 1981; Hayashi et al. 1981; Hayashi & Maclachlan 1984c).   The
xyloglucan can be assayed in reaction products and distinguished from
β-glucans that are formed concurrently by degrading the products with
endo-β-glucanase and a mixture of glycosidases and estimating the yield
of isoprimeverose (6-0-α-D-xylopyranosyl-D-glucopyranose) (Hayashi et al.
1981; Hayashi & Maclachlan 1984d).   The biosynthesis clearly takes place
by obligatory concurrent glycosyl transfer from both UDP-glucose and
UDP-xylose, and not by xylosyl transfer to a preformed 1,4-β-glucan.
Repeated tests were conducted in order to search for a lipid-linked

oligosaccharide corresponding to the heptasaccharide subunit of soybean
and pea xyloglucan (Fig. 1).   No such potential lipid intermediate has
been found.

However, the addition to pea membranes of UDP-(^{14}C)-xylose
(or UDP-(^{14}C)-arabinose) alone, without other sugar nucleotides, results
in formation of labelled dol-P-xylose and a highly charged glycolipid
containing labelled arabinose (Hayashi & Maclachlan 1984b).   The
membranes contain an active epimerase which rapidly interconverts the
pentose nucleotides.   It was demonstrated that the lipid-linked
arabinosyl oligosaccharide contained about 7 α-1,5-linked arabinofuranosyl units at the non-reducing terminal of an oligosaccharide with a
MW of 2300, i.e. there were up to 10 other unlabelled saccharides in the
molecule between the terminal arabinose units and the pyrophosphoryl-
lipid carrier (see Table 2).   It is technically difficult to establish
what these other saccharides may be, although they appear to contain
uronic acid residues since the oligosaccharide component is
electrophoretically mobile and not readily hydrolysed to free sugars.
Reduction of the oligosaccharide with borotritiide produces some
(^{3}H)-galactinol, but this does not necessarily mean that galactose was
the reducing terminal since it could have been derived from some other
unlabelled oligosaccharide that co-chromatographed with the ^{14}C-arabinosyl
oligosaccharide.

UDP-(^{14}C)-xylose also gave rise to electrophoretically-
mobile and -immobile (^{14}C)-arabinosyl-products which appeared to be high
MW glycoprotein and polysaccharide, respectively.   When the lipid-linked
(^{14}C)-arabinosyl oligosaccharide was provided back to fresh pea membranes
as a potential substrate, label was incorporated into products that
resembled the least electrophoretically mobile products that were formed
from UDP-(^{14}C)-xylose.   Thus, there are indications that the synthesis
of arabinose-containing products that may be cell-wall components, e.g.
arabinogalactans, in peas proceeds via a lipid-linked oligosaccharide.

In similar experiments with UDP-(^{14}C)-galactose as substrate
in the presence of ADP-ribose (which inhibits the epimerization of this
substrate to UDP-glucose), highly-charged lipid-linked oligosaccharide
was also formed which contained (^{14}C)-galactose as the only labelled
sugar (Hayashi & Maclachlan 1984a).   A product with similar properties
was reported (Mascara & Fincher 1982) to be formed from UDP-galactose by
rye grass membranes.   The product from peas was only partially (30-40%)

hydrolysed by dilute acid to a free oligosaccharide, and therefore only
a fraction could have been linked to polyisoprenyl phosphate.   The rest
may have represented a phytoglycolipid, e.g. phytophosphosphingolipid,
which is not readily susceptible to acid hydrolysis (Elbein 1980).   At
the same time, pea membranes supplied with UDP-(^{14}C)-galactose also
synthesized electrophoretically mobile protease-digestible glycoproteins
and immobile polysaccharides.   These contained (^{14}C)-galactose as the
labelled sugar and a preponderance of 1,6 linkages, with minor ammounts
of 1,3 and 1,4 linkages between galactose units.   Whether any part of
the lipid-linked galactose-containing oligosaccharides was a precursor
for these non-lipid products was not readily determined because of low
yields.

The glycoproteins and polysaccharides formed by pea membranes
from UDP-(^{14}C)-arabinose and UDP-(^{14}C)-galactose (Hayashi & Maclachlan
1984a,b) possess properties that resemble the most abundant class of
arabinogalactan-proteins and arabinogalactans, many of which contain a
1,6-linked galactosyl backbone or side chains, 1,5-α-linked arabinofuranosyl sequences and uronic acids (Clarke et al. 1979; Fincher et al.
1983).   One example of the structure of an allergenic arabinogalactan-
protein side branch from timothy pollen is shown in Figure 1.   The fact
that many such products contain sequences with a regular structure that
is repeated in the larger polysaccharide led Clarke et al. (1979) to
suggest that they derive from a pre-assembled precursor such as a lipid
intermediate by block transfer.   The above results with pea membranes
are consistent with such a proposal, but much more research needs to be
carried out with defined substrates, acceptors and products before the
mechanism can be considered as established.

CONCLUSIONS

There is abundant evidence that plant membranes, most
particularly the endoplasmic reticulum, are capable of synthesizing a
variety of polyisoprenyl monophosphoryl monosaccharides from the
corresponding sugar nucleotides (Table 1).   Dol-P-mannose and probably
dol-P-glucose, are almost certainly used as intermediates in
glycosyltransfer reactions leading to plant glycoprotein biosynthesis.
There is no clear or direct evidence, however, that such intermediates
are required for the elongation of cell-wall polysaccharides.

Plant membranes also readily assemble polyisoprenyl

pyrophosphoryl oligosaccharides containing N-acetylglucosamine, mannose
and glucose in a remarkably reproducible structure which appears to be
the universal precursor for N-linked glycoproteins.   We have also found
that UDP-arabinose and UDP-galactose are incorporated into charged lipid-
linked oligosaccharides (Table 2) whose complete structures are not yet
elucidated.   They may be intermediates in arabinogalactan protein
biosynthesis in instances where these products contain α-1,5-
arabinofuranosyl and 1,6-galactosyl linkages.   Much more analytical
and kinetic research is required in order to establish this possibility,
however.   Lipid-linked oligosaccharides containing other sugars which
could be precursors for repeating units in polysaccharides have been
proposed from time to time, but confirmation of their existence in plants
has not been forthcoming to date.   There are numerous technical
difficulties in establishing structures of such putative intermediates,
which can lead to misinterpretations of results.   It must be concluded
that the proposal that lipid-linked saccharides are precursors for
polysaccharide synthesis in plants remains speculative.

ACKNOWLEDGEMENTS

This review was prepared with helpful comment from A. Camirand,
R. Gordon and K. Torossian, which is gratefully acknowledged.   The
research results that are described from this laboratory derive from
projects supported by funds from the Natural Sciences and Engineering
Research Council of Canada and the Formation des Chercheurs et Action
Concertée du Québec.

REFERENCES

Alam, S.S. & Hemming, F.W. (1971). Betulaprenol phosphate as an acceptor
of mannose from GDP-mannose in Phaseolus aureus preparations.
FEBS Letters, 19, 60-62.
Alam, S.S. & Hemming, F.W. (1973). Polyprenyl phosphates and mannosyl
transferases in Phaseolus aureus. Phytochemistry, 12, 1641-
1649.
Albersheim, P. (1976). The primary cell wall. In Plant Biochemistry,
eds. J. Bonner & J.E. Varner, pp. 225-274. New York:
Academic Press.
Aloni, Y., Cohen, O.R., Benziman, M., Delmer, D. (1983). Solubilization
of the UDP-glucose: 1,4-β-Glucan 4β-glucosyltransferase
(cellulose synthase) from Acetobacter xylinum. Journal of
Biological Chemistry, 258, 4419-4423.
Bailey, D.S., Durr, M., Burke, J. & Maclachlan, G. (1979). The assembly
of lipid-linked oligosaccharides in plant and animal membranes.
Journal of Supramolecular Structure, 11, 123-138.

Banerji, N. & Rao, C.V.N. (1963). Structural studies on an arabinoxylan from pea skin. Canadian Journal of Chemistry, 41, 2844-2848.

Bligh, E.G. & Dyer, W.J. (1959). A rapid method of total lipid extraction and purification. Canadian Journal of Biochemistry and Physiology, 37, 911-917.

Bowles, D.J., Lehle, L. & Kauss, H. (1977). Glucosylation of sterols and polyprenolphosphate in the Golgi apparatus of Phaseolus aureus. Planta, 134, 177-181.

Brett, C.T. & Northcote, D.H. (1975). The formation of oligoglucans linked to lipid during synthesis of β-glucan fractions from peas. Biochemical Journal, 148, 107-117.

Brett, C.T. & Leloir, L.F. (1977). Dolichyl monophosphate and its sugar derivatives in plants. Biochemical Journal, 161, 93-101.

Brett, C.T. (1980). The isolation and characterization of polyprenyl-phosphate sugars. In Techniques in Carbohydrate Metabolism, B 305, 1-14. Elsevier/North Holland Science Publishers.

Carter, H.E., Betts, B.E. & Strobach, D.R. (1964). Biochemistry of the sphingolipids: the nature of the oligosaccharide component of phytoglycolipid. Biochemistry, 3, 1103-1107.

Catley, B.J. & McDowell, W. (1982). Lipid-linked saccharides formed during pullulan biosynthesis in Aureobasidium pullulans. Carbohydrate Research, 103, 65-75.

Chadwick, C.M. & Northcote, D.H. (1980). Glucosylation of phosphoryl polyisoprenol and sterol at the plasmamembrane of soya bean protoplasts. Biochemical Journal, 186, 411-421.

Clarke, A.E., Anderson, R.L. & Stone, B.A. (1979). Form and function of arabinogalactans and arabinogalactan-proteins. Phytochemistry, 18, 521-540.

Colvin, J.R. (1959). Synthesis of cellulose in ethanol extracts of Acetobacter xylinum. Nature, 183, 1135-1136.

Colvin, J.R. (1980). The biosynthesis of cellulose. In The Biochemistry of Plants, ed. J. Preiss, Vol. III, pp. 543-570. New York: Academic Press.

Daleo, G.R. & Pont-Lezica, R. (1977). Synthesis of dolichol phosphate by a cell-free extract from peas. FEBS Letters, 74, 247-250.

Delmer, D.P., Kulow, C. & Ericson, M.C. (1978). Glycoprotein synthesis in plants II. Structure of the mannolipid intermediate. Plant Physiology, 61, 25-29.

Delmer, D.P. (1982). Biosynthesis of cellulose. Advances in Carbohydrate Chemistry Biochem., 41, 105-153.

Delmer, D.P., Benziman, M., Klein, A.S., Bacic, A., Mitchell, B., Weinhouse, H., Aloni, Y. & Callaghan, T. (1983). A comparison of the mechanism of cellulose biosynthesis in plants and animals. Journal of Applied Polymer Science, 37, 1-16.

Duran, A. & Cabib, E. (1978. Solubilization and partial purification of yeast chitin synthetase. Journal of Biological Chemistry, 253, 4419-4425.

Durr, M., Bailey, D.S. & Maclachlan, G. (1979). Subcellular distribution of membrane-bound glycosyltransferases from pea stems. European Journal of Biochemistry, 97, 445-453.

Eggens, I., Chojnacki, T., Kenne, L. & Dallner, G. (1983). Biochimica et Biophysica Acta, 751, 355-368.

Eichenberger, W. (1977). Steryl glycosides and acylated steryl glycosides. In Lipids and Lipid Polymers in Higher Plants, pp. 169-182. Berlin: Springer-Verlag.

Elbein, A.D. (1979). The role of lipid-linked saccharides in the
    biosynthesis of complex carbohydrates. Annual Review of Plant
    Physiology, 30, 239-272.
Elbein, A.D. (1980). Glycolipids. In The Biochemistry of Plants, eds.
    P.K. Stumpf & E.S. Conn, 3, 571-587. New York: Academic Press.
Ericson, M.C. & Delmer, D.P. (1977). Glycoprotein synthesis in plants. 1.
    The role of lipid intermediates. Plant Physiology, 59, 341-347.
Ericson, M.C., Gafford, J.T. & Elbein, A.D. (1978). Evidence that the
    lipid carrier for N-acetylglucosamine is different from that
    for mannose in mung beans and cotton fibers. Plant
    Physiology, 61, 274-277.
Evans, P.J. & Hemming, F.W. (1973). The unambiguous characterization of
    dolichol phosphate mannose as a product of mannosyl
    transferase in pig liver endoplasmic reticulum. FEBS Letters,
    31, 335-338.
Fincher, G.B., Stone, B.A. & Clarke, A.E. (1983). Arabino-galactan
    proteins: structure, biosynthesis and function. Annual Review
    of Plant Physiology, 34, 47-70.
Forsee, W.T. & Elbein, A.D. (1973). Biosynthesis of mannosyl and glucosyl-
    phosphoryl polyprenols in cotton fibers. Journal of
    Biological Chemistry, 248, 2858-2867.
Furuya, M. & Galston, A.W. (1965). Flavonoid complexes in Pisum sativum.
    Phytochemistry, 4, 285-292.
Garcia, R.C., Recondo, E. & Dankert, M.A. (1974). Polysaccharide
    biosynthesis in Acetobacter xylinum. Enzymatic synthesis of
    lipid diphosphate and monophosphate sugars. European
    Journal of Biochemistry, 43, 93-105.
Grunwald, C. (1978). Functions of sterols. Philosophical Transactions of
    the Royal Society of London, Series B, 284, 541-558.
Haavik, S., Paulsen, B.S., Wold, J.K. & Grimmer, O. (1982).
    Arabinogalactans in a purified allergen preparation from
    pollen of timothy. Phytochemistry, 21, 1913-1919.
Hanover, J.A. & Lennarz, W.J. (1982). Transmembrane assembly of N-linked
    glycoproteins. Studies on the topology of saccharide-lipid
    synthesis. Journal of Biological Chemistry, 257, 2787-2794.
Haselbeck, A. & Tanner, W. (1981). A possible role of polyprenyl
    phosphates in membrane transport of cell wall precursors.
    In Plasmalemma and Tonoplast: Their Functions in the Plant
    Cell, eds. D. Marme et al., pp. 347-352. Holland: Elsevier.
Haselbeck, A. & Tanner, W. (1982). Dolichyl phosphate-mediated mannosyl
    transfer through liposomal membranes. Proceedings of the
    National Academy of Sciences of the United States of America,
    79, 1520-1524.
Hayashi, T., Kato, Y., Matsuda, K. (1980). Xyloglucan from suspension
    cultured soybean cells. Plant Cell Physiology, 21, 1405-1418.
Hayashi, T., Kato, Y. & Matsuda, K. (1981). Biosynthesis of xyloglucan in
    suspension cultured soybean cells. An assay method for
    xyloglucan xylosyltransferase. Journal of Biochemistry (Tokyo),
    89, 325-328.
Hayashi, T. & Matsuda, K. (1981). Biosynthesis of xyloglucan in suspension
    cultured soybean cells. Evidence that the enzyme system of
    xyloglucan synthesis does not contain β-1,4-glucan 4-β-D-
    glucosyltransferase activity. Plant & Cell Physiology, 22,
    1571-1584.
Hayashi, T. & Maclachlan, G.A. (1984a). Glycolipids and glycoproteins
    formed from UDP-galactose by pea membranes. Phytochemistry,
    23, 487-492.

Hayashi, T. & Maclachlan, G.A. (1984b). Biosynthesis of pentosyl lipids
    by pea membranes. Biochemical Journal, 217, 791-803.
Hayashi, T. & Maclachlan, G. (1984c). Pea cellulose and xyloglucan:
    biosynthesis and biodegradation. In Advances in Cellulose
    Chemistry and Technology, eds. R.A. Young & R.M. Rowell,
    in press. New York: Academic Press.
Hayashi, T. & Maclachlan, G. (1984d). Pea xyloglucan and cellulose:
    molecular organization. Plant Physiology, 75, part 3, in
    press.
Heftmann, E. (1977). Functions of steroids in plants. Progress in
    Phytochemistry, 4, 257-276.
Helsper, J.P.F.G. (1979). The possible role of lipid intermediates in
    the synthesis of β-glucans by a membrane fraction from pollen
    tubes of Petunia hybrida. Planta, 144, 443-450.
Hemming, F.W. (1977). The role of polyprenol-linked sugars in
    eukaryotic macromolecular synthesis. Biochemical Society
    Transactions, 5, 1682-1687.
Herscovics, A., Bugge, B. & Jeanloz, R.W. (1977). Glucosyltransferase
    activity in calf pancreas microsomes. Journal of Biological
    Chemistry, 252, 2271-2277.
Hopp, A.E., Romero, P.A. & Pont-Lezica, R. (1979). Subcellular
    localization of glucosyl transferases involved in
    glucosylation in Pisum sativum seedlings. Plant Cell
    Physiology, 20, 1063-1069.
Hori, H., James, D.W. & Elbein, A.D. (1982). Isolation and
    characterization of the Glc_3Man_9 $GlcNAc_2$ from lipid-linked
    oligosaccharides of plants. Archives of Biochemistry and
    Biophysics, 215, 12-21.
Hosel, W. (1981). Glycosylation and glycosidases. Biochemistry of Plants,
    7, 725-753.
Jourdan, R.S. & Mansell, R.L. (1982). Isolation and partial
    characterization of three glycosyltransferases involved in
    the biosynthesis of flavonol glycosides. Archives of
    Biochemistry and Biophysics, 213, 434-443.
Kahn, A.W. & Colvin, J.R. (1961a). Isolation of the precursor of
    bacterial cellulose. Journal of Polymer Science, 51, 1-9.
Kahn, A.W. & Colvin, J.R. (1961b). Synthesis of bacterial cellulose from
    labelled precursor. Science, 133, 2014-2015.
Kaul, K. & Lester, R.L. (1975). Characterization of inositol-containing
    phosphosphingolipids from tobacco leaves. Plant Physiology,
    55, 120-129.
Kauss, H. (1968). Enzymatische glucosylierung von pflanzlichen sterinen.
    Zeitschrift fuer Naturforschung, 23B, 1522-1526.
Kauss, H. (1969). A plant mannosyl lipid acting in reversible transfer of
    mannose, FEBS Letters, 5, 81-84.
Kemp, J. & Loughman, B.C. (1974). Cyclitol glucosides and their role in
    the synthesis of a glucan from UDPG in Phaseolus aureus.
    Biochemical Journal, 142, 153-159.
Kjosbakken, J. & J.R. Colvin (1973). Biosynthesis of cellulose by a
    particulate enzyme system from Acetobacter xylinum. In
    Biogenesis of Plant Cell Wall Polysaccharides, ed. F. Loewus,
    pp. 361-371. New York: Academic Press.
Lehle, L., Bowles, D.J. & Tanner, W. (1978). Subcellular site of mannosyl
    transfer to dolichyl phosphate in Phaseolus aureus. Plant
    Science Letters, 11, 27-34.

Lehle, L. (1980). Biosynthesis of the core region of yeast mannoproteins. Formation of a glucosylated dolichol-bound oligosaccharide precursor, its transfer to protein and subsequent modification. European Journal of Biochemistry, 109, 589-601.

Lehle, L. (1981). Plant cells synthesize glucose-containing lipid-linked oligosaccharides similar to those found in animals and yeast. FEBS Letters, 123, 63-66.

Lodish, H.F. & King, N. (1984). Glucose removal from N-linked oligosaccharides is required for efficient maturation of certain secretory glycoproteins from the rough endoplasmic reticulum in the Golgi complex. Journal of Cell Biology, 98, 1720-1729.

Maclachlan, G.A. (1982). Does β-glucan synthesis need a primer? In Cellulose and Other Natural Polymer Systems, ed. R.M. Brown, Jr., pp. 327-339. New York: Plenum Corp.

Maclachlan, G. & Singh, R. (1983). Transport and mobilization of asymmetrically-labelled sucrose in pea epicotyls. Recent Advances in Phytochemistry, 17, 153-172.

Mankowski, T., Jankowski, W., Chojnacki, T. & Franke, P. (1976). C55 Dolichol: Occurrence in pig liver and preparation by hydrogenation of plant undecaprenol. Biochemistry, 15, 2125-2130.

Mascara, T. & Fincher, G.B. (1982). Biosynthesis of arabinogalactan-protein in Lolium multiforum endosperm cells. In vitro incorporation of galactosyl residues from UDP-galactose into polymeric products. Australian Journal of Plant Physiology, 9, 31-45.

McCleary, B.V. (1979). Enzymatic hydrolysis, fine structure and gelling interaction of legume-seed D-galactomannans. Carbohydrate Research, 71, 205-230.

McNeil, M., Darvill, A.G. & Albersheim, P. (1973). Structure of plant cell walls. I. The macromolecular components of the walls of suspension cultured sycamore cells with a detailed analysis of the pectic polysaccharides. Plant Physiology, 51, 158-173.

McNeil, M., Albersheim, P., Taiz, L. & Jones, R.L. (1975). The structure of plant cell walls. VII. Barley aleurone cells. Plant Physiology, 55, 64-68.

Mellor, R.B., Roberts, L.M. & Lord, J.M. (1979). Glycosylation of exogenous protein by endoplasmic reticulum membranes from castor bean endosperm. Biochemical Journal, 182, 629-631.

Mellor, R.B. & Lord, J.M. (1979a). Involvement of a lipid-linked intermediate in the transfer of galactose from UDP(^{14}C)galactose to exogenous protein in castor bean endosperm homogenates. Planta, 147, 89-96.

Mellor, R.B. & Lord, J.M. (1979b). Subcellular localization of mannosyl transferase and glycoprotein biosynthesis in castor bean endosperm. Planta, 146, 147-153.

Michalczuk, L. & Bandurski, L.S. (1982). Enzymic synthesis of 1-0-indol-3-ylacetyl-β-D-glucose and indol-3-ylacetyl-myoinositol. Biochemical Journal, 207, 273-281.

Nagahashi, J. & Beevers, L. (1978). Subcellular localization of glycosyl transferases in glycoprotein biosynthesis in the cotyledons of Pisum sativum. Plant Physiology, 16, 451-459.

Nakano, K., Marahashi, A., Nohara, T., Tomimatsu, T., Imamura, N. & Kawasaki, T. (1983). A flavanol glycoside and a sesquiterpene cellobioside from Trillium tschonoskii. Phytochemistry, 22, 1249-1251.

Nevins, D.J., Yamamoto, R. & Huber, D.J. (1978). Cell-wall β-glucans of five grass species. Phytochemistry, 17, 1503-1505.

Paiement, J., Rachubinski, R.A., Ng Ying Kin, M.M.K., Sikstrom, R.A. & Bergeron, J.J.M. (1982). Membrane fusion and glycosylation in the rat hepatic Golgi apparatus. Journal of Cell Biology, 92, 147-154.

Palter, R., Haddon, W.F. & Lundin, R.E. (1972). Structure of a safflower steroid cellobioside. Phytochemistry, 11, 2327-2328.

Polonenko, D.R. & Maclachlan, G. (1984). Plasmamembrane sheets from pea protoplasts. Journal of Experimental Botany, 35, 1-8.

Pont-Lezica, R., Brett, C.T., Martinez, P.R. & Dankert, M.A. (1975). A glucose acceptor in plants with the properties of a saturated polyprenyl monophosphate. Biochemical and Biophysical Research Communications, 66, 980-987.

Pont-Lezica, R., Romero, P.A. & Dankert, M.A. (1976). Membrane bound UDP-glucose. Lipid glucosyltransferases from pea. Plant Physiology, 58, 675-680.

Pont-Lezica, R., Romero, P.A. & Hopp, H.E. (1978). Glucosylation of membrane-bound proteins by lipid-linked glucose. Planta, 140, 177-183.

Romero, P.A. & Pont-Lezica, R. (1976). Transfer of glucose-1-P from UDPG to lipid acceptors in plants. Acta Physiologica Latino Americana, 26, 364-370.

Romero, P., Garcia, R.C. & Dankert, M. (1977). Synthesis of polyprenol-monophosphate β-galactose by Acetobacter xylinum. Molecular and Cellular Biochemistry, 16, 205-212.

Ruiz-Herrera, J. & Bartnicki-Garcia, S. (1974). Synthesis of cell wall microfibrils in vitro by a soluble chitin synthetase from Mucor vouxii. Science, 186, 357-359.

Schmitt, R. & Sandermann, H. (1982). Specific localization of β-D-glucoside conjugates of 2,4-dichlorophenoxyacetic acid in soybean vacuoles. Naturforscher, 37c, 772-777.

Sharma, C.B., Lehle, L. & Tanner, W. (1981). N-glycosylation of yeast proteins. European Journal of Biochemistry, 116, 101-108.

Singh, R. & Maclachlan, G.A. (1983). Transport and metabolism of sucrose vs. hexoses in relation to growth in etiolated pea stem. Plant Physiology, 71, 531-535.

Snider, M.D. & Rogers, O.C. (1984). Transmembrane movement of oligosaccharide-lipids during glycoprotein synthesis. Cell, 36, 753-761.

Snider, M.D., Sultzman, L.A. & Robbins, P.W. (1982). Transmembrane location of oligosaccharide-lipid synthesis in microsomal vesicles. Cell, 24, 385-392.

Staneloni, R.J., Tolmansky, M.A., Petriella, C., Ugalde, R.A. & Leloir, L.F. (1980). Presence in a plant of a compound similar to the dolichyl diphosphate oligosaccharide in animal tissue. Biochemical Journal, 191, 257-260.

Staneloni, R.J., Tolmansky, M.E., Petriella, C. & Leloir, L.F. (1981). Transfer of oligosaccharide to protein from a lipid intermediate in plants. Plant Physiology, 68, 1175-1179.

Staudte, R.G., Woodward, J.R., Fincher, G.B. & Stone, B.A. (1983). Water-soluble 1,3/1,4 β-glucans from barley endosperm. Carbohydrate Polymers, 3, 299-312.

Sutter, A. & Grisebach, H. (1975). Free reversibility of the UDP-
        glucose-flavanol 3-0-glucosyl transferase reaction.
        Archives of Biochemistry and Biophysics, 167, 444-447.
Taguchi, R., Sakano, Y., Kikuchi, Y., Sakuma, M. & Kobayashi, T. (1973).
        Synthesis of pullulan by acetone-dried cells and cell free
        enzymes from Pullularia pullulans and the participation of
        lipid intermediate. Agricultural and Biological Chemistry,
        37, 1635-1641.
Waechter, B.M. & Lennarz, W.J. (1976). The role of polyprenol-linked
        sugars in glycoprotein synthesis. Annual Review of
        Biochemistry, 46, 95-112.
Wells, G.B., Turco, S.J., Hanson, B.A. & Lester, R.L. (1981). Separation
        of dolichylpyrophosphoryl-oligosaccharides by liquid
        chromatography. Analytical Biochemistry, 110, 397-406.
Whelan, W.H. (1976). On the origin of primer for glycogen synthesis.
        Trends in Biochemical Science, 1, 13-15.
Woodward, J.R., Fincher, G.B. & Stone, B.A. (1983). Water-soluble 1,3/1,4
        β-glucans from barley endosperm.   II. Fine structure.
        Carbohydrate Polymers, 3, 207-225.
Zatta, P., Zakim, D. & Vessey, D.A. (1975). The transfer of galactose
        from UDP-galactose to endogenous lipid acceptors in liver
        microsomes. Biochimica et Biophysica Acta, 392, 361-365.

9   BIOSYNTHESIS OF β-GLUCANS IN GROWING COTTON (<u>Gossypium</u>
<u>arboreum</u> L. and Gossypium <u>hirsutum</u> L.) FIBRES

A.J. Buchala
Institut de Biologie végétale, et de Phytochimie,
Université, CH 1700 Fribourg, Switzerland

H. Meier
Institut de Biologie végétale, et de Phytochimie,
Université, CH 1700 Fribourg, Switzerland

<u>Abstract</u>.   The biosynthesis <u>in vivo</u> and <u>in vitro</u> of
β-glucans in growing cotton fibres has been reviewed.   The
cell walls of cotton fibres contain two well-characterised
β-glucans i.e. callose and cellulose and other polysaccharides
containing β-D-glucosyl residues e.g. xyloglucan, which have
been recognised and partially characterised.   Synthesis <u>in</u>
<u>vivo</u> of callose and cellulose occurs at both the primary-
and secondary-cell-wall stage of development whilst that of
xyloglucan is limited to the former.   Most studies on
biosynthesis <u>in vitro</u> have been directed at the secondary-
cell-wall-forming stage where the products obtained depend
on the experimental system and the precursor used.   Damaged
fibres, fibre homogenates, and particulate fractions
readily synthesise callose from UDP-glucose, whilst intact
fibres synthesise both callose and cellulose from glucose,
fructose, or sucrose fed to the cell apoplast.

<u>Key words</u>:   β-Glucan, cellulose, callose, cotton fibre, cell
wall biosynthesis.

INTRODUCTION

    Mature cotton fibres have been long recognised as a source
of relatively pure cellulose and therefore seem apt for the study of its
biosynthesis.   Two basic approaches may be envisaged (i) <u>in vivo</u> with
feeding of labelled precursors to the cotton fruits via the plant's
vascular system or by treating the plants with $^{14}CO_2$ and allowing
photosynthesis to take place;   (ii) <u>in vitro</u> with feeding of radioactive
precursors to fibre homogenates or to the fibres (detached or intact)
themselves.   The former approach appears particularly attractive but
requires information concerning the fate of the precursor before it
attains the fibres.   Early work on feeding ^{14}C-glucose to the stem of
cotton plants was carried out by Greathouse (1953) and Shafizadeh &
Wolfrom (1955) but no kinetic work was attempted and few real conclusions
could be drawn.   The translocation of ^{14}C-labelled assimilate from the
main stem of the cotton plant to the cotton boll was studied by Brown

(1968) and in more detail by Benedict et al. (1973).   The second
approach has received more attention and yielded interesting results,
demonstrating in particular the synthesis of other polysaccharides
containing β-D-glucopyranosyl linkages such as glucomannan (Barber &
Hassid 1965) and callose (Heiniger & Delmer 1977).   Many of the
earlier conclusions drawn concerning cellulose synthesis thus need
reinterpretation.   In this overview, particular emphasis will be made
on the synthesis of callose and a possible relationship between such
synthesis and that of cellulose using both the in-vivo and in-vitro
approaches with cotton fibres.   For a comprehensive insight on the
field in general Delmer (1983) has written an excellent review.

GLUCOSE-CONTAINING POLYSACCHARIDES IN COTTON FIBRES
Cellulose
The linear extended chain structure of cellulose with its
(1→4)-β-D-glucopyranosyl residues and where the degree of polymerisation
varies with the source has long been accepted.   Marx-Figini & Schulz
(1966) showed that biosynthesis proceeds in two distinct stages - a slow
initial process yielding primary-wall cellulose with a non-uniform
degree of polymerisation of about 2000 - 6000, followed by a more rapid
process yielding secondary-wall cellulose with a high and uniform degree
of polymerisation of about 14000.   It has also been suggested that
cotton cellulose is a glycoprotein (Nowak-Ossorio et al. 1976).
Strong hydrogen bonding occurs between the D-glucosyl residues giving
rise to a highly insoluble, partially crystalline, fibrillar structure.
When these criteria are fulfilled then a product of biosynthesis may be
referred to as "cellulose".   Otherwise one should use the term "(1→4)-β-
D-glucan" - insolubility in strong alkali is not really sufficient to
define a product as cellulose.   Some workers use insolubility in the
Updegraff reagent (1969) as a criterion of the cellulosic nature of a
product, but Delmer reports (1983) that although native (in vivo)
cellulose is insoluble, (1→4)-β-D-glucan synthesised in vitro is nearly
always soluble.

Callose
The presence of other polysaccharides containing glucose
residues occurring in vivo was recognised by Meinert & Delmer (1977)
and Huwyler et al. (1978), and that such polysaccharides were produced

in vitro, by Barber & Hassid (1965) and Delmer et al. (1977).   One of
these polysaccharides was partially characterised (Maltby et al. 1979)
and isolated and characterised by Huwyler et al. (1978) as a (1→3)-β-D-
glucan which is hereafter referred to as "callose".   These and later
results (unpublished) also show the presence of some (1→6)-β-linkages
arising from chain branching by single-residue side chains.   The
repeating unit has been found to vary from about 10 to 20 residues
depending on the actual source of the material studied.   We have little
precise information concerning its molecular weight.   Fractionation
gives material which still contains amino acids but proof that callose
is a glycoprotein is lacking.   It may be solubilised from the fibres
with boiling water but hot dimethyl sulphoxide (DMSO) is the most
effective solvent.   Waterkeyn (1981) has shown that the callose occurs,
independently of fibre age, in the innermost wall layer bordering the
cell lumen.

Xyloglucan and other polysaccharides

Xyloglucan type of molecules are also present in the cell
wall of cotton fibres (Buchala & Meier, unpublished).   Fractionation
gave material containing glucose, xylose, galactose and fucose (moles
ratio 2.4:1:0.3:0.1) which appeared homogeneous.   Methylation analysis
showed that the structure was typical of that known for xyloglucans i.e.
a main chain with (1→4)-β-D-linked glucopyranosyl residues with side
chains containing xylosyl-, galactosylxylosyl-, and fucosylgalactosyl-
xylosyl units.   The similarity of the main chain with cellulose
molecules must not be overlooked.

Cotton dust (the source of the dust is not clear) contains
a glycosaminoglycan with glucose, galactose, mannose and glucosamine
residues (ratio ca. 1:1:1:0.5) of undetermined structure (Mohammed et al.
1971) which may be a glycoprotein.   Mannose residues have also been
detected in hydrolysates of cotton fibres (Meinert & Delmer 1977;
Huwyler et al. 1979).   It is not clear whether they constitute a β-
linked glucomannan or an α-linked glycoprotein-type molecule.   Since
these macromolecules may contain glucosyl residues, their presence should
not be entirely forgotten when interpreting results concerning
incorporation of glucose.   β-Glucans containing (1→3)-, and (1→4)-
linked residues of the cereal endosperm type and starch are known to be
absent.

β-GLUCAN CONTENT OF FIBRES DURING DEVELOPMENT

The absolute amount of cellulosic and non-cellulosic glucose-containing polysaccharides was determined by both Meinert & Delmer (1977) and Huwyler et al. (1979).   The former workers used a direct trifluoroacetic-acid hydrolysis method which in our opinion does not give reliable results, whilst the latter used classical extraction methods followed by mineral-acid hydrolysis.   These results, expressed on a constant-cell-number basis, i.e. with respect to the fibres of one seed, showed that the amount of cellulose laid down during the primary-wall stage was low and that after ca. 21-25 days post-anthesis (DPA) the cellulose content increased linearly until fibre maturity (ca. 60 DPA) to give a final weight of ca. 50 mg/fibres of one seed.   The non-cellulosic glucan content showed a maximum (ca. 1 mg) at the beginning (ca. 25 DPA) of secondary-wall formation.   These results were confirmed by Jaquet et al. (1982) who used an exo-(1→3)-β-glucanase for the determination of the glucan and thereby demonstrated that nearly all of the non-cellulosic glucose derived from a callose type of molecule. Fig. 1 shows the results obtained for G. arboreum (Jaquet et al. 1982). The period at which the maximum occurs has been found to depend on the species and the conditions under which the cotton plants are cultivated. It should be noted that fibres which develop on ovules cultured in vitro mature more rapidly than those cultured normally in vivo.

No reliable method has been found to determine the xyloglucan content of the fibres, but the amount is evidently small and the polysaccharide probably derives from the primary wall.

β-GLUCAN PRECURSORS

As previously pointed out, not all of the experimental systems are suitable for a study of the normally accepted β-glucan precursors, e.g. sugar nucleotides per se cannot normally penetrate the plasma membrane and incorporation must be studied using fibre homogenates, enzyme extracts or detached fibres.   Therefore, when intact cells appear to incorporate sugars from sugar nucleotides this may be taken as evidence for the presence of damaged cells (Brett 1978;   Delmer et al. 1977).   The uptake of neutral sugars, however, readily occurs with intact cells.

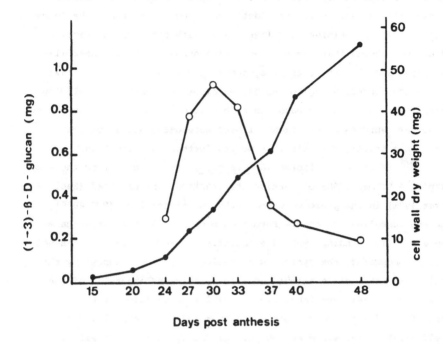

Fig. 1.   Changes in the cell wall dry weight (—●—)
and in the callose content (—O—) of cotton G. arboreum
fibres after treatment with 80% methanol.   The values
are expressed in mg per fibres of one ovule (adapted from
Jaquet et al. 1982).

Incorporation with non-intact cells

Membrane fractions.   Most of the earlier work has been
carried out with non-intact cotton fibres, particulate membrane fractions
or soluble (although not in a strict sense) fractions.   Using a
particulate fraction derived from fibres at essentially the primary-wall
stage, Barber & Hassid (1965) obtained an alkali-insoluble product from
only GDP-glucose and incorporation was stimulated in the presence of GDP-
mannose.   The product was not characterised but Hassid (1967) later
interpreted the results to indicate that cellulose had been synthesised.
The product could also have been an insoluble glucomannan (cf. Heller &
Villemez 1972).   Similar results were later obtained with detached
fibres (G. hirsutum) by Delmer et al. (1974) who showed that the
incorporation from GDP-glucose diminished markedly with fibre
development and was almost negligible at the stage of secondary-wall

formation whilst incorporation from UDP-glucose into alkali-insoluble
material (lipids and polysaccharides) increased.   Later work by Delmer
et al. (1977) and Heiniger & Delmer (1977) with particulate fractions
and UDP-Glc showed that the polysaccharide produced was a linear (1→3)-
β-D-glucan and that the rate of synthesis showed a well-defined maximum
at a period corresponding to the beginning of secondary-wall cellulose
synthesis.   On the other hand, no such well defined maximum was observed
by the same workers when detached fibres were used instead of the
particulate fractions - this fact merits further investigation.

         When cotton fibres (G. hirsutum) are detached from the seeds
(Carpita & Delmer 1980) or particulate fractions are prepared (Bacic &
Delmer 1981) in the presence of polyethylene glycol 4000 (PEG 4000),
membrane resealing or vesicle formation occurs.   In the latter case,
when a positive (inside) membrane potential is induced in the presence of
certain ionophores, the synthesis of β-glucan from UDP-glucose by the
particulate fraction is stimulated 4 to 12 fold and about 10% of the
product is a (1→4)-β-D-linked glucan, albeit solubilised by the
Updegraff (1969) procedure (Bacic & Delmer 1981).   Recently, Delmer
et al. (1984) re-examined the UDP-glc:glucan synthetase activity in
particulate membrane fractions and found that, by modifying the isolation
method, the products contained a mixture of (1→3)-β- and (1→4)-β-glucosyl
residues.   The incubation media in these experiments nearly always
contained cellobiose since oligosaccharides have been found to
significantly stimulate β-glucan synthesis.   If glucose-1-P and UTP
are fed instead of UDP-glucose, then similar products are obtained
(Delmer et al. 1977), suggesting that these metabolites are converted
into UDP-glucose.   Other possible precursors, e.g. glucose, sucrose, are
very poorly, or not at all, incorporated into β-glucans under the above
conditions.

         Detached fibres.   When the cotton fibres (G. hirsutum) are
removed from the seeds in the absence of PEG 4000 the cell contents spill
out and cellular organisation is disrupted.   In the presence of PEG 4000,
or less well in the presence of other osmotica, e.g. mannitol, over 50%
of the detached fibres remained essentially similar, under the optical
microscope, to undetached fibres (Carpita & Delmer 1980).   With
unprotected fibres the results obtained on feeding radioactive
precursors are quite similar to those with particulate fractions, i.e.
glucose is only really incorporated from UDP-glucose to give callose and

no (1→4)-β-glucan.   Such fibres at the stage of primary-wall
formation do, however, incorporate glucose better from GDP-glucose
(Delmer et al. 1974) and we have been able to confirm this fact in our
own laboratory (Allenbach, Buchala & Meier, unpublished).   The identity
of the products, however, is less clear although both (1→3)-, and (1→4)-
β-glucosyl linkages are formed.   The major product obtained from UDP-
glucose is sucrose, but when labelled sucrose is used as precursor the
radioactivity incorporated into β-glucan is negligible.

When the fibres are protected with PEG 4000 the results
obtained are different.   Carpita & Delmer (1980) have reported that UDP-
glucose is less well incorporated into (1→3)-β-glucan and radioactive
glucose itself is incorporated into both (1→3)-β-, and (1→4)-β-glucan
in the ratio ca. 3:1, i.e. in the same ratio obtained with intact fibres
not protected with PEG.   The results obtained correspond well with the
idea that a part of the fibres behaved as if they were intact (see below).
However, in our laboratory (Meier 1983) the PEG stimulated the
incorporation from both glucose and UDP-glucose.   For both precursors
the absolute amounts of callose synthesised were much higher than those
of cellulose.

Incorporation with intact cells

Three different systems have been described in the literature:
(i) feeding to the ovules (G. hirsutum) via the culture medium, cf Maltby
et al. (1979);   (ii) feeding to the apoplast of isolated seed clusters
(G. arboreum), cf Pillonel et al. (1980);   and (iii) feeding via the
fruit petiole to the intact fruit (G. arboreum), cf Meier et al. (1981).

Cotton seed ovules.   When glucose (Maltby et al. 1979;
Francey & Meier 1984), fructose or sucrose (Francey & Meier 1984) is fed
to the culture medium, synthesis of callose and cellulose occurs
(glucose seems to be the best precursor) and the only significant
difference lies in the absolute incorporation of the radioactivity.
UDP-glucose has also been reported to be incorporated by this system
(Dugger & Palmer 1980) to give callose and short-chain (1→4)-β-linked
glucose-containing molecules.   These authors also reported that the
presence of borate in the culture medium had an effect on the ratio of
the products synthesised.   However, since sugar nucleotides are not
normally directly transported across the plasma membrane, it appears that
the UDP-glucose was first metabolised in the cell-wall apoplast or by

damaged cells.   Carpita & Delmer (1981) carried out a kinetic study of
the products, both of low-molecular-weight sugars and of β-glucans,
when glucose was fed to the culture medium.   They found, as did Franz
(1969), that UDP-glucose was the main sugar nucleotide at the stage of
secondary-wall formation and that the labelling of the UDP-glucose pool
during a pulse-chase experiment with glucose indicated that UDP-glucose
was probably a precursor for both secondary-wall callose and cellulose.

    Intact seed clusters.   Another approach which we have
exploited in our laboratory involves the feeding, by vacuum infiltration,
of the precursors to the apoplast of isolated intact seed clusters
(Pillonel et al. 1980).   Several precursors have been studied (glucose,
fructose, sucrose, UDP-glucose, and GDP-glucose) at both primary- and
secondary-wall forming stages of fibre development, but only the products
derived from glucose, fructose and sucrose, at the secondary-cell-wall
stage, have been characterised (Pillonel et al. 1980; & unpublished
results).   Both callose (ca. 70%) and cellulose (ca. 30%) are
synthesised - the factors which influence the proportions of the two
products are discussed later on.   The products formed at the primary-
cell-wall stage and from the nucleotide sugars are less well known,
mainly due to the relatively poor rates of incorporation.   As noted
above, if the seed clusters are intact, then no incorporation is to be
expected from the sugar nucleotides and thus such incorporation, e.g.
Pillonel et al. (1980), may be taken to indicate that some of the fibres
were indeed damaged during the removal of the seed cluster from the fruit.
It is of interest that when intact seed clusters are mechanically
disturbed, e.g. by teasing out the fibres slightly, then incorporation
from sucrose is much diminished whilst that from UDP-glucose is improved.

    One of the reasons for adopting this approach was the idea
that the sugar precursor arrived to the fibres via the apoplast and was
most likely glucose, fructose or sucrose since these free sugars
predominate.   Our work (Jaquet et al. 1982) showed that at the early
stages of secondary-wall formation, glucose and fructose exhibited a
maximum which was closely followed by a maximum in the callose content and
in the sugar phosphates.   On the other hand, the sucrose content
increased regularly until fibre maturity.   At the time when glucose and
fructose were at a maximum the total invertase activity was also
maximal (Buchala & Meier, unpublished).   The probable route for the
sugars necessary for fibre cell-wall synthesis is via the phloem in the

form of sucrose to the cotton fruit.   The sucrose may then pass into
the fibre cell-wall free space (apoplast) through the carpels or into
the cotton seeds through the funiculi.   In the latter case, due to the
apparent lack of plasmodesmata (Ryser, personal communication) connecting
fibre cells to the neighbouring seed cells, sugars (hexose or sucrose?)
must be directed to the fibres via the apoplast.   The role of the
invertase could then be to assure a sufficient supply of hexose for
cell-wall synthesis.   However, whether sucrose may be taken up directly,
or only after hydrolysis by a cell-wall-bound, or plasma-membrane-
associated invertase is not yet known.   The incorporation from sucrose,
but not that from UDP-glucose, is strongly inhibited by inhibitors of
respiratory ATP formation such as KCN or dinitrophenol (DNP), but it is
not known whether the transport of sucrose or its further metabolism in
the cell is inhibited (Pillonel et al. 1980).

Intact fruits.   When ^{14}C-labelled sucrose is fed to the
petioles of whole fruit capsules at the stage of secondary-wall
formation both callose and cellulose are synthesised but, contrary to
the seed-cluster experiments, cellulose is the major product (Meier et al.
1981).   The results are somewhat similar to those obtained when glucose
is fed to cotton-seed ovules (see above) and to those obtained when $^{14}CO_2$
is fed to intact cotton plants and incorporation occurs from the
labelled assimilates after photosynthesis (Meier et al. 1981).   At the
stage of primary-wall formation, feeding of $^{14}CO_2$ gives the same products,
but in addition other polysaccharides, including xyloglucan, are labelled.

Relationship between precursor and product

Any attempt to relate a precursor to its product implies a
knowledge of the intervening steps.   The situation with homogenates or
particulate fractions is less complicated than that with non-intact or
intact cells.   The problems associated with the presence of different
pools do not exist but the results obtained do not necessarily reflect
the situation in vivo since certain enzymes may be stimulated, whereas
others may be deactivated due to membrane disruption.   Once the cell
compartmentation is destroyed then the appropriate enzymes or substrates
are no longer present in the correct proportions, and due to the
reversibility of most enzymic processes the reaction in vitro may proceed
in the opposite direction to that in vivo.

When the precursor has crossed the plasma membrane, most
workers seem to agree that the first step to active cell-wall formation is

the formation of nucleotide sugars.   For the synthesis of β-glucans
both UDP-glucose and GDP-glucose have been reported to be effective.
The case for UDP-glucose as the precursor for callose does not seem in
doubt, but the only clear evidence which shows that UDP-glucose may also
be the precursor of cellulose comes from the work of Carpita & Delmer
(1981), who carried out a series of experiments tracing the flow of
carbon in ovules cultured in vitro when ^{14}C-glucose was fed to the
culture medium.   Their analysis of the various sugar pools led them to
conclude that UDP-glucose was probably a precursor for both callose and
cellulose and that no other nucleotide sugar displayed an appropriate
pattern of labelling consistent with precursor properties.   The scheme
proposed does not take into account metabolic paths involving sucrose or
fructose which are major free sugars in developing cotton fibres
(Jaquet et al. 1982).   Both sucrose, via sucrose synthetase, or
glucose-1P, via UDP-glucose pyrophosphorylase, are supposed to furnish
the UDP-glucose.   However, when UDP-glucose is fed to non-intact cells,
sucrose is the main product (Buchs 1982) and attempts to reverse the
reaction were unsuccessful.   UDP-glucose pyrophosphorylase activity has
been found in cotton fibres at all stages of development (Delmer 1977),
whereas that of GDP-glucose pyrophosphorylase was not detected.   For the
moment no other viable path of glucosylation has been demonstrated and the
results obtained in vivo on feeding sucrose or glucose to intact cells
do not invalidate the hypothesis.   The latter sugars may therefore
produce UDP-glucose, but in lower concentrations and under metabolic
regulation.   It has been shown for other higher plants that where UDP-
glucose is used to synthesise both (1→3)-β-, and (1→4)-β-linkages
(the products were not necessarily callose or cellulose;   see citations
in Delmer 1983), high concentrations favour synthesis of (1→3)-β-
linkages.   However, with non-intact cotton fibres the effect of varying
the UDP-Glc concentration on the synthesis of (1→4)-glucan is difficult
to estimate since normally more than 95% of the labelled product is callose.

        The steps following the synthesis of UDP-glucose are less
well documented in higher plants.   Apart from sucrose, glycosyl lipids
are also synthesised from UDP-glucose, and these have been shown to be
mainly of the sterol type (Forsee et al. 1974) or dolichol (Dol) type
(Forsee & Elbein 1973).   Similar lipids are also synthesised when
sucrose is fed to the apoplast of intact seed clusters (Buchala & Meier,
unpublished).   There is no convincing evidence for the implication of

such lipids as intermediates in β-glucan synthesis;   several inhibitors
of lipid glycosylation are without effect on β-glucan synthesis
(Pillonel 1983) and in turn inhibitors of cellulose synthesis, e.g. 2,6-
dichlorobenzonitrile (DCB), do not give rise to accumulation of such
lipids (Montezinos & Delmer 1980).   Plant and animal membrane fractions
catalyse the formation of lipid-linked oligosaccharides containing 3 to
14 sugar moieties to give a final product, namely $Glc_3Man_9GlcNAc_2$-PP-Dol,
which is used in lipid-dependent protein glycosylation.   However, the
terminal glucosyl residues, which are probably α-linked, and some of the
mannosyl residues are normally trimmed off before the glycosylation
step.   Although mannose residues are found in polysaccharide fractions
from cotton fibres, there has been no demonstration of the presence of
mannose- or glucose-containing glycoproteins.   Callose is in fact the
only macromolecule other than cellulose labelled at the stage of
secondary-wall formation and there is no real evidence to show that it
is a glycoprotein.   The possibility that callose may function as an
intermediate in cellulose synthesis will be discussed below.

        In conclusion, although glucose, fructose or sucrose are
used by intact cotton cells to synthesise both callose and cellulose,
the first step in glycosylation appears to be the formation of UDP-
glucose.

        Factors that influence the ratio of callose to cellulose
        synthesised in short-time feeding experiments
        From the foregoing discussion it is clear that only feeding
of radioactive glucose or sucrose to intact fibre systems need be
considered.   When a radioactive precursor (^{14}C-glucose or ^{14}C-sucrose)
is fed to intact fibres in vitro during periods ranging from 10 min
to several hours, the ratio of the radioactivity in newly formed callose
to that in newly formed cellulose may vary from ca. 0.7 to 7 depending
on the experimental system.   Some results are given in Table 1 which
shows that:
(i) temperature has a large influence.   Low temperature strongly favours
the synthesis of callose compared with cellulose, but the total
incorporation into both products is diminished.
(ii) the relative incorporation of radioactivity into callose is highest
in the experimental system where the precursor is fed by vacuum
infiltration to isolated seed clusters, obtained from fruits grown in the

greenhouse, and lowest in the system where the precursor is fed via the
nutrient solution to the fibres of ovules cultured in vitro.
(iii) disturbing the plasma membrane by plasmolysis with 1M mannitol or
by the ionic detergent sodium dodecylsulphate (SDS) increases the ratio
of radioactivity found in newly formed callose to cellulose. When the
feeding time is extended over several hours the ratios observed slowly
diminish compared to those given in Table 1.   A number of inhibitors
such as DNP, KCN, valinomycin or bacitracin inhibit both callose and
cellulose synthesis to a similar extent, whereas DCB and coumarin inhibit
the synthesis of callose much less than that of cellulose (Pillonel 1983).
No inhibitor was found that specifically inhibited callose formation and
not cellulose formation.

 The reasons for the different ratios of newly formed callose
to cellulose in the three experimental systems mentioned in Table 1 are
not clear.   If the precursor pools for callose and cellulose are
different then one could argue that these pools attain different
specific activities depending on the incubation system.   It may well be
that some of the callose formed is a result of stress and that the
extent to which the values obtained are affected depends on the amount of
such stress.   Cellulose synthesis seems to be a much more labile
metabolic process than callose synthesis.   The former is supposed to take
place in synthesising complexes in the plasma membrane.   When the latter
is dissociated from the cell wall by e.g. plasmolysis, or is rendered
permeable by treatment with detergents, it appears evident that cellulose
synthesis should be perturbed.   Callose synthesis can occur in cell-free
systems where such rigorously controlled conditions do not appear
necessary for its synthesis.

THE TURNOVER OF CALLOSE

 When radioactive precursors are fed to cotton fruits via
the petioles or by treating the plants with $^{14}CO_2$ and allowing
photosynthesis to take place or in vitro to cultured ovules via the
seeds or fibre surface it is observed, at any stage of secondary-wall
formation, that after a short time 30 to 40% of the radioactivity
incorporated into β-glucans is in callose.   This indicates a turnover of
callose since, at any stage during secondary-wall formation except during
the transition phase from primary- to secondary-wall deposition, the
absolute amount of callose is only about 0.5 to 1.5% of the total β-glucan

Table 1.   Ratio of the radioactivity incorporated into callose and cellulose after 1 h of feeding <u>in vitro</u> with ^{14}C-glucose or ^{14}C-sucrose to intact cotton fibres at the stage of secondary-wall formation.

radioactive precursor and special conditions	experimental procedure		
	intact seed cluster fed by vacuum infiltration §	cultured ovules fed by vacuum infiltration §	cultured ovules fed via the culture medium*
sucrose, 30°C, Triton X-100 (0.01%)	1.7	1.2	-
sucrose, 15°C Triton X-100 (0.01%)	3.8	-	-
sucrose, 10°C, Triton X-100 (0.01%)	7.3	-	-
glucose, 30°C, Triton X-100 (0.01%)	1.7	1.0	-
glucose, 34°C, culture medium	-	-	0.7
sucrose, 34°C, culture medium	-	-	0.7
sucrose, 30°C, Triton X-100 (0.01%) + mannitol (1 M)	3.7	-	-
sucrose, 30°C, SDS (175 mM)	3.0	-	-

The incorporation ratio was determined thus: dpm callose/dpm cellulose. § Vacuum infiltration was carried out in HEPES buffer (20 mM; pH 7.5) according to Pillonel <u>et al.</u> (1980).   * The culture medium used was that described by Beasely & Ting (1974) and only the aerial fibres were analysed.

(Huwyler et al. 1979, Jaquet et al. 1982).   Turnover of callose is
clearly demonstrated by pulse-chase experiments (Meier et al. 1981).
From Figs. 2 (results obtained by feeding ^{14}C-sucrose to the petiole of
intact fruits) and 3 (results obtained by treating an intact branch with
$^{14}CO_2$) it can be seen that the total radioactivity in callose as well as
in the methanol-soluble sugar pool reaches a maximum about 10 h and 24 h
respectively after the pulse and then slowly decreases.   The radioactivity
in the callose finally attains a value of 2 to 3% of the total
incorporation into β-glucan, and this value corresponds well to the
actual absolute amount of callose present.   The apparent turnover of
callose is obviously so slow because of the very large vacuolar sugar

Fig. 2.   Pulse-chase feeding of sucrose to petioles of fruit
capsules (G. arboreum) harvested 35 DPA. Pulse: ^{14}C-sucrose
(1 mM, 20 μl, 1.85 MBq, 20 min).   Cellulose (—■—),
callose (—●—), 80% methanol-soluble material (—○—).
In parentheses: %radioactivity in callose- and cellulose-
type glucans respectively.   Results from Meier et al. (1981).

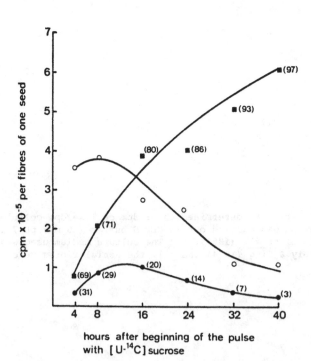

hours after beginning of the pulse
with [U-^{14}C] sucrose

pool (methanol-soluble) of the fibre cells.   The radioactivity in this
sugar pool and that in the callose pool drop in parallel.   From the
quantitative analyses of the three main sugars (glucose, fructose, and
sucrose) in the fibres (Jaquet et al. 1982) it can be calculated that
the sugar pool of a fibre in the early stage of secondary-wall
formation is sufficient for the synthesis of secondary-wall material
deposited during a three- to four-day period.   It is therefore not
astonishing that in a pulse-chase experiment, the total radioactivity
in the sugars and in callose, after having reached a maximum, decreases
only slowly.

Fig. 3.   Pulse-chase feeding of $^{14}CO_2$ to branches of
intact plants (G. arboreum) bearing fruit capsules at
the age of 35-40 DPA.   Pulse: (7.4 MBq, 30 min);
cellulose (——■——), callose (——●——), 80% methanol-
soluble material (——○——).   In parentheses, % radioactivity
in callose- and cellulose-type glucans respectively.
From Meier et al. (1981).

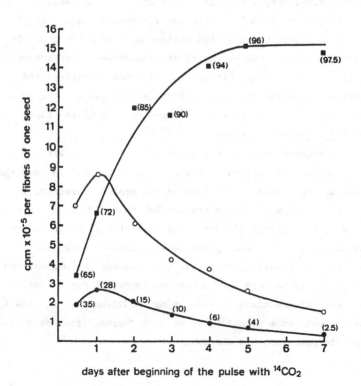

days after beginning of the pulse with $^{14}CO_2$

The continuous breakdown of callose, which obviously occurs in all the experimental systems studied, is probably due to the action of one or several (1→3)-β-D-glucanases.   It has been found (Dürr 1981, and Bucheli et al. 1984) that an exo-(1→3)-β-glucanase, bound to the cell walls of cotton fibres, has a high activity during secondary-wall formation.   As this enzyme can also function as a transglucosylase it might be argued that it could transfer glucose residues from callose to growing cellulose chains.   This, however, is speculation for the moment. It is also possible that the enzyme merely hydrolyses callose or that it transfers glucose derived from callose to an acceptor other than cellulose.   In this way, glucose residues incorporated into callose could be returned to the general sugar pool.

In experiments with continuous feeding (i.e. without chase) of ^{14}C-sucrose (Fig. 4) to the petioles of intact cotton fruits, it was observed that the steady-state level of radioactivity in callose was attained after somewhat less than one day.   It can be seen from Fig. 4 that callose behaves very similarly to the low-molecular-weight (methanol-soluble) sugar pool.   From the analyses of Jaquet et al. (1982) it can be calculated that the callose pool of a fibre should undergo turnover two (at the beginning of secondary-wall formation) to eight (at later stages) times within 24 h in order to deliver the glucose residues necessary for the production of cellulose during the same period (ca. 2 mg of cellulose is formed in 24 h by the fibres of one ovule of Gossypium arboreum).

A precursor role of callose for cellulose is, after all, feasible.   However, as mentioned above, it is also possible that the callose pathway represents a side loop in the general carbohydrate metabolism of the fibres, but the reasons for this side loop are not yet evident.   Waterkeyn (1981) discussed the role of callose in cotton fibres and postulated that most plant cells possess a normally "permanent" non-cellulosic glycan layer just outside the plasmalemma which is supposed to perform many important functions during cell development and differentiation, e.g. it may function as a lubricant matrix for the cellulose microfibrils or as a "screen" against diffusion and loss of the precursors of the latter.

Fig. 4.   Incorporation of radioactive sucrose (20 mM),
fed to the petioles of unripe capsules (G. arboreum)
during secondary-wall formation, into β-glucans.
From Meier (1981).

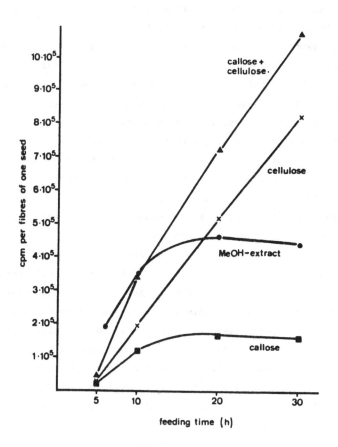

FINAL REMARKS

One should be aware that direct evidence for UDP-glucose as a
precursor for cellulose in cotton fibres is lacking.   In a large number
of experiments (unpublished) where cotton   (G. hirsutum) fibres
detached in PEG-containing buffer were incubated with radioactive UDP-
glucose, the β-glucans formed consisted of 96 to 99% callose and perhaps
(it is technically difficult to be sure) 1 to 4% of (1→4)-β-glucan.
When the detached fibres were incubated with ^{14}C-glucose between 10 and
20% of the radioactivity in β-glucans was found in (1→4)-β-glucan.

When ^{14}C-glucose was fed together with unlabelled UDP-glucose to
detached fibres, the added UDP-glucose had no influence on the
incorporation of radioactive glucose into β-glucans. This indicates
that (i) UDP-glucose is not on the pathway from glucose to cellulose, or
(ii) glucose is only incorporated by the intact protoplasts in the assay
mixture after having been taken up through the plasma membrane. UDP-
glucose added to the assay mixture, on the other hand, is not taken up by
the intact protoplasts, but is incorporated into callose by the content
of broken protoplasts - these are not able to synthesise cellulose.
The very small amounts of putative (1→4)-β-glucan formed from the UDP-
glucose can be explained by a slight hydrolysis of the sugar nucleotide,
after which the radioactive glucose is taken up by intact protoplasts.
The bulk of the UDP-glucose is directed to sucrose synthesis.

We really seem to be in a dilemma at present: the most
likely precursor of cellulose, namely UDP-glucose, cannot penetrate the
cell membrane of intact protoplasts and cannot, therefore, serve as a
cellulose precursor since normally, intact protoplasts are necessary for
cellulose production, although Delmer et al. (1984) have preliminary
evidence for such synthesis with a membrane fraction. Broken protoplasts
of cotton fibres use UDP-glucose only for the synthesis of sucrose and
callose and for the glycosylation of lipids. To obtain direct evidence
for the precursor role of UDP-glucose in cellulose synthesis one should
succeed in preparing plasma membranes, which are not sealed to form
vesicles, and which are still able to produce cellulose. The same is
true for showing unequivocally whether or not callose is a precursor for
cellulose.

ACKNOWLEDGEMENTS

The authors wish to thank their colleagues and students
(A. Allenbach, P. Bucheli, L. Buchs, U. Christ, Y. Francey, T. Homewood,
J.P. Jaquet, Y. Jenny, Ch. Pillonel, Ph. Rime, & U. Ryser) whose results
have been liberally quoted in this article. The research project was
financed by the Swiss National Science Foundation.

REFERENCES

Bacic, A. & Delmer, D.P. (1981). Stimulation of membrane-associated
    polysaccharide synthetases by a membrane potential in
    developing cotton fibers. Planta, 152, 346-351.

Barber, G.A. & Hassid, W.Z. (1965). Synthesis of cellulose by enzyme preparations from the developing cotton boll. Nature, 207, 295-296.

Beasley, C.A. & Ting, I.P. (1974). Effects of plant growth substances on in vitro fiber development from unfertilized cotton ovules. American Journal of Botany, 61, 188-194.

Benedict, C.R., Smith, R.H. & Kohel, R.J. (1973). Incorporation of ^{14}C-photosynthate into developing cotton bolls, Gossypium hirsutum L.. Crop Science, 13, 88-91.

Brett, C.T. (1978). Synthesis of β-(1→3)-glucan from extracellular uridine diphosphate glucose as a wound response in suspension cultured soybean cells. Plant Physiology, 62, 377-382.

Brown, K.J. (1968). Translocation of carbohydrate in cotton: movement to the fruiting bodies. Annals of Botany, 32, 703-713.

Bucheli, P., Buchala, A.J. & Meier, H. (1984). β-Glucanases in developing cotton (Gossypium hirsutum L.) fibres. Presented at the Third Cell Wall Meeting, University of Fribourg, Switzerland.

Buchs, L. (1982). Einbau radioactiver Vorläufer in (1→3)-β-D-Glucane (Callose) und in die Cellulose der Samenhaare der Baumwolle (Gossypium arboreum L.). Doctoral thesis, University of Fribourg, Switzerland.

Carpita, N.C. & Delmer, D.P. (1980). Protection of cellulose synthesis in detached cotton fibres by polyethylene glycol. Plant Physiology, 66, 911-916.

Carpita, N.C. & Delmer, D.P. (1981). Concentration and metabolic turnover of UDP-glucose in developing cotton fibers. Journal of Biological Chemistry, 256, 308-315.

Delmer, D.P., Beasley, C.A. & Ordin, L. (1974). Utilization of nucleoside diphosphate glucoses in developing cotton fibers. Plant Physiology, 53, 149-153.

Delmer, D.P. (1977). The biosynthesis of cellulose and other plant cell wall polysaccharides. Recent Advances in Phytochemistry, 11, 45-77.

Delmer, D.P., Heiniger, U. & Kulow, C. (1977). UDP-glucose:glucan synthetase from developing cotton fibers. I. Kinetic and physiological properties. Plant Physiology, 59, 713-718.

Delmer, D.P. (1983). Biosynthesis of cellulose. Advances in Carbohydrate Chemistry and Biochemistry, 41, 105-153.

Delmer, D.P., Thelen, M. & Alexander, D. (1984). Regulation of β-glucan synthesis in cotton fibers. Presented at the Third Cell Wall Meeting, University of Fribourg, Switzerland.

Durr, M. (1981). Studies on differentiating cotton fibres: β-glucanases and β-glucosidases. Presented at the Second Cell Wall Meeting, University of Gottingen, FRG.

Dugger, W.M. & Palmer, R.L. (1980). Effect of boron on the incorporation of glucose from UDP-glucose into cotton fibers grown in vitro. Plant Physiology, 65, 266-273.

Forsee, W.T. & Elbein, A.D. (1973). Biosynthesis of mannosyl- and glucosyl-phosphoryl-polyprenols in cotton fibers. Journal of Biological Chemistry, 248, 2858-2867.

Forsee, W.T., Laine, R.A. & Elbein, A.D. (1974). Solubilization of a particulate UDP-glucose:sterol β-glucosyl-transferase in developing cotton fibers and seeds and characterization of steryl 6-acyl-D-glucosides. Archives of Biochemistry and Biophysics, 161, 248-259.

Francey, Y. & Meier, H. (1984). Glucan synthesis by cotton fibre ovules cultured in vitro. Presented at the Third Cell Wall Meeting, University of Fribourg, Switzerland.

Franz, G. (1969). Soluble nucleotides in developing cotton hair. Phytochemistry, 8, 737-741.

Greathouse, G.A. (1955). Biosynthesis of C^{14}-specifically labeled cotton cellulose. Science, 117, 553-554.

Heiniger, U. & Delmer, D.P. (1977). UDP-glucose:glucan synthetase in developing cotton fibers. II. Structure of the reaction product. Plant Physiology, 59, 719-723.

Huwyler, H.R., Franz, G. & Meier, H. (1978). β-Glucans in the cell walls of cotton fibres (Gossypium arboreum L.). Plant Science Letters, 12, 55-62.

Huwyler, H.R., Franz, G. & Meier, H. (1979). Changes in the composition of the cotton fibre cell walls during development. Planta, 146, 635-642.

Jaquet, J.-P., Buchala, A.J. & Meier, H. (1982). Changes in the non-structural carbohydrate content of cotton (Gossypium spp.) fibres at different stages of development. Planta, 156, 481-486.

Maltby, D., Carpita, N.C., Montezinos, D., Kulow, C. & Delmer, D.P. (1979). β-1,3-Glucan in developing cotton fibers. Plant Physiology, 63, 1158-1164.

Marx-Figini, M. & Schulz, G.V. (1966). Ueber die Kinetik und den Mechanismus der Biosynthese der Cellulose in den höheren Pflanzen (nach Versuchen an den Samenhaaren der Baumwolle). Biochimica et Biophysica Acta, 112, 81-101.

Meier, H. (1981). In "Cell Walls '81", eds. D.G. Robinson & H. Quader, Wissenschaftliche Verlagsgesellschaft, Stuttgart, 75-83.

Meier, H. Buchs, L., Buchala, A.J. & Homewood, T. (1981). (1→3)-β-D-Glucan (callose) is a probable intermediate in biosynthesis of cellulose of cotton fibres. Nature, 289, 821-822.

Meier, H. (1983). Biosynthesis of (1→3)- and (1→4)-β-D-glucans in cotton fibers (Gossypium arboreum and Gossypium hirsutum). Journal of Applied Polymer Science, 37, 123-130.

Meinert, M.C. & Delmer, D.P. (1977). Changes in biochemical composition of the cell wall of the cotton fiber during development. Plant Physiology, 59, 1088-1097.

Mohammed, Y.S., El-Gazzar, R.M. and (in part) Adamyova, K. (1971). Carbohydrate Research, 20, 431-435.

Montezinos, D. & Delmer, D.P. (1980). Characterization of inhibitors of cellulose synthesis in cotton fibers. Planta, 148, 305-311.

Nowak-Ossorio, M., Gruber, E. & Schurz., J. (1976). Untersuchungen zur Cellulosebildung in Baumwollsamen. Protoplasma, 88, 255-263.

Pillonel, Ch., Buchala, A.J. & Meier, H. (1980). Glucan synthesis by intact cotton fibres fed with different precursors at the stages of primary and secondary wall formation. Planta, 149, 306-312.

Pillonel, Ch. (1983). La biosynthèse in vitro de β-glucanes (callose et cellulose) par des fibres intactes de coton (Gossypium arboreum L.). Doctoral thesis, University of Fribourg, Switzerland.

Shafizadeh, F. & Wolfrom, M.L. (1955). Biosynthesis of C^{14}-labeled cellulose from D-glucose-1-C^{14} and D-glucose-6-C^{14}. Journal of the American Chemical Society, 77, 5182-5183.

Updegraff, D.M. (1969). Semi-micro determination of cellulose in biological materials. Analytical Biochemistry, 32, 420-424.

Waterkeyn, L. (1981). Cytochemical localization and function of the
          3-linked glucan callose in the developing cotton fibre cell
          wall. Protoplasma, 106, 49-67.

10 BIOSYNTHESIS OF 1,4-β-D-GLUCAN IN CELL-FREE
    PREPARATIONS FROM MUNG BEANS

M. Benziman
Department of Biological Chemistry, Institute of Life
Sciences, The Hebrew University of Jerusalem, 91904
Jerusalem, Israel

T. Callaghan
Department of Biological Chemistry, Institute of Life
Sciences, The Hebrew University of Jerusalem, 91904
Jerusalem, Israel

Abstract. Conditions have been found for an extremely
efficient transfer of glucose from UDP-glucose to a
cellulosic 1,4-β-D-glucan product, using enzyme preparations
derived from Phaseolus aureus seedlings. Membrane fractions
obtained in the presence of 20% (w/v) polyethyleneglycol
(PEG)-4000 exhibit UDP-glucose: 1,4-β-D-glucan synthase
activity 15-20 times higher than those previously reported.
The alkali-insoluble reaction product has been characterized
as 1,4-β-D-glucan by the following criteria: complete
conversion to glucose upon acid hydrolysis, susceptibility to
digestion either by highly purified endo-1,4-β-D-glucanase or
exocellobiohydrolase; methylation analysis yields
exclusively the derivative expected for 1,4-linked glucosyl
moieties; and periodate oxidation yields primarily
erythritol. Product formation is linear with time up to
30 min and is not affected by excess of UDP-xylose or GDP-
glucose; enzyme activity is inhibited by coumarin and
dichlorobenzonitrile. The enzyme has an absolute
requirement for both Mg^{2+} and Ca^{2+}. Centrifuged PEG-
membrane fractions exhibit low activity. Activity is
restored in a synergistic manner upon re-addition of the
supernatant fraction (SF). Addition of cellobiose (K_a =
0.75 mM) to centrifuged membranes results in a ten-fold
increase in activity; however, in the presence of SF, only
a two-fold increase is observed. Unlike the 1,4-β-D-glucan
synthase of Acetobacter xylinum, the plant enzyme has not been
found to respond to the addition of GTP. SF, however,
interferes with GTP activation of the bacterial enzyme. The
possibility that the in vitro system described here is
already in a fully activated state, is compatible with its
ability to form 1,4-β-D-glucan at rates which approach those
of cellulose synthesis observed in vivo during primary cell-
wall formation.

INTRODUCTION

Progress in the study of biosynthesis of cellulose
(1,4-β-D-glucan) has been peristently hampered by the low rates of
in vitro synthesis obtained using cell-free preparations derived from

plant or bacterial cells capable of ample cellulose synthesis in vivo
(Delmer 1983).   We recently reported success in achieving high rates
of synthesis of 1,4-β-D-glucan from UDP-glucose, using membrane
preparations obtained from the bacterium Acetobacter xylinum.   The
key to the success lay in the discovery of a complex GTP-Ca^{2+}-protein
factor-mediated regulatory system for the A. xylinum UDP-glucose:1,4-
β-D-glucan 4-β-D-glucosyl transferase (cellulose synthase) (Aloni et al.
1982, 1983).   In this chapter we describe conditions found for an
extremely efficient transfer of glucose from UDP-glucose to a cellulosic
1,4-β-D-glucan product, using enzyme preparations derived from a higher
plant, namely seedlings of mung beans (Phaseolus aureus).   The rates of
enzyme activity achieved approach those of cellulose synthesis observed
in vivo.

METHODS AND MATERIALS
Enzyme preparation
        Phaseolus aureus seeds were allowed to germinate in the dark
at 25°C in vermiculite saturated with distilled water.   The shoots were
harvested when the length of hypocotyls reached 2 to 3 cm, which usually
required 3 days.   The hypocotyls were separated from the remainder of
the seedlings and placed in a pre-weighed beaker, containing buffer
(50 mM N-(-2-hydroxyethyl)-piperazine-N'-3-propane sulphonic acid (EPPS),
20% w/v polyethyleneglycol-4000(PEG) and 50 μM dithiothreitol, pH 7.5),
and fresh weight was determined.   The tissue was ground at 4°C, using
a pre-cooled mortar and pestle in 2 ml of buffer per gram of tissue, and
the resultant homogenate was strained through several layers of gauze.
The filtrate was placed in small Eppendorf tubes, frozen in liquid
nitrogen and stored at -70°C.   When required, the homogenate was
centrigued at 17,000 xg for 10 min, and the resulting membrane fraction
was resuspended in the original volume of 50 mM buffer.   This fraction
typically contained 0.8 mg protein per ml.   The supernatant fraction
(SF) was saved and kept separately.

Enzyme assay
        Standard assays performed at 20°C contained, in a final
volume of 0.2 ml:50 mM buffer, pH 7.5, 5 μM UDP-[^{14}C] glucose
(320 cpm/pmol), 6 mM MgCl$_2$, 4 mM CaCl$_2$, 5mM cellobiose and enzyme.
Reactions were terminated by addition of 2 ml 24% KOH.   Approximately

20 mg of carrier cellulose was added, and the mixture incubated for 12 h
at 40°C (Hayashi & Matsuda 1981).   The mixtures were then filtered onto
Whatman GF/A glass fiber filters, and washed consecutively with water
and methanol.

The filters containing alkali-insoluble product were dried
at 60°C, and the radioactivity was counted and corrected for quenching
as described (Swissa et al. 1980).

Analysis of reaction products

The linkages of (radioactive)neutral-sugar residues were
determined by methylation analysis, according to the method of
Hakamori (1964), as modified by Klein et al. (1981).   Separation of
the derivatized sugars was performed in a Packard model 438 gas
chromatograph, equipped with an effluent stream splitter.   Collected
samples were assayed for radioactivity.   Peaks were identified by
comparison with permethylated derivatives of laminarin and cellobiose.
Enzyme digestion of the alkali-insoluble product was performed at 40°C
for 15 h in 0.05 M sodium acetate buffer (pH 5.0).   Purified exocellobiohydrolase and endo-1,4-β-glucanase were used at a final concentration
of 0.1 mg/ml.   After incubation, reactions were brought to 66% ethanol,
chilled and filtered onto Whatman GF/A glass fiber filters.   Radioactivity in the dried filters and in an aliquot of the filtrate was
counted to determine the percent digestion.   A portion of the filtrate
from exocellobiohydrolase digestion was concentrated and chromatographed
on Whatman No. 4 paper, by using propan-1-ol/ethyl-acetate/water
(7/1/2 v/v) as solvent.   Sugar standards were detected by spraying with
alkaline silver nitrate.   Periodate oxidation was performed as
described by Heiniger & Delmer (1977).

Chemicals

UDP-[^{14}C]glucose (240 mCi/mmol) was obtained from
Radiochemical Centre, Amersham, England.   Nucleotides, sugar nucleotides
and polyethyleneglycol were purchased from Sigma Chemical Co., St. Louis,
Missouri, U.S.A.   Coumarin and dichlorobenzonitrile were from Fluka AG,
Buchs, Switzerland.   Purified exocellobiohydrolase and endo-1,4-β-
glucanase from Trichoderma reesei were a generous gift of Dr. S. Shoemaker,
Cetus Corp., Emeryville, California, U.S.A.   These purified enzymes were
free of β-glucosidase and endo-1,3-β-D-glucanase activities.   The

exocellobiohydrolase was also free of endo-1,4-β-D-glucanase activity.

RESULTS

Characterization of alkali-insoluble product

In all the studies described below, data are presented only
for the radioactive 24% KOH-insoluble glucan product formed from
UDP-[^{14}C]glucose, which has been characterized as 1,4-β-D-glucan by the
following criteria:- (i) Total acid hydrolysis converted > 90% of the
radioactivity to [^{14}C]glucose;   (ii) The product was more than 95%
digested by either highly purified Trichoderma reesei exocellobiohydrolase or by highly purified Trichoderma endo-1,4-β-glucanase.
Chromatography of the exocellobiohydrolase digest showed that > 80% of
the radioactivity solubilized, co-chromatographed with a cellobiose
standard.   The remainder was found at the origin and in glucose.
(iii) Methylation analysis of the products formed either from 5 μM,
500 μM or 1 mM UDP-glucose yielded exclusively 1,4,5-tri-0-acetyl-
2,3,6-tri-0-methyl-[^{14}C]glucitol, the derivative expected for 1,4-
linked glucosyl moieties.   (iv) Periodate oxidation, followed by
borohydride reduction, complete acid hydrolysis and paper chromatography
yielded primarily erythritol (indicative of 1,4-linkages), a trace of
glycerol and some material which remained at the origin.   In no case
could a significant peak of glucose (indicative of 1,3-linkages) be
detected.   The small peak of glycerol detected amounted to 1-3% of the
labelling of the major erythritol peak.   Assuming that this glycerol
arises from terminal glucose residues, it indicates that the enzyme is
producing long chains and is not a chain-terminating enzyme adding just
single glucose residues to pre-existing polysaccharides.

Effect of PEG on isolation and activity of β-1,4-glucan synthase

If cells were ruptured in an extracting medium containing
20% PEG-4000, the membranes and all associated glucan synthase activity
(when assayed under standard assay conditions) pelleted at relatively
low centrifugal forces (17,000 xg for 10 min).   In contrast, in the
absence of PEG, higher centrifugal forces (150,000 xg for 60 min) were
needed to sediment membranes and enzyme activity.   A similar effect of
PEG has been observed with other plant glucan synthases (Delmer 1983)
and with A. xylinum 1,4-β-D-glucan synthase (Aloni et al. 1982), and may

be attributed to the known effect of PEG in promoting membrane fusion
and protein coagulation (Boni et al. 1981).   In addition, the inclusion
of PEG in the extracting medium enhanced 4-5-fold the rate of product
formation by the membrane preparation.   Activity was retained in PEG
preparations for two months when these were frozen in liquid nitrogen
and kept at -70°C.

Characteristics of the 1,4-β-glucan synthase

Under standard assay conditions (pH 7.5), product
formation was linear with time, at 20°C for up to 30 min, and proportional
to enzyme concentration up to 100 µg protein (Fig. 1).   At pH 8.6,
activity was considerably lower, and linearity with time ceased after 2
to 5 min.   Increasing the temperature to 30°C did not affect activity
at pH 7.5, but linearity with time was impaired after 15 min.   The
increase in temperature caused a further decrease in activity at pH 8.6.

Requirements for the reaction are shown in Table 1.   The
enzyme has an absolute requirement for both Mg^{2+} and Ca^{2+}.   Whereas the
dependence on Mg^{2+} was demonstrable with regular enzyme preparations,
the effect of Ca^{2+} was much more pronounced when EGTA was included in
the enzyme extracting medium.   A synergistic effect by Mg^{2+} and Ca^{2+}
in promoting cellulose synthesis in detached cotton fibers has been
reported by Carpita & Delmer (1980).   In A. xylinum, the 1,4-β-glucan
synthase has an absolute requirement for Mg^{2+} and is further activated
by Ca^{2+}.   In this system the effect of Ca^{2+} appears to be specifically
involved in the GTP-mediated regulation of the enzyme (Aloni et al. 1983).

The particulate membrane fraction, sedimenting from a
centrifuged crude homogenate, exhibits low activity.   Activity is
restored upon re-addition of the supernatant fraction (SF) which by
itself was found to be devoid of synthase activity.   Product formation
is markedly enhanced by the addition of cellobiose to the reaction
mixture.   However, the extent of the effect varies with enzyme
preparation.   Whereas with centrifuged fractions the addition of
cellobiose results in a 10-fold increase in activity, with uncentrifuged
preparations only a 2-fold increase is observed.   Activity of the
particulate fraction increases with increasing cellobiose concentrations
and with varying levels of SF (Figs. 2 and 3).   Half-maximal
stimulations are obtained with 0.75 mM cellobiose and 1.5 µl of SF (an
amount of supernatant derived from 0.8 mg fresh weight of tissue).   The

Fig. 1.   Time course of reaction: effect of pH and
temperature.

Standard assay mixtures contained 50 mM EPPS buffer at
pH 7.5 or 8.6.   Mixtures were incubated at 20°C or 30°C,
and the reaction was stopped at the times indicated.
Centrifuged membrane fraction, derived from 10 mg fresh
weight of tissue, was used per assay.

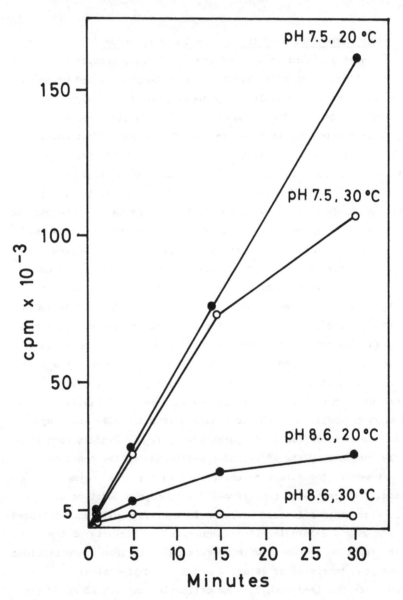

Table 1   Factors affecting UDP-glucose : 1,4-β-D-glucan
            synthase activity.

Incubation mixture	Activity pmol product/min/g tissue
Complete system	1,700
Minus Mg^{2+}	50
Minus Ca^{2+}	40
Minus SF minus cellobiose	170
Minus cellobiose	850
Minus SF	1,700
Plus DCB* (6 µM)	120
Plus coumarin (2 mM)	40

The complete system contains: crude homogenate derived
from 10 mg of tissue, in the presence of 5 mM EGTA;
6 mM $MgCl_2$; 4 mM $CaCl_2$; 5mM cellobiose; and 5 µM
UDP-[^{14}C]glucose.

*DCB = 2,6-dichlorobenzonitrile

Fig. 2.   Effect of varying the concentration of cellobiose
on enzyme activity.

Conditions were as in the standard assay, except that the
concentration of cellobiose was varied, as indicated.
Centrifuged membrane fraction from 10 mg fresh weight of
tissue, was used per assay.

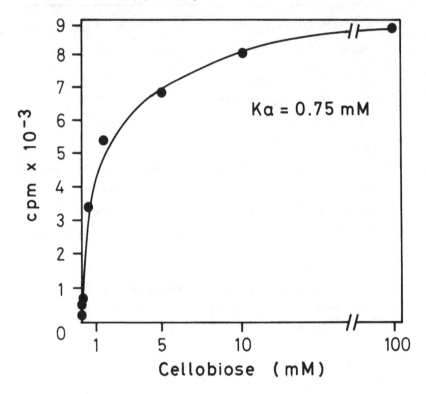

enzyme displays normal Michaelis-Menten kinetics with respect to UDP-
glucose (Fig. 4).   The presence of saturating amounts of cellobiose or
SF does not alter the apparent K_m for the substrate (100 μM) but the
V_{max} is increased in both cases about 5-fold.   Stimulation by cellobiose
has been reported for β-glucan synthases from various plant sources
(Henry & Stone 1982; Maclachlan 1982; Delmer 1983).   In the case of
1,3-β-glucan synthase of detached cotton fibers, which shows substrate
activation by UDP-glucose, the effect of cellobiose is to shift the range
in which this activation occurs to lower concentrations of UDP-glucose
(Delmer et al. 1977).   The 1,4-β-glucan synthase of A. xylinum, on the
other hand, is not affected at all by cellobiose (Aloni & Benziman,
unpublished results).

Preliminary characterization of the stimulatory factor in SF, based on stability studies, indicate that it is an organic, dialyzable, alkali-labile compound which is absorbed to charcoal.   It is resistant to heat and acid treatment (0.1 N HCl, 10 min at 100°C) and to digestion by exocellobiohydrolase, endo-1,4-β-glucanase, β-glucosidase, crude cellulase, and alkaline and acid phosphatases.   Unlike the 1,4-β-glucan synthase of A. xylinum, the plant enzyme has not been found to respond to the addition of GTP.   SF, however, strongly interferes with GTP activation of the bacterial enzyme (Weinhouse & Benziman, unpublished results).

Fig. 3.   Effect of supernatant fraction (SF) on enzyme activity in the absence of cellobiose.

Conditions were as in the standard assay, except that cellobiose was omitted and supernatant fraction (SF) was added in varying amounts, as indicated.   Centrifuged membrane fraction, derived from 10 mg fresh weight of tissue, was used per assay.

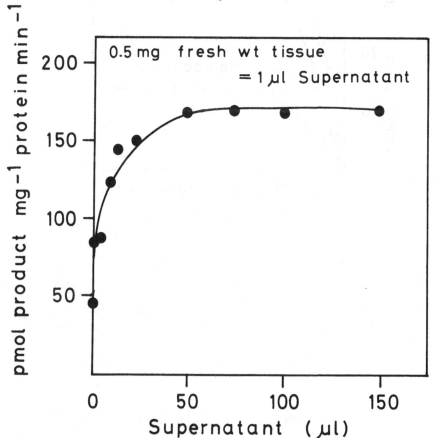

Fig. 4.   Effect of varying the concentration of UDP-glucose
on enzyme activity.

UDP-glucose concentrations were as indicated, and centrifuged
membrane fraction derived from 10 mg fresh weight of tissue,
was used per assay.   Additions indicated were: 5 mM
cellobiose (+CB) and 25 µl of supernatant fraction (+SF).

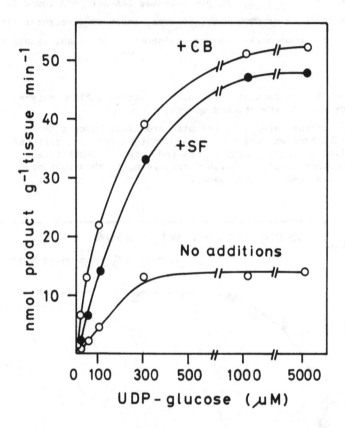

In the light of the reports that UDP-xylose stimulates
incorporation from UDP-glucose into insoluble glucan products and the
possibility that 1,4-β-glucan synthesized in some cases may be a
backbone for other non-cellulosic polymers (Ray 1975, 1980; Villemez &
Hinman 1975; Hayashi & Matsuda 1981), we assayed our preparation in the
presence of UDP-xylose.   Incorporation of [^{14}C]glucose from UDP-
[^{14}C]glucose into the cellulosic product was neither stimulated nor
inhibited by the addition of up to a 100-fold excess of unlabelled UDP-
xylose.   Similar results were obtained upon addition to the reaction
mixture of a same excess of unlabelled GDP-glucose (c.f., Hayashi &
Matsuda 1981).

Coumarin and dichlorobenzonitrile - specific inhibitors of
in-vivo cellulose synthesis in plants (Montezinos & Delmer 1980) -
strongly inhibit our enzyme (Table 1).   It is noteworthy that these
compounds inhibit GTP activation of the A. xylinum 1,4-β-glucan synthase
(Aloni & Benziman, unpublished results).

DISCUSSION

Because the reaction product is evidently a 1,4-β-D-glucan,
we presume that the enzyme studied here is a true cellulose synthase;
however, since we lack information on the crystalline and microfibrillar nature of the product, we refer to it as a 1,4-β-glucan
synthase.   The rates of synthesis of 1,4-β-glucan in vitro reported
here far exceed those previously reported for cell-free preparations
from various higher plants (Barber et al. 1964;   Villemez et al. 1967;
Shore et al. 1975; Abdul-Baki & Ray 1971; Carpita & Delmer 1980).   The
reason for the much higher activity is still not clear, and probably lies
in the method used for isolating the membranes and associated enzyme
systems from the cells.

The presence of PEG during cell rupture increased by 4-5-fold
the synthase activity of the membrane preparations.   This effect could
involve protection of the enzyme during cell breakage or the promotion
of interactions between components of the enzyme system.   A similar
effect of PEG in promoting synthase activity has been observed with the
1,4-β-glucan synthase of A. xylinum (Aloni et al. 1982).   In the
bacterial system, however, the presence of PEG has an additional and more
pronounced effect in conjunction with the activation by GTP.   The
ability to demonstrate GTP activation of the synthase depends on the

presence of a protein factor which easily dissociates from the membrane-
bound enzyme.   The factor was not detected previously because it was
lost during conventional membrane preparation procedures used by others.
However, we found that membranes prepared in the presence of PEG-4000
retain the protein factor, and the enzyme in such preparations shows
marked activation by GTP (Aloni et al. 1982; Benziman et al. 1983).
In more recent studies we discovered that the protein factor is
actually an enzyme, reacting with GTP.   The product of this reaction
is a low molecular weight quanosine derivative (GX) which is the actual
activator of the synthase reaction.   The mung bean enzyme described
here has thus far not been found to be affected by GTP nor by the above
mentioned GX derivative produced in A. xylinum.   Its activity, however,
is significantly enhanced by a factor present in supernatant fractions
of crude homogenates of mung beans.   Unlike the factor of A. xylinum,
the stimulatory factor is probably not a protein, being of low molecular
weight as well as both heat and acid resistant.   The possibility that
this factor is structurally related to GX is intriguing, especially since
SF interferes with GX formation from GTP in the A. xylinum system.
Moreover, the latter reaction is also inhibited by DCB and coumarin which,
as reported here, also inhibit 1,4-ß-glucan formation by the plant
enzyme.   On the other hand, the various effects of SF could be
completely unrelated and due to different compounds present in this
crude fraction.

Unlike the bacterial 1,4-ß-glucan synthase, the mung bean
enzyme is activated by cellobiose.   The effect of cellobiose is
markedly more prominent in the absence of SF, but is still significant
in the presence of saturating amounts of SF.   The mode of action of SF
and its relationship to the effect of cellobiose require further
exploration.   At this stage, the present data may be of some interest
in relation to the studies of Maclachlan's group with peas (Chao &
Maclachlan 1978) and Delmer's group with cultured soy bean cells
(Delmer et al. 1983), which indicate the possible existence in plant
extracts of dissociable, heat stable factors which can influence the
capacity for ß-glucan synthesis in vitro.

A major issue which at this stage is still not clear to us,
is the relationship between our present findings and the highly active
UDP-glucose:1,3-ß-D-glucan synthase reportedly present in almost all
plant extracts, including mung bean seedlings (Feingold et al. 1958).

The physiological significance of this enzyme activity and the possible
role of its product in cellulose biosynthesis have been widely discussed
(Delmer 1977, 1983; Meir et al. 1981; Fincher & Stone 1981).

The rates of synthesis of 1,4-β-glucan in vitro described here
exceed those previously reported.   At concentrations of UDP-glucose in
the range that may occur in vivo, the in vitro rates of 1,4-β-glucan
formation are within the range of those of cellulose synthesis observed
in vivo during primary cell wall formation (12-30 nmol/min/g tissue)
(Shore et al. 1975; Meinert & Delmer 1977; Abdul-Baki & Ray 1971).   This
system should be of great value for exploring further the details of the
mechanism and regulation of cellulose biosynthesis.

REFERENCES

Abdul-Baki, A.A. & Ray, P.M. (1971). Regulation by auxin of carbohydrate
    metabolism involved in cell-wall synthesis by pea-stem
    tissue. Plant Physiology, 47, 537-544.
Aloni, Y., Delmer, D.P. & Benziman, M. (1982). Achievement of high
    rates of in vitro synthesis of 1,4-β-D-glucan: Activation
    by cooperative interaction of the Acetobacter xylinum
    enzyme system with GTP, polyethyleneglycol, and a protein
    factor. Proceedings of the National Academy of Sciences of
    the United States of America, 79, 6448-6452.
Aloni, Y., Cohen, R., Benziman, M. & Delmer, D.P. (1983). Solubilization
    of the UDP-glucose:1,4-β-D-glucan 4-β-D-glucosyltransferase
    (cellulose synthase) from Acetobacter xylinum: A comparison
    of regulatory properties with those of the membrane-bound
    form of the enzyme. Journal of Biological Chemistry, 258,
    4419-4423.
Barber, G., Elbein, A.D. & Hassid, W.Z. (1964). Synthesis of cellulose
    by enzyme systems from higher plants. Journal of Biological
    Chemistry, 239, 4056-4061.
Benziman, M., Aloni, Y. & Delmer, D.P. (1983). Unique regulatory
    properties of the UDP-glucose:1,4-β-glucan synthetase of
    Acetobacter xylinum. Journal of Applied Polymer Science, 37,
    131-143.
Boni, L.T., Stewart, T.P., Alderfer, J.L. & Hui, S.W. (1981). Lipid-
    polyethylene glycol interactions: I. Induction of fusion
    between liposomes. Journal of Membrane Biology, 62, 65-70.
Carpita, N. & Delmer, D.P. (1980). Protection of cellulose synthesis in
    detached cotton fibers by polyethylene glycol. Plant
    Physiology, 66, 911-916.
Chao, H.-Y. & Maclachlan, G.A. (1978). Soluble factors in Pisum extracts
    which moderate Pisum β-glucan synthetase activity. Plant
    Physiology, 61, 943-948.
Delmer, D.P., Heiniger, U. & Kulow, C. (1977). UDP-glucose:glucan
    synthetase in developing cotton fibers. Plant Physiology, 59,
    713-718.
Delmer, D.P. (1977). The biosynthesis of cellulose and other plant cell
    wall polysaccharides. In Recent Advances in Phytochemistry,
    Vol. 11, eds. F.A. Loewus & V.C. Runeckles, pp. 45-77.
    New York: Plenum Press.

Delmer, D.P. (1983). Cellulose Biosynthesis. Advances in Carbohydrate
    Chemistry and Biochemistry, 41, 105-153.
Delmer, D.P., Benziman, M., Klein, A.S., Bacic, A., Mitchell, B.,
    Weinhouse, H., Aloni, Y. & Callaghan, T. (1983). A
    comparison of the mechanism of cellulose biosynthesis in
    plants and bacteria. Journal of Applied Polymer Science:
    Applied Polymer Symposium, 37, 1-16.
Feingold, D.S., Neufeld, E.F. & Hassid, W.Z. (1958). Synthesis of a
    β-1,3-linked glucan by extracts of Phaseolus aureus
    seedlings. Journal of Biological Chemistry, 233, 783-786.
Fincher, G. & Stone, B.A. (1981). Metabolism of non-cellulosic wall
    polysaccharides. In Encyclopaedia of Plant Physiology,
    New Series 13B, eds. F.A. Loewus & W. Tanner, pp. 68-132.
    Springer Verlag.
Hakamori, S. (1964). A rapid permethylation of glycolipid and poly-
    saccharides catalyzed by methylsulfinyl carbanion in
    dimethylsulphoxide.   Journal of Biochemistry, 55, 205-208.
Hayashi, T. & Matsuda, K. (1981). Biosynthesis of xyloglucan in
    suspension cultured soybean cells: Evidence that the
    enzyme system of xyloglucan synthesis does not contain
    β-1,4-glucan 4-β-D-glucosyl-transferase activity
    (E.C.2.4.1.12).   Plant and Cell Physiology, 22, 1571-1584.
Heiniger, U. & Delmer, D.P. (1977). UDP-glucose:glucan synthetase in
    developing cotton fibers. Plant Physiology, 59, 719-723.
Henry, R.J. & Stone, B.A. (1982). Factors influencing β-glucan
    synthesis by particulate enzymes from suspension cultured
    Lolium multiflorum endosperm cells. Plant Physiology, 69,
    632-636.
Klein, A.S., Montezinos, D. & Delmer, D.P. (1981). Cellulose and
    1,3-glucan synthesis during the early stages of wall
    regeneration in soybean protoplasts. Planta, 152, 105-114.
Maclachlan, G.A. (1982). Does β-glucan synthesis need a primer? In
    Cellulose and Other Natural Polymer Systems: Biogenesis,
    Structure and Degradation, ed. R.M. Brown, pp. 327-337.
    New York: Plenum Press.
Meier,H., Buchs, L., Buchala, A.J. & Homewood, T. (1981). (1→3)-β-D-
    glucan (callose) is a probable intermediate in biosynthesis
    of cellulose of cotton fibres. Nature, 289, 821-822.
Montezinos, D. & Delmer, D.P. (1980). Characterization of inhibitors
    of cellulose synthesis in cotton fibers. Planta, 148,
    305-311.
Ray, P.M. (1975). Golgi membranes form xyloglucan from UDP-glucose and
    UDP-xylose. Plant Physiology, 56, Supp. No. 84.
Ray, P.M. (1980). Cooperative action of β-glucan synthetase and
    UDP-xylose xylosyl-transferase of Golgi membranes in the
    synthesis of xyloglucan-like polysaccharides. Biochimica
    et Biophysica Acta, 629, 431-444.
Shore, G., Raymond, Y. & Maclachlan, G.A. (1975). The site of cellulose
    synthesis.   Cell surface and intracellular 1,4-β-glucan
    (cellulose) synthetase activities in relation to the stage
    and direction of cell growth. Plant Physiology, 56, 34-38.
Swissa, M., Aloni,Y., Weinhouse, H. & Benziman, M. (1980). Intermediary
    steps in cellulose synthesis in Acetobacter xylinum:
    Studies with whole cells and cell-free preparations of the
    wild type and a celluloseless mutant. Journal of
    Bacteriology, 143, 1142-1150.

Villemez, C.L. & Hinman, M. (1975). UDP-glucose stimulated formation
    of xylose containing polysaccharides. Plant Physiology,
    56, Supp. No. 79.
Villemez, C.L., Franz, G. & Hassid, W.Z. (1967). Biosynthesis of alkali-
    insoluble polysaccharide from UDP-glucose with particulate
    enzyme preparations from Phaseolus aureus. Plant Physiology,
    42, 1219-1223.

11 STRUCTURE AND FUNCTION IN LEGUME-SEED POLYSACCHARIDES

J.S.G. Reid
Department of Biological Science, University of Stirling
Stirling FK9 4LA, Scotland

INTRODUCTION

Although starch is the best-known legume-seed polysaccharide
it will not be discussed here.   It is accepted that the structures of
the starch components are known in reasonable detail, and that they are
reserve carbohydrates.   This article concerns the other polysaccharides
which occur in large amounts in leguminous seeds: the galactomannans,
the xyloglucans and the complex galactose-rich polysaccharide materials
which are referred to here for convenience as the "galactans".   All
three are cell-wall polysaccharides.   They are also reserves.

The galactomannans, xyloglucans and "galactans" have been
reviewed recently in the wider contexts of non-starch polysaccharides in
higher plants (Meier & Reid 1982) and of cell-wall storage carbohydrates
in seeds (Reid 1984).   This article specifically examines the overall
biological functions of these molecules.

GALACTOMANNANS, XYLOGLUCANS AND "GALACTANS" AS RESERVE POLYSACCHARIDES

The biosynthesis and mobilisation of galactomannans and the
mobilisation of xyloglucans and "galactans" have been reviewed (Meier &
Reid 1982;   Reid 1984).   This section briefly summarises and supplements
the published information concerning the post-germinative utilisation of
these polysaccharides.

Galactomannans

The galactomannans (Fig. 1) are characteristic of the
leguminous seed endosperm.   Consequently the amount of galactomannan
stored relative to cotyledonary reserves is dependent upon the relative
size of the endosperm.   For example, the soya bean (Glycine max) has
a rudimentary endosperm and contains less than 1% galactomannan

(Whistler & Saarnio 1957) whereas the carob seed has a massively developed endosperm and contains 38% galactomannan by weight (Anderson 1949).

Fig. 1. **Structural features of leguminous seed galactomannans**

A linear (1→4)-β-linked D-mannan backbone is substituted by single-unit (1→6)-β-linked D-galactopyranosyl side-chains. Mannose/galactose ratio varies from about 1 to about 4. The distribution of side-chains along the backbone is not regular (Dea & Morrison 1975).

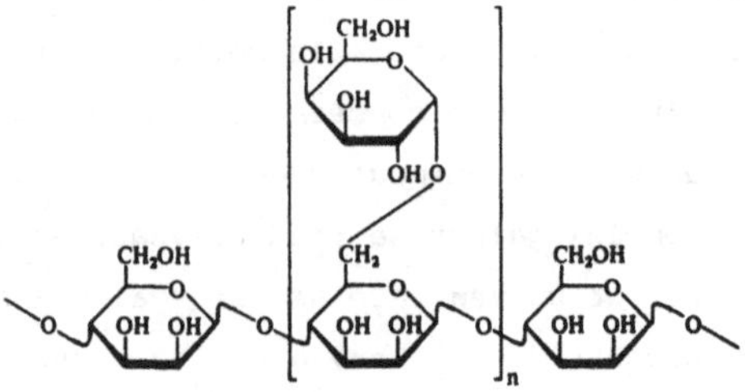

The post-germinative mobilisation of galactomannan has been studied in detail in the seeds of three species: fenugreek, guar and carob. Their botanical classifications and the amounts and compositions of the galactomannans in their endosperms are listed in Table 1.   The anatomical features of the carob and fenugreek endosperms are illustrated in Fig. 2.   The carob endosperm is composed of thick-walled cells, all of which have living protoplasts (Seiler 1977); the galactomannan is present in the cell-walls.   By contrast the fenugreek endosperm consists mainly of galactomannan-"filled" storage tissue (Reid 1971). The only living cells are those of the aleurone layer which forms the outer periphery of the endosperm (Reid & Meier 1972).   [Galactomannan formation in fenugreek has been studied (Reid & Meier 1970, 1973a; Meier & Reid 1977; Campbell & Reid 1982) and it is a cell-wall polysaccharide.]

Table 1. Characteristics of three galactomannan-containing species.

Species	Sub-family & tribe	Ratio man/gal in galactomannan	% by weight galactomannan in seed	Endosperm anatomy
Trigonella foenum-graecum	Faboideae - trifolieae	~ 1 [1]	18 [2]	see Fig. 2a
Cyamopsis tetragonoloba	Faboideae - indigoferae	~ 2 [1]	35 [1]	As Fig. 2a, but aleurone layer several cells thick (McClendon et al. 1976)
Ceratonia siliqua	Caesalpinioideae	~ 4 [1]	38 [1]	see Fig. 2b

(1) Dea & Morrison (1975) Table 1

(2) Reid & Meier (1970)

Fig. 2.   <u>Cross-sections of fenugreek (Trigonella foenum-</u>
<u>graecum) and carob seeds (Ceratonia siliqua)</u>

Cryostat sections, stained with periodic acid and Schiff's
reagent.
C = cotyledons; E = endosperm; T = testa; A = aleurone layer
a. <u>Fenugreek</u> - Note the absence of cell lumina in most of
the endosperm, with the exception of the one-cell-thick
aleurone layer.
b. <u>Carob</u> - Cell lumina are present throughout the endosperm
and there is no obvious aleurone layer (after Seiler 1977).

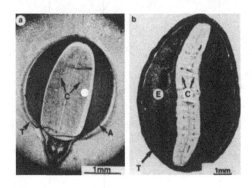

In all three seeds galactomannan breakdown occurs by the
concerted action of three enzymes: α-galactosidase (Reid & Meier 1973b;
McCleary & Matheson 1974;   Seiler 1977), <u>endo</u>-β-mannanase (McCleary &
Matheson 1975;   Reid <u>et al</u>. 1977;   Seiler 1977) and β-mannosidase (Reid &
Meier 1973b;   McCleary & Matheson 1975;   Seiler 1977) or <u>exo</u>-β-mannanase
(McCleary 1982).   In fenugreek and guar the α-galactosidase and <u>endo</u>-β-
mannanase are synthesised <u>de</u> <u>novo</u> and secreted by the aleurone layer into
the storage tissue (Reid & Meier 1972;   Reid <u>et al</u>. 1977;   McCleary 1983).
In guar it has been shown that the <u>exo</u>-β-mannanase is present in an
active form in the endosperm of the resting seed (McCleary 1983).   <u>De</u>
<u>novo</u> synthesis of α-galactosidase has been demonstrated in carob (Seiler
1977).   Presumably this occurs in all the cells of the endosperm.
          The end-products of galactomannan breakdown are galactose and
mannose (Reid 1971;   Seiler 1977).   They are very rapidly taken up from

the endosperm into the embryo (Reid 1971) by passive diffusion (Uebelmann 1978) and converted to sucrose and transitory starch (Reid 1971).

No positive, hormonal control of galactomannan breakdown by the embryo has been detected.

Xyloglucans and "Galactans"

The xyloglucans and "galactans" are cotyledonary reserve polysaccharides.   The xyloglucans (Fig. 3) are present as massive wall-thickenings in many nonendospermic leguminous species.   Most belong to the sub-family Caesalpinioideae, but a few species in the Faboideae also contain xyloglucan (Kooiman 1960;   Meier & Reid 1972).   Xyloglucans are not restricted to the Leguminosae.   They are also found in the seeds of many non-leguminous species (Kooiman 1960;   Meier & Reid 1972).   The best known xyloglucan-containing leguminous seed is that of tamarind (Tamarindus indica).   The "galactans" appear to be restricted to the leguminous genus Lupinus.

Fig. 3.   Structural features of seed xyloglucans

A linear (1→4)-β-linked D-glucan backbone is substituted by D-xylopyranosyl and β-galactopyranosyl(1→2)-D-xylopyranosyl side-chains.   Both types of side chain are attached (1→6)-α to the backbone.

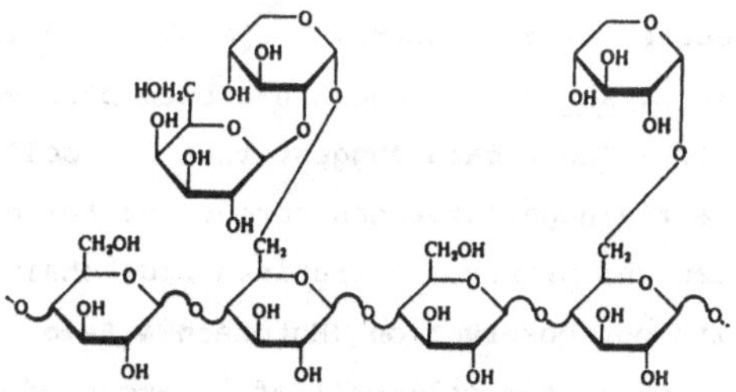

Although xyloglucan mobilisation has been observed in the seeds of several species (Reiss 1889), it has been studied in detail only in the seed of nasturtium (Tropaeolum majus).   Nasturtium xyloglucan (Le Dizet 1972) is almost identical in composition to that of tamarind (Kooiman 1961).   Following germination, nasturtium xyloglucan is completely broken down by a hydrolytic process involving three enzymes: endo(1→4)-β-glucanase, α-xylosidase and β-galactosidase (Edwards et al. 1984).   Of these the first two are of particular interest with respect to their substrate specificities.   The endo-β-glucanase shows a very high degree of specificity for xyloglucans, having no effect upon the carboxymethyl and hydroxyethylcelluloses which are normally used to assay endo(1→4)-β-glucanases (Edwards et al. 1984). The α-xylosidase is also inactive towards the conventional substrate. It liberates xylose from xyloglucans, but does not hydrolyse p-nitrophenyl-α-D-xylopyranoside (Edwards & Reid, unpublished).   Similar enzymes are possibly involved in xyloglucan mobilisation in tamarind and other xyloglucan-containing leguminous seeds.

The formation (Parker 1984a) and mobilisation (Parker 1984b) of "galactan" have been observed cytologically in seeds of Lupinus angustifolius and the biochemistry of mobilisation has been studied in the same system (Crawshaw & Reid 1984).   Clearly (Fig. 4) most of the galactose residues and some of the arabinose residues in the cell-wall of L. angustifolius are mobilised, whereas the other sugar residues remain relatively unchanged.   Structural studies on the cell-walls of L. angustifolius cotyledons are incomplete, but a detailed investigation of the polysaccharide in L. albus cotyledons has been carried out by Carré & Brillouet (1982).   Their data suggest that the cell-wall poly-saccharides consist of a rhamnogalacturonan core substituted by linear (1→4)-β-linked D-galactan and branched L-arabinan side-chains.   This is certainly consistent with our observation (Hutcheon & Reid, mss in prep) that the galactose residues in the cell-wall of L. angustifolius are liberated by an exo-β-D-galactanase.

GALACTOMANNANS AND WATER

Because of their essentially linear yet highly-branched structures (Fig. 1), the galactomannans are hydrophilic.   They imbibe water and can be dispersed to form mucilaginous solutions of high viscosity.   The interactions of galactomannans with water and other polysaccharides have been reviewed (Dea & Morrison 1975).

Fig. 4.   Changes in the total non-starch polysaccharide
and in the major component sugar residues in the cotyledon
pairs of light-grown (open-symbols) and dark-grown (filled
symbols) lupins (L. angustifolius cv. Unicrop) during
germination and establishment of the seedling (after
Crawshaw & Reid 1984).   T: Total;   G: Galactose;
A: Arabinose;   U: Uronic acid;   X: Xylose.

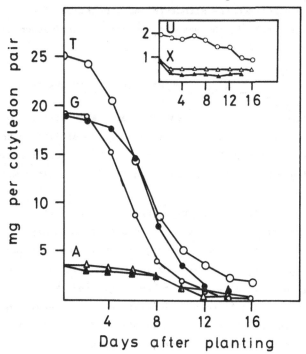

In vivo the galactomannan in the endosperm of the fenugreek
seed imbibes 62% of the water taken up by the seed on hydration.   Yet
it constitutes only 30% of the seed's dry weight (Reid & Bewley 1979).
In the endospermic legumes, the embryo is completely enclosed by
endosperm tissue (Fig. 2).   Consequently the water associated with the
galactomannan is interposed between the embryo and its environment.   It
has been shown experimentally (Reid & Bewley 1979) that the imbibed
endosperm is able to "buffer" or protect the embryo against desiccation
during periods of post-imbibition drought.   The physical basis of this
is the galactomannan's ability to lose a large proportion of its
imbibed water with little change in its water-potential (Ψ_w) (Reid &
Bewley 1979).

In fenugreek, the endosperm has three biological functions which it fulfils in succession.   It first acts as an imbibition-tissue for the seed.   It is then able to buffer the germinating embryo against water-loss during post-imbibition drought.   Finally, following successful germination, it acts as a food reserve for the embryo.   Only a full ecological study will reveal the relative importance of these roles in the seed's germination strategy.   There is, however, an obvious relationship between the structure and properties of the galactomannan and its role in the water-relations of germination (Reid & Bewley 1979).

STRUCTURE-FUNCTION RELATIONSHIPS OF XYLOGLUCANS AND "GALACTANS"

In structure the xyloglucans are remarkably similar to the galactomannans (compare Figs. 1 & 3).   They are also similar in their properties (Dea & Morrison 1975).   Perhaps their biological function in non-endospermic leguminous seeds is analogous to that of galactomannan in endospermic types.

If the "galactans" of Lupinus species have a biological function beyond that of a food reserve, it is almost certainly unconnected with water-imbibition.   Storage mesophyll cell-walls isolated from L. angustifolius cotyledons imbibe but little water (Hutcheon & Reid, unpublished).   It has been suggested (Matheson & Saini 1977) that the removal of galactose residues from the cell-walls of L. luteus following germination is a modification of the cell-wall to allow cotyledonary expansion.   Some cotyledon expansion does take place during "galactan" mobilisation in L. angustifolius, but the extent of it is limited.   The "galactan" mobilised, however, accounts for 33% by weight of the seed's reserves (Crawshaw & Reid 1984).   Perhaps within the genus Lupinus a wall turnover process associated with the regulation of wall plasticity has been adapted to carbohydrate storage.

CONCLUSIONS

It is evident that there is a striking structural resemblance between the cell-wall "storage" polysaccharides of seeds and the non-cellulosic polysaccharides of "normal" primary and secondary cell-walls in vegetative tissues.   Work of the type described in this paper is therefore important in two respects.   It contributes to our understanding

of seed physiology and it also provides excellent model systems for the
study of the enzymatic mechanisms which are probably responsible for
cell-wall turnover in vegetative tissues.

ACKNOWLEDGEMENTS

I am grateful to the Agriculture & Food Research Council and
to Unilever (U.K.) Ltd. for financial support.

REFERENCES

Anderson, E. (1949). Endosperm mucilages of legumes. Industrial and
    Engineering Chemistry, 41, 2887-2890.
Campbell, J. McA., Reid, J.S.G. (1982). Galactomannan formation and
    guanosine 5'-diphosphate-mannose: galactomannan mannosyl-
    transferase in developing seeds of fenugreek (Trigonella
    foenum-graecum L., Leguminosae). Planta, 155, 105-111.
Carré, B., Brillouet, J.M. (1982). Composition and methylation analysis
    of cell wall material from white lupin (Lupinus albus L.)
    cotyledons. II. International Lupin Conference,
    Torremolinos (Malaga). Abstracts, pp. 243-246. International
    Lupine Association, Cordoba.
Crawshaw, L.A., Reid, J.S.G. (1984). Changes in cell-wall polysaccharides
    in relation to seedling development and the mobilisation of
    reserves in the cotyledons of Lupinus angustifolius cv.
    Unicrop. Planta, 160, 449-454.
Edwards, M., Bulpin, P.V., Dea, I.C.M., Reid, J.S.G. (1984). Xyloglucan
    (amyloid) mobilisation in the cotyledons of Tropaeolum majus
    seeds following germination. Planta, in press.
Kooiman, P. (1960). On the occurrence of amyloids in plant seeds. Acta
    Botanica Neerlandica, 9, 208-219.
Kooiman, P. (1961). The constitution of Tamarindus-amyloid. Recueil des
    Travaux Chimiques des Pays-Bas, 80, 849-865.
Le Dizet, P. (1972). Quelques précisions sur la structure de l'amyloïde
    de capucine. Carbohydrate Research, 24, 505-509.
Matheson, N.K., Saini, H.S. (1977). Polysaccharide and oligosaccharide
    changes in germinating lupin cotyledons. Phytochemistry, 16,
    59-66.
McCleary, B.V. (1982). Purification and properties of a β-D-mannoside
    mannohydrolase from guar. Carbohydrate Research, 101, 75-92.
McCleary, B.V. (1983). Enzymic interactions in the hydrolysis of
    galactomannan in germinating guar: the role of exo-β-mannanase.
    Phytochemistry, 22, 649-658.
McCleary, B.V., Matheson, N.K. (1974). β-Galactosidase activity and
    galactomannan and galactosylsucrose oligosaccharide
    depletion in germinating legume seeds. Phytochemistry, 13,
    1747-1757.
McCleary, B.V., Matheson, N.K. (1975). Galactomannan structure and β-
    mannanase and β-mannanosidase activity in germinating legume
    seeds. Phytochemistry, 14, 1187-1194.
McClendon, J.H., Nolan, W.G., Wenzler, H.F. (1976). The role of the
    endosperm in the germination of legumes: galactomannan,
    nitrogen and phosphorus changes in the germination of guar
    (Cyamopsis tetragonoloba, Leguminosae). American Journal of
    Botany, 63, 790-797.

Meier, H., Reid, J.S.G. (1977). Morphological aspects of the
    galactomannan formation in the endosperm of Trigonella
    foenum-graecum L. (Leguminosae). Planta, 133,
    234-248.

Meier, H., Reid, J.S.G. (1982). Reserve polysaccharides other than
    starch in higher plants. F.A. Loewus & W. Tanner, eds.
    Encyclopedia of Plant Physiology New Series, Vol 13A.
    Plant Carbohydrates I, pp. 418-471.

Parker, M.L. (1984a). Cell wall storage polysaccharides in cotyledons
    of Lupinus angustifolius L. I. Deposition during seed
    development. Protoplasma, 120, 224-232.

Parker, M.L. (1984b). Cell wall storage polysaccharides in cotyledons
    of Lupinus angustifolius L. II. Mobilization during
    germination and seedling development. Protoplasma, 120, 233-
    241.

Reid, J.S.G. (1971). Reserve galactomannan metabolism in germinating
    seeds of Trigonella foenum-graecum L. (Leguminosae).
    Planta, 100, 131-142.

Reid, J.S.G. (1984). Cell-wall storage carbohydrates in seeds:
    biochemistry of the seed "gums" and "hemicelluloses".
    Advances in Botanical Research, 11, in press.

Reid, J.S.G., Bewley, J.D. (1979). A dual role for the endosperm and
    its galactomannan reserves in the germinative physiology
    of fenugreek (Trigonella foenum-graecum L.)
    an endospermic leguminous seed. Planta, 147, 145-150.

Reid, J.S.G., Meier, H. (1970). Formation of reserve galactomannan in
    the seeds of Trigonella foenum-graecum. Phytochemistry, 9,
    513-520.

Reid, J.S.G., Meier, H. (1972). The function of the aleurone layer
    during galactomannan mobilisation in germinating seeds of
    fenugreek (Trigonella foenum-graecum L.), crimson clover
    (Trifolium incarnatum L.) and lucerne (Medicago sativa L.):
    a correlative biochemical and ultrastructural study.
    Planta, 106, 44-60.

Reid, J.S.G., Meier, H. (1973a). Formation of the endosperm
    galactomannan in leguminous seeds: preliminary communications.
    Caryologia Supplement, 25, 219-222.

Reid, J.S.G., Meier, H. (1973b). Enzymic activities and galactomannan
    mobilisation in germinating seeds of fenugreek
    (Trigonella foenum-graecum L. Leguminosae).
    Secretion of ß-galactosidase and ß-mannosidase by the
    aleurone layer. Planta, 112, 301-308.

Reid, J.S.G., Davies, C., Meier, H. (1977). Endo-ß-mannanase, the
    leguminous aleurone layer and the storage galactomannan in
    germinating seeds of Trigonella foenum-graecum L. Planta,
    133, 219-222.

Reiss, R. (1889). Ueber die Natur der Reservecellulose und über ihre
    Auflösunsweise bei der Keimung der Samen. Landwirtschaftliche
    Jahrbuecher, 18, 711-765.

Seiler, A. (1977). Galaktomannanabbau in keimenden Johannisbrotsamen
    (Ceratonia siliqua L.). Planta, 134, 209-221.

Uebelmann, G. (1978). Samenkeimung bei Trigonella foenum-graecum L.:
    Aufnahme der beim Galaktomannanabbau im Endosperm
    freiwerdenden Zucker durch den Embryo. Zeitschrift für
    Pflanzenphysiologie, 88, 235-253.

Whistler, R.L., Saarnio, J. (1957). Galactomannan from soy bean hulls.
    Journal of the American Chemical Society, 79, 6055-6057.

12 CELL-WALL CHANGES DURING AUXIN-INDUCED CELL EXTENSION.
MECHANICAL PROPERTIES AND CONSTITUENT POLYSACCHARIDES OF
THE CELL WALL.

Y. Masuda
Department of Biology, Faculty of Science, Osaka City
University, Sumiyoshi-ku, Osaka 558, Japan

R. Yamamoto
Department of Biology, Faculty of Science, Osaka City
University, Sumiyoshi-ku, Osaka 558, Japan

Abstract. Auxin-induced changes in mechanical properties of
the cell wall which are thought to govern primarily cell
extension, have been measured by the stress-relaxation
technique, and the minimum stress-relaxation time (T_o) has
been found to represent cell-wall loosening. The
underlying modifications in cell-wall polysaccharides leading
to cell-wall loosening are also described and the shift of
xyloglucans towards a lower molecular weight is suggested to
be involved in auxin-induced cell-wall loosening. For
continued cell extension, new cell-wall synthesis is
necessary. Galactose inhibition of auxin-induced long-term
extension, particularly in monocotyledonous coleoptile
segments, is shown to be due to its inhibition of UDP-
glucose formation and thus of cell-wall synthesis.

Key words: Auxin, Cell-wall loosening, Cell-wall poly-
saccharides, Dicotyledonous species, Galactose inhibition,
Gibberellin, Mechanical properties of the cell wall,
Monocotyledonous species, Stress-relaxation, UDP-glucose,
Xyloglucan.

INTRODUCTION

Auxin induces rapid elongation of an organ segment which is
floated on dilute buffer solution containing auxin. The cell extension
is thought to be driven by turgor, but unless the cell wall encasing the
protoplast becomes extensible or loosened, the cell cannot enlarge by
water absorption. Thus, cell extension in excised organs in the presence
of added auxin seems to be regulated primarily by the osmotic concentration
of the vacuolar sap and by the mechanical properties of the cell wall,
although hydraulic conductivity is said to contribute about as much as
cell-wall extensibility to control of the growth rate of the auxin-
sensitive hypocotyl (Boyer & Wu 1978).

Our laboratory has been mainly concerned with the changes in
mechanical properties and polysaccharide components of the cell wall
during auxin-induced cell extension. We developed a stress-relaxation

technique to measure the mechanical properties of the cell wall in order
to evaluate them in rheological terms (Masuda et al. 1972).   Cell-wall
loosening should be brought about by biochemical modifications of polysaccharide molecules constituting the cell wall.   The extended cell wall
should then become fixed by new cell-wall synthesis so that the
extension becomes irreversible.   The synthesis also seems to prevent
thinning of the extended cell wall, rendering it capable of extending
further.

The present paper first discusses measurements of stress-
relaxation parameters and changes in the parameters caused by auxin and
their significance in cell-wall loosening, and then describes studies of
changes in cell-wall polysaccharides during auxin-induced cell extension,
their degradation and synthesis.

MECHANICAL PROPERTIES OF THE CELL WALL

Since Heyn and van Overbeek first measured the deformability
of coleoptiles using a bending method with a load in 1931, several methods
such as plasmolysis (Burström 1942;   Ketellapper 1953;   Cleland 1958,
1959;   Masuda 1961), bending (Tagawa & Bonner 1957;   Cleland 1958;   Yoda
& Ashida 1960;   Black et al. 1967;   Kohji et al. 1981), and stretching
(Brauner & Hasman 1952;   Preston & Hepton 1960;   Yoda & Ashida 1960;
Masuda & Wada 1966;   Black et al. 1967), have been adopted to measure
mechanical properties of the cell wall by load-extension analyses.   In
these methods a force is applied, either as a load or as turgor by
plasmolysis and deplasmolysis, to a cell-wall specimen such as
deproteinized Avena coleoptile segments, in order to measure its
deformation which consists of reversible (elastic) and irreversible
(plastic) components.   Based on the findings from many investigations
using these methods, it has been postulated that auxin causes an increase
in "extensibility" (elastic and/or plastic) or "plasticity".   An
automatic Instron-type tensile-tester was first adopted for load-
extension analyses using Avena coleoptiles by Olson et al. in 1965.   The
use of this technique has prevailed since then, although the parameters
obtained still represent reversible and irreversible extensibilities or
compliance (Cleland 1967 a,b;   Masuda 1968).   Polymers like those in the
primary cell wall in plants, which consists of cellulose microfibrils
associated with noncellulosic polysaccharide molecules, usually show a
viscoelastic property (Lockhart 1965;   Cleland 1971 a) such as "creep",

which appears in load-extension curves obtained even by classical
techniques (Tagawa & Bonner 1957; Preston & Hepton 1960; Masuda & Wada
1966).   Thus, a more physically defined parameter rather than the
conventionally used "extensibility" was needed to understand the
mechanical property of the cell wall and to know what happens when the
cell wall is "loosened" by auxin.   Creep was simulated by a viscoelastic
model called a "Voigt" element, consisting of a spring (elastic component)
and a dash-pot (viscosity component) connected in parallel fashion, and
some attempts have been made to analyze creep in the plant cell wall
(Cleland 1971 b; Jaccard and Pilet 1975, 1977, 1979).   However, rather
serious technical difficulties are encountered when making quantitative
creep measurements.   The cell-wall specimen fixed between upper and
lower clamps usually shows a "slack", otherwise it will be slightly
stretched; thus it is practically impossible to fix a specimen at zero
load without slack.

The easiest method by which to analyze viscoelastic
properties of the cell wall using an Instron-type tensile-tester is
"stress-relaxation" which was first adopted by Yamamoto et al. (1970),
then by Cleland and Haughton (1971).   A cell-wall specimen, usually
obtained by boiling coleoptile or stem segments in methanol for 5 min
followed by treatment with pronase (200 ppm + 5% ethanol, at 37° C for
18 h) to remove proteins, is fixed between the upper and lower clamps of
a tensile-tester.   When the specimen is quickly stretched by lowering
the bottom clamp, a stress is produced by the load-cell of the machine.
If the clamp is stopped at a preset position, usually when the stress
has reached 5 - 20 g, the stress starts to relax in an exponential
manner, first rapidly, then gradually to a certain value.   Hence the
plot of stress (S) against log time gives a straight line.

Stress-relaxation is usually simulated by a "Maxwell" model
consisting of a spring and a dash-pot connected in series and expressed
in the equation (1).

$$S = \gamma G \exp(-t/\tau) = So \exp(-t/\tau) \qquad (1)$$

where S: stress, γ: strain, t: time in sec, So: stress at t = O,
τ: relaxation time, G: elastic modulus.

This simple Maxwell model, however, is not really applicable
to the stress-relaxation curve of the plant cell wall.   A rather
generalized Maxwell model consisting of an infinite number of components

connected in parallel can represent the stress-relaxation process of the
cell wall (Yamamoto & Masuda 1971; Tanimoto & Masuda 1971; Masuda et al.
1972).   The stress-relaxation process of the cell wall is thus
represented well by the following semi-empirical equation.

$$S = R \log \left[(t + T_m)/(t + T_o) \right] + c \qquad\qquad (2)$$

where S: stress, t: time, R and c are constants, T_o: minimum stress-
relation time, T_m: maximum stress-relaxation time.

In stress-relaxation analysis, the cell wall specimen is
assumed to be stretched instantaneously, i.e., pre-extension time should
be zero.   However, a pre-extension period actually is needed before the
initial stress (S_o) is produced.   Fujihara et al. (1978 a,b) attempted
to correct for the pre-extension by assuming that the elements with a
relaxation time shorter than the pre-extension time must have relaxed
during the pre-extension process and may not have contributed to the
subsequent stress-relaxation process.   To equation (2), they introduced
a new parameter to represent the influence of pre-extension on stress-
relaxation, which was in fact equal to the pre-extension time
(Fujihara et al. 1979).   The data of stresses changing with time at
millisecond intervals can be recorded with a minicomputer system
connected with the tensile-tester and again put into a computer (e.g.
Hewlett Packard Model 85) which in turn does a simulation with the
equation using the least squares method, and finally produces the stress-
relaxation parameters, T_o, T_m, R and c.   An example of the data
obtained by the procedure is shown in Fig. 1.   Fujihara et al. (1978 a,b)
compared the parameters calculated from data obtained using different
pre-extension times and concluded that the pre-extension period needed
for the actual manipulation had very little effect on the parameters and
thus zero pre-extension time can be assumed for equation (2).   Fujihara
et al. (1978 c) further attempted to simulate the stress-relaxation curve
for compliance determinations to a generalized Maxwell model and
concluded that mathematical formulations for an experiment with a
constant rate of extension are closely related to those for a stress-
relaxation experiment.   The T_o value is small and the T_m is large in the
elongating region of stems (Yamamoto et al. 1974 a; Nishitani & Masuda
1979; Masuda et al. 1981; Capesius et al. 1981), coleoptiles (Fujihara
et al. 1972; Yamamoto et al. 1974 a; Sakurai & Masuda 1978 a,b;
Sakurai et al. 1982; Zarra & Masuda 1979 a), or roots (Masuda & Pilet

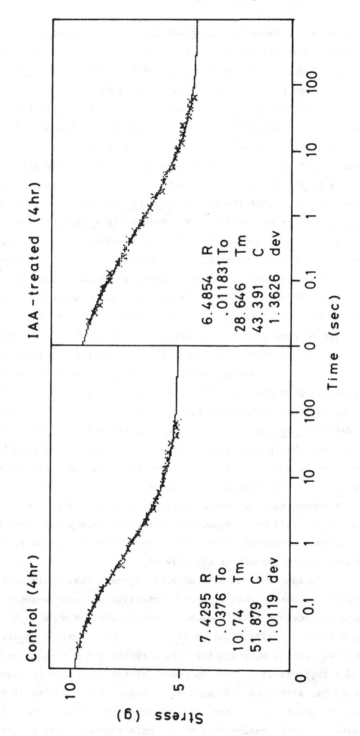

Fig. 1. Stress-relaxation process of the epidermal cell wall of *Vigna angularis* epicotyls: simulation by computer (see text).

Epicotyl segments were treated with or without 10^{-5} M IAA for 2 h, then the epidermis was peeled off, killed in boiling methanol, treated with Pronase-P, and subjected to stress-relaxation analysis.

1983), the former becoming larger and the latter smaller towards the
basal, non-elongating regions.   The R value is small in the elongating
region of coleoptiles (Yamamoto et al. 1974 a; Sakurai et al. 1982).
It has also been reported that auxin causes a decrease in the T_o value
in the cell wall of Avena coleoptiles (cf. Yamamoto et al. 1970;
Yamamoto & Masuda 1971; Shen-Miller & Masuda 1973; Masuda et al. 1974 b),
pea epicotyl epidermis (cf. Tanimoto & Masuda 1971), pea hook epidermis
(Nakamura et al. 1975), and Vigna angularis epicotyl epidermis
(Nishitani et al. 1979; Nishitani & Masuda 1981).   Auxin also causes a
decrease in the R value (former b) in Avena coleoptiles (Yamamoto et al.
1974 a; Sakurai et al. 1982) but an increase in Vigna epicotyl segments
(Kiyota et al. in preparation).   There are several features regarding
auxin-induced changes in mechanical properties which differ between
monocotyledonous (monocot) and dicotyledonous (dicot) species.   This
point will be discussed later.   In general, changes due to auxin in the
parameters obtained by stress-relaxation analysis or other techniques
are thought to represent auxin-induced cell-wall loosening.   However,
the changes may reflect extension of the cell wall which already has
occurred.   In fact, several studies have pointed out that extensibility
(strain/stress) or plastic compliance (DP) may mainly represent the
degree of cell-wall extension which already has occurred (cf. Masuda
1968; Masuda et al. 1974).   By using mannitol solution which inhibits
osmotic water absorption, the direct effect of auxin on the cell wall
property can be seen, and in fact, auxin was shown to cause a decrease
in the T_o value, suggesting that this value represents the degree of
cell-wall loosening, i.e. the capacity of the cell wall to extend
(Masuda et al. 1974 b).   Also, a linear relationship was found between
the T_o value at a certain incubation time and the extension rate at the
subsequent period (Masuda et al. 1974 b).

        Evidence from many studies suggests that auxin primarily
acts on the epidermis when it causes extension of organ segments
(Thimann & Schneider 1938; Tanimoto & Masuda 1967; Masuda et al. 1972;
Masuda & Yamamoto 1972; Yamamoto et al. 1974 b,c; Masuda et al. 1974 a;
Penny et al. 1972), although there is a report which is against this view
(Durand & Rayle 1973).   The importance of the epidermis in segment
extension has also been indicated by evidence that the rate of extension,
or that of curvatures in tropistic reactions, of stem segments is
dependent on the extension rate of the outermost cell layers (Firn &

Digby 1977).   Thus, the question arises of whether only the epidermis
responds to auxin or whether the inner tissue also plays a role in segment
extension.   The isolated epidermis of pea internodes cannot extend in
response to auxin alone but can when GTP, ATP, ITP or UTP is also present
(Yamagata & Masuda 1976).   Since the nucleotides are not needed when
hydrogen ions and fusicoccin induce extension of the isolated epidermis
(Durand & Rayle 1973; Yamagata & Masuda 1975), they appear to be
required for auxin action on the epidermis to activate proton pumps.
One of the roles of the inner tissue may be to provide the epidermis with
factors such as nucleotides to make it responsive to auxin.   The
epidermal strip shrinks about 5% in length after it is peeled off, while
peeled segments expand rapidly when floated on water, indicating that the
inner tissue is compressed by the physical strength of the epidermis when
the stem segments are intact.   In fact, the thickness of the outer
epidermal cell wall is more than ten times the thickness of the
parenchymatous cell wall (Masuda & Yamamoto 1972).

There are certain differences in the auxin-induced changes in
cell-wall properties between monocots and dicots, as already pointed out.
An auxin-induced decrease in the T_o value can be seen when deproteinized
whole segments of coleoptiles are subjected to stress-relaxation analysis.
This is not always seen when whole segments of dicot stems are used for
the analysis.   However, if the epidermis is peeled off from stem
segments, which have been incubated on buffer solution containing auxin
(IAA: indole-3-acetic acid), and then subjected to stress-relaxation
analysis, IAA causes a decrease in the T_o value.   Coleoptiles have two
epidermal layers occupying about 75% of the whole volume of the organ,
and the inner parenchyma consists of only a few cell layers.   Thus, even
though IAA has little effect on the cell wall of the inner tissue, changes
in the property of the epidermal cell wall due to IAA may be observed
even if whole coleoptile segments are subjected to the analysis.   On the
other hand, in dicot stems, the epidermis occupies a relatively small
part of the organ, and if whole or half segments are used, the auxin
effect on the epidermis can barely be found.   According to quantitative
determination of noncellulosic and cellulosic polysaccharides in Vigna
epicotyls, the epidermal cell wall accounts for approximately 50% of the
total cell wall of the epicotyl (Nishitani et al. 1979).   Therefore, in
the case of coleoptiles, the amount of the epidermal cell wall well

exceeds 50% of the total cell wall.

Hormonal control of the stress-relaxation property of the cell wall also differs between dicots and monocots. With dicot stems, auxin and gibberellin often show a synergistic promotive effect on segment extension. Unless the segments are excised from an extremely young part of the stem, gibberellin shows very little growth-promotive effect when given alone (Purves & Hillman 1958; Katsumi & Kazama 1978), but with segments which rapidly elongate in response to added auxin, gibberellin synergistically enhances auxin-induced extension (Tanimoto et al. 1967; Ockerse & Galston 1967; Ockerse 1970; Katusmi & Kazama 1974; Shibaoka 1972; Hogetsu et al. 1974 a; Nishitani & Masuda 1982).   If segments are pretreated with gibberellin, then with auxin, the gibberellin enhances the subsequent effect of auxin on extension, but not if the segments are treated first with auxin.   Thus, gibberellin apparently sensitizes the dicot stems to auxin.   Similar interactions between gibberellin and auxin have been found in the expansion growth of tuber tissue slices from Jerusalem artichoke (Helianthus tuberosus) (Setterfield 1963; Masuda 1965).

As shown in Table 1, gibberellin alone has no effect on the T_o value of the epidermal cell wall of Vigna epicotyl segments, but IAA causes it to decrease.   Addition of gibberellin to IAA solution causes no difference in the T_o value from the IAA treatment of the epidermal cell wall.   On the other hand, with segments cut longitudinally in half, even though IAA alone exerts no significant effect on the T_o value, it causes a decrease in the presence of gibberellin.   Thus, gibberellin appears to sensitize the inner tissue to make it responsive to auxin and this seems to be part of its synergistic effect on auxin action on segment elongation.

In Helianthus tuberosus tuber slices, the enhancing effect of gibberellin on the subsequent effect of auxin is inhibited if the slices are treated with inhibitors of nucleic acid synthesis such as 5-fluorouracil or fluorodeoxyuridine (Setterfield 1963; Masuda 1966; Kamisaka & Masuda 1971).   It has therefore been suggested that nucleic acid metabolism is involved in the sensitization of the tissue to auxin by gibberellin.   On the other hand, Shibaoka (1974) observed by electron microscopy that cell-wall microtubules ran transverse to the cell axis of Vigna epicotyls treated with gibberellin and IAA whereas they were randomly oriented in segments treated with IAA only.   This effect of

Table 1.   Effect of IAA and gibberellic acid on the T_o
value of the cell wall of <u>Vigna</u> <u>angularis</u> epicotyl
segments.

Treatment	T_o, ms	
	Epidermis	Halved segments
Initial	18.9+1.9	19.2+1.5
[-Sucrose] Control	15.5+0.9	20.4+0.9
IAA	11.3+1.3	19.5+0.9
GA_3	16.1+1.7	19.5+0.9
IAA + GA_3	11.1+1.7	16.4+1.1
[+50mM Sucrose] Control	17.1+1.2	18.2+0.7
IAA	10.5+0.9	18.6+0.8
GA_3	16.3+1.5	19.5+1.3
IAA + GA_3	10.5+0.8	15.6+1.2

Segments were incubated for 4 h in K-phosphate buffer
solution, pH6.0, containing or not containing 10^{-5}M IAA
and/or 10^{-5}M gibberellic acid (GA_3) in the presence and
absence of 50 mM sucrose.   After incubation, the epidermis
was peeled off, or segments were longitudinally halved with
a razor blade, killed in boiling methanol, treated with
Pronase-P, then subjected to stress-relaxation analysis.

gibberellin was negated when colchicine was simultaneously supplied;
the wall which formed consisted of microfibrils running in the same
direction (Takeda & Shibaoka 1981).   Coumarin and 2,6-dichlorobenzo-
nitrile, which inhibit cellulose synthesis, reverse the gibberellin
promotion of auxin-induced epicotyl elongation (Hogetsu et al. 1974 a,b).
Gibberellin seems to enhance auxin-induced extension by modifying the
orientation of microtubules, and hence that of cellulose microfibrils.
However, these modifications are observed in epidermal cells where changes
in mechanical properties due to auxin are not modified by gibberellin,
as mentioned previously.

BIOCHEMICAL SIGNIFICANCE OF THE PARAMETERS

In general, the stress-relaxation process is determined by
the equation (see equation 1) which is a function of η (viscosity) and
G (elastic modulus) $(\tau=\eta/G)$.   If stress relaxation time decreases, this
is caused either by a decrease of η and/or an increase in G.   There is
little evidence that auxin influences the elastic modulus of the cell
wall.   Burström et al. (1967, 1970) using a resonance method with pea
stems reported that auxin rapidly decreases Young's modulus.   On the
other hand, auxin has been shown to cause a decrease in the η of
hemicellulose B which has been isolated and purified from Avena
coleoptile segments treated with IAA (Sakurai et al. 1979).   In a
generalized Maxwell model the stress-relaxation is represented as
follows:

$$S = \int_0^\infty H(\tau) \cdot \gamma \cdot \exp(-t/\tau) \cdot d\tau \qquad (3)$$

where S: stress, τ: stress-relaxation time = η/G, γ: strain,
$H(\tau) = G(\tau)d\tau$: distribution function of stress-relaxation
(Yamamoto & Masuda 1971).   The viscosity η can be represented by the
following equation derived from equation (3).

$$\eta = \int_0^\infty H(\tau)d\tau$$

The hemicellulose B of Avena or other monocot coleoptile cell walls
consists of $\beta(1-3),\beta(1-4)$glucans and glucuronoarabinoxylans (Nevins et al.
1977, 1978; Wada & Ray 1978; Yamamoto & Nevins 1978).   Calculations
according to Flory's theory (1953) give 37 nm as the effective radii for
β-glucans and 20 nm for arabinoxylans based on the molecular
distributions of these polymers on Sepharose 4B (Sakurai et al. 1979).
If these molecules assume random coils, the volume of a β-glucan

molecule is much larger than that of an arabinoxylan molecule, suggesting
that β-glucan degradation contributes to the decrease in viscosity of the
cell-wall matrix.   However, the kinetics of the auxin-induced decrease
in η that occurs after 1 hour of incubation are not really comparable to
that of auxin-induced decrease in the T_o value that is usually found as
early as 10 min after auxin treatment in Avena coleoptile segments
(Sakurai et al. 1979).   In addition, although IAA rapidly causes a
decrease in the T_o value of the epidermal cell wall of dicot stem
segments as mentioned above, the dicot cell-wall matrix contains no
β-glucans or arabinoxylans but other polymers such as arabinogalactans or
galactans and pectic substances (Labavitch & Ray 1978 b; Nishitani &
Masuda 1981; Masuda 1980).

An approximation from an equation of stress-relaxation
formulated by Tobolsky & Eyring (1943), based upon chemical kinetics,
gives (Yamamoto et al. 1981 a):

$$R = 2Nk\ T/\lambda \qquad\qquad (5)$$

where N: number of flowing factors per unit area, λ: flow distance of
the factors, k: Boltzmann constant, T: absolute temperature.   Therefore,
R represents the concentration of the stress-relaxation components.   In
the case of Avena coleoptiles, R may also represent the degradation of
matrix components of the cell wall.   However, the physical and
chemical significance of the parameters R and Tm still remains unclear
and needs further investigation.   Parameter R may represent the amount
of the cell-wall polysaccharide per unit length.

THE CELL-WALL MATRIX COMPONENTS RESPONSIBLE FOR CELL-WALL
LOOSENING

After extracting pectic fractions by acidic oxalate buffer
solution, the cell-wall holocellulose of Vigna epicotyls is extracted
with dilute alkali solution (4% KOH) to obtain hemicellulose 1 (HC-1),
then with concentrated alkali (24% KOH) to obtain hemicellulose 2 (HC-2),
the residue being cellulose.   Monocot plant cell walls contain only a
small amount of pectic substances but dicot cell walls contain a large
amount (Masuda 1980).   HC-1 contains arabinogalactans and galactans and
HC-2 contains mainly xyloglucans (Nishitani & Masuda 1972, 1973).   In
Avena coleoptile segments Loescher & Nevins (1972, 1973) found that the
glucose content of noncellulosic polysaccharides of the wall decreased

during auxin-induced cell extension.   These effects of auxin on
elongation, the glucose content and the T_o value are all inhibited by a
potent inhibitor of glucanase, nojirimycin (5-amino-5-deoxy-D-
glucopyranose) which also inhibits auxin-induced extension (Nevins 1975;
Sakurai & Masuda 1977; Masuda 1977).   Exogenously applied β-1,3-
glucanase causes elongation of Avena coleoptile segments and mediates
cell-wall loosening (Masuda 1968).   Good correlations exist in Avena
coleoptile segments between the decrease in the glucose content, the
decrease in the T_o value and cell extension, which are all caused by
auxin (Sakurai et al. 1977).   The same situation was found with barley
and rice coleoptiles (Sakurai & Masuda 1978 b; Zarra & Masuda 1979 b).
In Vigna epicotyl segments, on the other hand, IAA enhances the
degradation of galactans in both pectin and hemicellulose fractions in
the absence of sucrose, although the hormone promotes polymerization of
the polymers (Nishitani & Masuda 1980; Nishitani et al. 1979).   When
segments are treated with IAA in the presence of sucrose, IAA increases
the amount of polysaccharides and enhances the polymerization of
galactans, xylans and glucans in the hemicellulose fraction.   Thus, the
changes due to IAA in the polysaccharide compositions are quite different
between Avena and Vigna, probably due to difference in the matrix
composition (Fig. 2).

The acid growth theory has been proposed to explain the
mechanism of auxin action on rapid cell extension (cf. Rayle & Cleland
1970, 1972; Rayle et al. 1970; Hager et al. 1971; Marre et al. 1973;
Yamamoto et al. 1974 b,c; Cleland & Rayle 1978; Cleland 1973, 1983).
According to this theory, auxin activates proton pumps located on the
plasma membrane to cause hydrogen-ion secretion into the cell wall
compartment and thereby acidifies the wall leading to cell-wall
loosening and hence cell extension.   Acid-induced cell-wall loosening
is thought to be brought about by some modifications of cell-wall
polysaccharides.   An acid pH solution as well as auxin causes a
decrease in the T_o value but β-glucan degradation in the Avena
coleoptile cell wall is not brought about by protons (Sakurai et al.
1977).   Thus, partial degradation of β-glucans in response to IAA in
Avena coleoptile segments is not likely to be responsible for the
decrease in the T_o value which may be due to some secondary or
accompanying changes in the cell wall.   Darvill et al. (1977, 1978)
presented data from maize coleoptiles that auxin and protons induced,

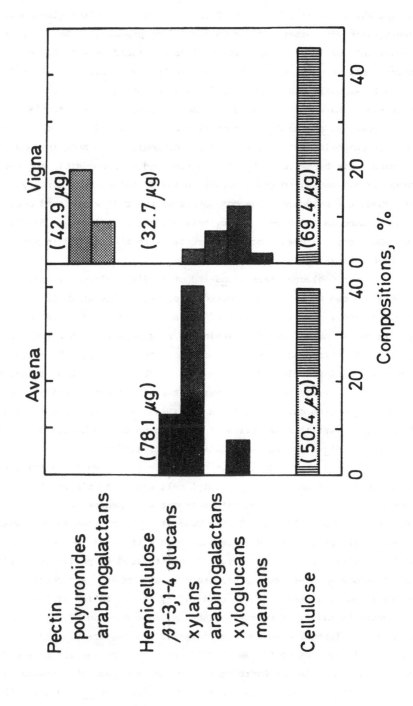

Fig. 2.   Cell wall polysaccharide compositions of <u>Avena</u> coleoptile and <u>Vigna</u> epicotyl cell walls.

perhaps enzymatically, breakage of the linkages connecting glucuronoarabinoxylan with other similar molecules and proteins.   However, in
primary cell walls glucuronoarabinoxylans are found in significant
amounts only in monocots.   In Vigna epicotyl segments, when IAA induces
cell extension, it causes partial degradation of arabinogalactans
(Nishitani & Masuda 1981), which are not components of the matrix
polysaccharides of the Avena coleoptile cell wall.   Thus, some other
polymers composing the cell-wall matrix commonly found both in monocots
and dicots may be degraded by auxin action and be responsible for the
change in η, leading to the decrease in the T_o value.   As already
mentioned, the polysaccharide components of the cell wall found commonly
in both monocots and dicots are xyloglucans and cellulose.   This is why
our laboratory has been concerned recently with the changes in molecular
structures of xyloglucans due to hormones.

Xyloglucans from Vigna and other dicot plants are water-
soluble and can be easily determined colorimetrically at OD 625 nm
(Kooiman 1960).   In the absence of exogenous sugars the amount of
xyloglucans is not affected by auxin and/or gibberellin when the hormones
promote cell extension growth (Nishitani & Masuda 1982).   Molecular
distribution of xyloglucans on Sepharose CL-4B has been found to shift
to a lower molecular weight, the weight average molecular weight (MW)
being changed from ca. 1.2 to 1.0 X 10^6 (Nishitani & Masuda 1981, 1983)
(Table 2).   Partial degradation of xyloglucans caused by auxin was also
reported for Pisum epicotyls by Gilkes & Hall (1977).

Acidic pH also induces a decrease in the MW of xyloglucans
in the HC-2 fraction of Vigna epicotyl cell wall and this effect is
completely reversed by subsequent treatment with neutral pH solution
(Nishitani & Masuda 1983).   In the case of Avena coleoptile cell walls,
noncellulosic polysaccharides are extracted with 17.5% KOH solution
(Wada & Ray 1978), then the extract is neutralized to pH 4.5 to obtain
hemicellulose A (HC-A) as a precipitate.   The supernatant contains
hemicellulose B (HC-B) which is obtained by precipitation with alcohol.
HC-A consists mainly of water-insoluble xyloglucans and HC-B of water-
soluble xyloglucans, glucuronoarabinoxylans and β-glucans.   IAA
substantially decreases the MW of HC-A and acidic pH has the same effect.
However, neither IAA nor hydrogen ions cause any remarkable changes in
the MW of water-soluble xyloglucans in HC-B (Inoue et al. 1984).   The
effect of acidic pH on HC-A is reversed quickly when segments are

Table 2. Effect of IAA and acid pH on the weight average
molecular weight (MW) of the water-soluble xyloglucans
extracted from Vigna angularis epicotyl segments.

Treatment	MW, $\times 10^{-6}$	%
Initial	1.20	100
pH 6.5	1.22	102
pH 6.5 + IAA	0.92	77
pH 4.0	0.86	72

Segments were treated with K-phosphate buffer solution at
pH 6.5 in the presence or absence of 10^{-5}M IAA and at pH
4.0 for 4 h.   Hemicelluloses were extracted and
chromatographed on Sepharose CL 4B.   The MW was then
calculated using the formula MW = ΣMi wi/Σwi
(wi:xyloglucan content of the ith fraction; Mi: the MW of
the ith fraction estimated from the calibration curve using
standard dextrans).

subsequently treated with neutral buffer solution.   The MW decrease in
HC-A caused by IAA and acidic pH becomes apparent as rapidly as the
decrease in the T_o value caused by these agents in Avena coleoptile
segments, as in the case of Vigna epicotyl segments.   Xyloglucans in
Avena coleoptile cell walls are found in two forms, a water-insoluble
one in HC-A and a water-soluble one in HC-B, whereas xyloglucans in Vigna
epicotyl cell walls are found only in HC-2 which is water-soluble.
Water-soluble xyloglucans have galactose and fucose on the side chains
connected to the backbone of β-1,4-glucans via xylose, having glucose,
xylose, galactose and fucose in the ratio of 10:7:2.5:1 (Albersheim 1974;
Kato & Matsuda 1976), whereas water-insoluble xyloglucans in the monocot
cell wall have no such neutral sugar moieties attached to xylose side
chains (Kato et al. 1981, 1982).   Xyloglucans in HC-A of the Avena coleoptile cell wall have a MW of ca. 1.2 X 10^6 (6.4 µg/10 mm segment) which
decreases to ca. 1.0 X 10^6 in response to IAA or acid pH, as in the case
of water-soluble xyloglucans found in Vigna cell walls (18.6 µg/10 mm
segment).   On the other hand, water-soluble xyloglucans in the HC-B of

Avena cell walls (3.3 μg/10 mm segment) have a MW of 0.02 X 10^6, which
is much smaller than those in HC-A.   With etiolated mung bean hypocotyls,
the molecular weight of cell wall xyloglucans is 160,000 and that of
water-soluble ones released when the organ is homogenized is 20,000
(Kato & Matsuda 1981).   Although there are some differences in the
properties of xyloglucans between Avena and Vigna as described, IAA and
acid pH cause a shift from larger to smaller molecular weight in
xyloglucans and these changes are very likely to be responsible for the
change in the T_o value of the cell wall due to auxin and protons.

Endoglucanases have been suggested to be responsible for
xyloglucan degradation (Koyama et al. 1981; Hayashi & Maclachlan 1984;
Hayashi et al. 1984).   It appears likely that auxin and protons activate
the endoglucanases in the coleoptiles and stem cells, but direct evidence
is needed to confirm this.   Some reports support the view that auxin
stimulates the release of xyloglucans in dicotyledonous plants
(Labavitch & Ray 1974; Terry & Bonner 1980; Terry et al. 1981) and
Avena coleoptile segments (Inoue et al. 1984).

LONG-TERM GROWTH AND CELL-WALL SYNTHESIS

An extensive analysis by Vanderhoef et al. (1976), with plots
of the extension rates of segments from a number of monocot coleoptiles
and dicot stems against time taken from the literature, yielded evidence
that there are two separable responses of organ segments to auxin.   They
reported that average lag times for the first and second responses were
12.4 and 35.4 min, and the maximum rates of extension were 0.57 and 0.54
mm/h, respectively.   Their evidence also indicated that the acid
response approximated the first but not the second response.   Kazama
and Katsumi (1976, 1978) also presented data using cucumber hypocotyl
segments indicating that there are two phases for auxin-induced
extension, the second being metabolism-dependent.   Thus, when organ
segments are floated on buffer solution containing auxin, they show
first rapid extension, followed by a second extension;   the first is due
to acid growth and the second to growth requiring nucleic acid and
protein syntheses.

A number of metabolic inhibitors can be used to analyze the
cell-extension process caused by added hormones and there have been many
reports indicating that various metabolic pathways are involved in
hormone-induced cell extension.   Recently, Jacobs and Taiz (1980) studied

the auxin-induced extension of Avena coleoptile and Pisum epicotyl
segments using an inhibitor of plasma-membrane-associated hydrogen-ion-
pumping ATPase, sodium ortho-vanadate. They showed experimental results
that vanadate inhibits IAA-induced extension of segments and IAA-
enhanced medium acidification within about 5 min after its application
to IAA-treated segments.   The inhibition is reversed within 10 min after
vanadate removal, showing the effect is reversible.   Thus, the
inhibitory effects of vanadate on acidification and elongation appear in
parallel, indicating that short-term auxin-induced growth is due to cell-
wall acidification which may be caused by plasma-membrane-associated
ATPase acting as a proton pump.   Vanadate does not inhibit respiration
or protein synthesis, but since it strongly inhibits leucine uptake, they
concluded that cell-wall acidification is important also for long-term
elongation.

        Nucleic acid and protein syntheses have been shown to be
required for auxin-induced growth (Key & Ingle 1964; Noodén & Thimann
1966; Masuda & Wada 1966; Masuda et al. 1967; Masuda & Kamisaka 1969).
These studies used metabolic inhibitors, such as actinomycin D,
cycloheximide, and puromycin, and showed that usually a relatively long
time is needed for them to be effective in inhibiting auxin-induced
growth.   The inhibitors also inhibit auxin-induced changes in
mechanical properties of the cell wall (Masuda & Wada 1966; Coartney
et al. 1967; Black et al. 1967; Masuda 1969).   Cycloheximide inhibits
auxin-induced extension and auxin-induced changes in extensibility or
compliance in Avena coleoptile and soybean hypocotyl segments.   It also
inhibits medium acidification when added to buffer solution containing
IAA on which Avena coleoptile segments are floated.   Thus, although the
situation appears rather complicated, we can generally say that auxin-
induced short-term rapid extension is mediated by wall acidification and
long-term, continued growth accompanies a variety of metabolic activities
including nucleic acid and protein synthesis, and also cell-wall
synthesis.

        When segments of Vigna epicotyls are floated on buffer
solution containing IAA in the presence of sucrose, the sucrose enhances
IAA-induced extension (Nishitani et al. 1980).   Sucrose or glucose added
also promote IAA-induced extension and wall synthesis in Avena
coleoptile segment (Baker & Ray 1965 a,b; Ray 1973 a,b).   Promotion of
growth in isolated organ segments by various sugars has been extensively

reported (cf. Göring & Gerlach 1966).   In the absence of added sugars,
synthesis of new cell-wall materials is small, resulting in a decrease
in the weight of cell-wall materials per unit length during the period of
IAA-induced coleoptile extension (Ray 1962).   In the presence of sugar,
new cell-wall synthesis occurs, but thinning of the wall during
extension is still found, although its degree is much smaller than that
found in the absence of sugars.   In dicots, such as Vigna, the amount
of pectic substances per segment decreases during IAA-induced
extension in the absence of added sucrose and that of hemicelluloses and
cellulose changes very little or increases slightly (Nishitani & Masuda
1982).   In the presence of sucrose, all components of the cell wall,
particularly HC-2 and cellulose, increase during IAA-induced extension.
The polysaccharides in HC-2 which increase are xyloglucans.   Here again,
xyloglucans appear to be a key polymer in cell-wall extension.   These
results indicate that cell-wall synthesis is necessary for enhancement of
auxin-induced cell extension and contributes to continued growth.   In
fact, there are reports indicating that inhibitors of cell-wall synthesis
such as coumarin and dichlobenil, inhibit growth of intact plants
(Kawamura et al. 1976) and hormone-controlled growth of organ segments
(Hogetsu et al. 1974 a,b).

Galactose has been known to inhibit auxin-induced extension
of Avena coleoptiles at 3 - 10 mM (Ordin & Bonner 1957;   Ray 1962;
Baker & Ray 1965 b).   This sugar also inhibits IAA-induced extension of
other monocot coleoptile segments, but has very little effect on that of
dicot stem segments (Yamamoto et al. 1981 b) (Table 3).   With Avena
coleoptile segments, galactose does not inhibit IAA uptake by the tissue,
respiration, water permeability of the tissue, IAA-induced decrease in
the T_o value and IAA-induced molecular weight shift of xyloglucans
(Yamamoto et al. 1984 a,b).   In addition, galactose does not inhibit
acid-induced extension of Avena coleoptile segments.   And its
inhibition of auxin-induced extension of Avena coleoptile segments
requires a lag period of about two hours at the early period of extension
and another of one hour at the later stage of extension (Yamamoto et al.
1981 b).   The inhibitory effect of galactose is immediately reversed by
its removal.   These results suggest that galactose inhibits long-term,
continued elongation due to auxin but not the short-term, probably acid-
mediated one.   Galactose at 10 mM completely inhibits incorporation of
[14]C-labelled glucose into noncellulosic and cellulosic polysaccharide

Table 3.   Effect of galactose on IAA-induced elongation of segments
excised from different organs of different plants.

| | Elongation, % of the initial length | | | |
| | - Galactose | | + Galactose | |
Plant & organs	-IAA	+IAA	-IAA	+IAA
Avena sativa				
coleoptiles	4.7±0.3	22.0±1.0	4.0±0.3	10.6±0.8
Triticum aestivum				
coleoptiles	2.4±0.1	21.0±2.0	2.0±0.1	8.0±0.8
Oryza sativa				
coleoptiles	1.0±0.1	16.5±1.5	1.0±0.1	6.0±0.7
Phaseolus aureus				
hypocotyls	10.5±0.5	30.5±2.0	9.5±0.5	29.0±2.0
Pisum sativum				
epicotyls	4.5±0.5	18.5±2.5	3.0±0.2	17.5±2.0

Segments were incubated for 6 hours in 10 mM K-phosphate buffer
solution, pH 6.5, containing and not containing 10^{-5}M IAA in the
presence and absence of 10 mM galactose.   Initial length: 10 mm.

fractions of the Avena coleoptile cell wall (Yamamoto et al. 1984 a,b).
This result suggests again that cell-wall synthesis is needed for IAA-
induced long-term, continued extension growth and galactose specifically
and reversibly inhibits this process.   In fact, the inhibitory effect of
low concentrations of galactose, e.g. 1 mM, on auxin-induced extension
can be overcome by the addition of sucrose.

Auxin-induced extension of dicot stem segments is much less
sensitive to galactose and more than tenfold concentrations are needed
to inhibit such extension in monocot coleoptile segments.   Galactose
inhibits not only shoot elongation but also root growth more effectively
in monocots than in dicots (Göring & Reckin 1968 a,b).   As mentioned
previously, cell-wall composition in monocot coleoptiles is different
from those in dicot stems, the former in general being rich in glucose
and the latter in galactose in noncellulosic polysaccharides.   When

Avena coleoptile segments are incubated in buffer solution containing
galactose and then noncellulosic polysaccharides are extracted, the
relative amounts of glucose, arabinose, xylose and galactose do not
change, indicating that galactose does not modify the composition of the
cell wall (Yamamoto et al. 1981 b).

    Göring and Reckin (1968 a,b) observed that labelled galactose
is respired much more strongly by root segments of Cucumis sativus
(a dicot) than by those of Zea mays (a monocot); galactose-phosphate
accumulates in the latter but not in the former.   Based on these
results, they proposed a hypothesis (Fig. 3) that galactose, as
galactose-1-phosphate, after being taken up into the monocot tissue,
competitively inhibits the conversion of glucose-1-phosphate into UDP-
glucose by blocking the activity of an enzyme UTP:α-D-glucose-1-phosphate
uridylyltransferase (Yamamoto & Masuda 1984 b), leading to the
inhibition of new cell-wall components, and thereby inhibiting cell
extension.   We examined this hypothesis by conducting experiments:

Fig. 3.   Hypothetical representation of the galactose
           inhibition of cell-wall synthesis.

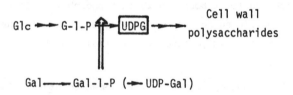

(i) to see the in-vitro effect of galactose on UTP:α-D-glucose-1-
phosphate uridylyltransferase using glucose-1-phosphate as substrate
and enzyme preparations extracted from Avena coleoptiles;   (ii) to
extract UDP-hexoses and to see the effect of galactose on their amounts
after the segments were incubated on IAA solution in the presence and
absence of galactose;   (iii) to compare the effect of galactose on the
amount of UDP-hexoses in coleoptiles and dicot stem segments.

    Coleoptile segments were homogenized with water and
centrifuged at 3,000 rpm for 10 min.   The supernatant was then dialyzed
against water overnight.   The dialyzate (the final concentration of
proteins was 66 µg/ml) was incubated with various concentrations of

glucose-1-phosphate, 10 mM galactose-1-phosphate, and 4mM UTP for 1 h
at 37°C.   The solvent of the reaction mixture was 20 mM sodium phosphate
at pH 7.5 containing 5 mM $MgCl_2$.   After incubation, the reaction mixture
was heated for 5 min at 100°C to inactivate the enzymes and then
diluted five times.   Next, 50 µl each of NAD solution (12.5 mg/ml),
glycine buffer solution (1 M, pH 8.7), and UDPG dehydrogenase solution
(0.016 µmol unit/ml, Sigma No. 600-5000) was added to 1.25 ml of the
diluted reaction mixture and its absorbance at 340 nm was determined.
The effect of galactose-1-phosphate on the enzyme activity was then
examined by double-reciprocal analysis.

Galactose-1-phosphate, in fact, competitively inhibits the
enzyme activity extracted from Avena coleoptiles, suggesting that
galactose inhibits the formation of UDP-glucose and hence cellulosic and
noncellulosic polysaccharides of the cell wall (Yamamoto & Masuda 1984 b).
We adopted the chromatography technique using DEAE-Sephadex A-25 of
Carpita and Delmer (1981) to separate UDP-sugars from the tissue extracts.
Coleoptile segments treated with IAA + galactose were frozen and dipped
into 80% ethanol at 80°C for 30 min and extracted twice, then the
segments were squeezed.   The combined ethanol extracts were chilled to
-20°C and centrifuged at 1,500 xg for 30 min.   The supernatant was
evaporated to dryness and the residue was resuspended in 10 ml of water
and extracted twice with diethyl ether.   The aqueous phase was
collected and subjected to chromatography.   The aqueous phase was
loaded onto a DEAE-Sephadex A-25 (Pharmacia) anion-exchange column (0.9 x
4.0 cm), and the column was washed with 10 ml water.   Anions were
eluted with 10 ml of 0.5 M NH_4HCO_3 which was then removed by evaporation.
The residue containing anions was resuspended in water and loaded onto a
DEAE-Sephadex A-25 anion-exchange column (0.9 x 50.0 cm) equilibrated
with 0.1 M NH_4HCO_3 (pH 8.0).   The nucleotides and nucleoside-sugars
were eluted with a 1 1 linear gradient of 0.1 to 0.4 M NH_4HCO_3 (pH 8.0)
and the elution profile was monitored by UV absorption at 254 nm.   The
UDP-sugar fraction was then collected, frozen and lyophilized.   The
fraction was subjected to neutral sugar analysis by gas-liquid
chromatography and the amounts of UDP-glucose and UDP-galactose were
estimated.   Authentic radioactive compounds were added to the extract
in order to estimate the yield.   As shown in the Fig. 4, when Avena
coleoptile segments are incubated for two hours in the presence and
absence of 10^{-5} M IAA with and without 10 mM galactose, galactose causes

Fig. 4.   Effect of galactose on the formation of UDP-glucose in Avena coleoptile and Pisum epicotyl segments.

Segments were incubated for 2 h in the K-phosphate buffer solution, pH 6.5, with and without 10^{-5} M IAA in the presence and absence of 10 mM galactose. Segments were then extracted with 80% ethanol, and the extract was then subjected to DEAE-Sephadex A-25 chromatography to separate UDP-sugars which were then subjected to gas-liquid chromatography to determine glucose.

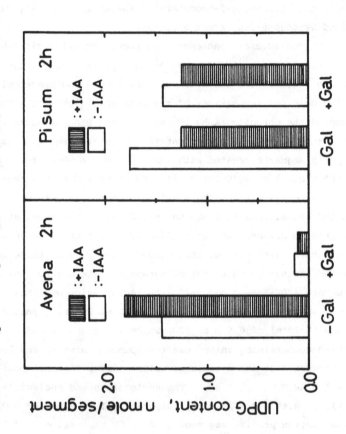

a large decrease in the amount of UDP-glucose and an increase in the amount
of UDP-galactose.   If galactose is removed and the segments are incubated
further, the level of UDP-glucose in the cell is restored slowly.   Therefore, galactose inhibition of UDP-glucose formation as well as of auxin-
induced extension of Avena coleoptile segments is reversible.   These
results provide partial evidence to support the above hypothesis.   It is
interesting that galactose at this concentration shows very little effect
on the amounts of these nucleotide sugars in the Pisum epicotyl segments.
The results offer an explanation of why dicots are much less sensitive
to the inhibitory effect of galactose.

CONCLUSION

Auxin induces cell extension in a biphasic manner when
segments of monocot coleoptiles or dicot stems are floated on buffer
solution containing IAA, the first phase of cell extension also being
induced by hydrogen ions.   IAA rapidly causes a decrease in one of the
parameters obtained by the stress-relaxation method, T_o, which represents
cell-wall loosening probably in the first phase.   Hydrogen ions also
cause a decrease in the T_o value.   Partial degradation of a polysaccharide
component of the cell wall commonly found in monocots and dicots, xyloglucan, seems to have some role in the first phase of cell-wall loosening.
β-Glucan degradation in the monocot cell wall and galactan degradation in
the dicot cell wall also occur, but their direct involvement in cell-wall
loosening seems unlikely.   The cell-wall synthesis, besides other
metabolic activities, required for continued cell extension, might be
represented by the other stress-relaxation parameter, R, which increases
in the dicot cell wall in response to auxin.   However, the biochemical
significance of the parameters, R and T_m, is not clear yet.   For the
second phase of cell extension, i.e. the long-term, continued growth,
cell-wall synthesis is necessary.   Galactose, a specific inhibitor for
auxin-induced long-term extension in monocot coleoptile segments, inhibits
cell-wall synthesis by blocking the formation of UDP-glucose from glucose-
1-phosphate.   Galactose has very little effect on auxin-induced extension
of dicot stem segments and on the formation of UDP-glucose.   In Avena
coleoptiles, galactose, which probably is converted to galactose-1-
phosphate in the cell, inhibits the formation of UDP-glucose by
inhibiting the enzyme UTP:α-D-glucose-1-phosphate uridylyltransferase
in a competitive manner with glucose-1-phosphate.   All of this leads

to the conclusion that for auxin-induced cell extension, partial
degradation of polysaccharides constituting the cell wall, probably
xyloglucans, occurs first, leading to cell-wall loosening; and then for
continued growth, new cell-wall synthesis occurs by continuous formation
of UDP-glucose as a precursor for various kinds of cell-wall polysaccharides.

ACKNOWLEDGEMENTS
        We wish to thank our colleague, Dr. S. Kamisaka, for his
invaluable criticism and discussions.

REFERENCES

Albersheim, P. (1974). The primary cell wall and central control of
        elongation growth. In Plant Carbohydrate Chemistry, ed.
        J.B. Pridham, pp. 145-164. London: Academic Press.
Baker, B. & Ray, P.M. (1965a). Direct and indirect effects of auxin on
        cell wall synthesis in oat coleoptile tissue. Plant
        Physiology, 40, 345-352.
Baker, B. & Ray, P.M. (1965b). Relation between effects of auxin on cell
        wall synthesis and cell elongation. Plant Physiology, 40,
        360-368.
Black, M., Bullock, C., Chatler, E.N., Hanson, A.D. & Jolley, G.M. (1967).
        Effect of inhibitors of protein synthesis on the plastic
        deformation and growth of plant tissues. Nature, 215, 1289-
        1290.
Boyer, J.S. & Wu, G. (1978). Auxin increases the hydraulic conductivity
        of auxin-sensitive hypocotyl tissue. Planta, 139, 227-237.
Brauner, L. & Hasman, M. (1952). Weitere Untersuchungen uber den
        Wirkungsmechanismus des Heteroauxins bei der Wasser-
        aufnahme von Pflanzenparenchymen. Protoplasma, 41, 302-326.
Burström, H. (1942). The influence of heteroauxin on cell growth and
        root development. Annals of the Agricultural College,
        Sweden, 10, 211-230.
Burström, H.G., Uhrström, I. & Wurscher, R. (1967). Growth, turgor, water
        potential and Young's modulus in pea internodes. Physiologia
        Plantarum, 20, 213-231.
Burström, H.G., Uhrström, I. & Olausson, B. (1970). Influence of auxin on
        Young's modulus in stems and roots of Pisum and the theory of
        changing the modulus in tissues. Physiologia Plantarum, 23,
        1223-1233.
Capesius, I., Bopp, M. & Masuda, Y. (1981). Effect of 5-FdUrd on the cell
        wall composition of Sinapis alba hypocotyls. Zeitschrift für
        Pflanzenphysiologie, 103, 87-93.
Carpita, N.C. & Delmer, D.P. (1981). Concentration and metabolic
        turnover of UDP-glucose in developing cotton fibers.
        Journal of Biological Chemistry, 256, 308-315.
Cleland, R. (1958). A separation of auxin-induced cell wall loosening
        into its plastic and elastic components. Physiologia
        Plantarum, 11, 599-609.

Cleland, R. (1967a). Auxin and the mechanical properties of the cell wall. Annals of the New York Academy of Sciences, 144, 3-18.

Cleland, R. (1967b). Extensibility of isolated cell walls: Measurement and changes during cell elongation. Planta, 74, 197-209.

Cleland, R. (1959). Effect of osmotic concentration on auxin-action and on irreversible and reversible expansion of the Avena coleoptile. Physiologia Plantarum, 12, 809-825.

Cleland, R. (1971a). Cell wall extension. Annual Review of Plant Physiology, 22, 197-222.

Cleland, R. (1971b). The mechanical behaviour of isolated Avena coleoptile walls subjected to constant stress. Plant Physiology, 47, 805-811.

Cleland, R. & Haughton, P.M. (1971). The effect of auxin on stress relaxation in isolated Avena coleoptiles. Plant Physiology, 47, 812-815.

Cleland, R. (1973). Auxin-induced hydrogen ion excretion from Avena coleoptiles. Proceedings of the National Academy of Sciences of the United States of America, 70, 3092-3093.

Cleland, R. & Rayle, D.L. (1978). Auxin, H-excretion and cell elongation. Bot. Mag. Tokyo, Special Issue 1, 125-139.

Cleland, R. (1983). The capacity for acid-induced wall loosening as a factor in the control of Avena coleoptile cell elongation. Journal of Experimental Botany, 34, 676-680.

Coartney, S., Morre, D.J. & Key, J.L. (1967). Inhibition of RNA synthesis and auxin-induced cell wall extensibility and growth by actinomycin D. Plant Physiology, 42, 434-439.

Darvill, A.G., Smith, C.J. & Hall, M.A. (1977). Auxin induced proton release, cell wall structure and elongation growth: a hypothesis. In Regulation of Cell Membrane Activities in Plants, ed. E. Marre & O. Ciferri, pp. 275-282. Amsterdam: Elsevier/North Holland Biomedical Press.

Darvill, A.G., Smith, C.J. & Hall, M.A. (1978). Cell wall structure and elongation growth in Zea mays coleoptile tissue. New Phytologist, 80, 503-516.

Durand, H. & Rayle, D.L. (1973). Physiological evidence for auxin-induced hyrogen-ion secretion and the epidermal paradox. Planta, 114, 185-193.

Firn, R. & Digby, J. (1977). The role of the peripheral cell layers in the geotropic curvature of sunflower hypocotyls: a new model of shoot geotropism. Australian Journal of Plant Physiology, 4, 337-347.

Flory, P.J. (1953). Principles of Polymer Chemistry. Cornell University Press.

Fujihara, S. Yamamoto, R. & Masuda, Y. (1978a). Viscoelastic properties of plant cell walls. I. Mathematical formulation for stress relaxation with consideration for pre-extension rate. Biorheology, 15, 63-75.

Fujihara, S., Yamamoto, R. & Masuda, Y. (1978b). Viscoelastic properties of plant cell walls. II. Effect of pre-extension rate on stress relaxation. Biorheology, 15, 77-85.

Fujihara, S., Yamamoto, R. & Masuda, Y. (1978c). Viscoelastic properties of plant cell walls. III. Hysteresis loop in the stress-strain curve at constant strain rate. Biorheology, 15, 87-97.

Fujihara, S., Yamamoto, R. & Masuda, Y. 1979). Viscoelastic properties
    of plant cell walls. IV. Physical meaning of parameter K
    in mathematical formulation for stress-relaxation and load-
    extension processes. Biorheology, 16, 387-396.
Furuya, M., Masuda, Y. & Yamamoto, R. (1972). Effect of environmental
    factors on mechanical properties of the cell wall in rice
    coleoptiles. Developmental Growth and Differentiation, 14,
    95-105.
Gilkes, N.R. & Hall, M.A. (1977). The hormonal control of cell wall
    turnover in Pisum sativum L. New Phytologist, 78, 1-15.
Göring, H. & Gerlach, I. (1966). Das Wachstum kurzfristig isolierter
    pflanzlicher Gewebe in Abhängigkeit von der exogenen
    Kohlenstoffquelle. Zeitschrift für Pflanzenphysiologie,
    55, 429-44.
Göring, H. & Reckin, R. (1968a). Einfluss der D-Galaktose auf den
    Kohlenhydratstoffwechsel pflanzlicher Gewebe. Flora A159,
    82-103.
Göring, H. & Reckin, R. (1968b). Galaktose-1-phosphat als Inhibitor von
    Polysaccharid-Synthese höherer Pflanzen. Naturwissenschaften,
    55, 40-41.
Hager, A., Menzel, H. & Kaus, A. (1971). Versuche und Hypothese zur
    Primärwirkung des Auxins beim Streckungswachstum. Planta,
    100, 47-75.
Hayashi, T. & Maclachlan, G.A. (1984). Pea xyloglucan and cellulose.
    I. Macromolecular organization. Plant Physiology (in press).
Hayashi, T., Wong, Y.S. & Maclachlan, G.A. (1984). Pea xyloglucan and
    cellulose. II. Hydrolysis by pea endo-1,4-glucanase. Plant
    Physiology (in press).
Heyn, A.N.J. & van Overbeek, J. (1931). Weiteres Versuchsmaterial zur
    plastischen und elastischen Dehnbarkeit der Zellmembran.
    Kon. Acad. Wet. Amsterdam 34, 1190-1195.
Hogetsu, T., Shibaoka, H. & Shimokoriyama, M. (1974a). Involvement of
    cellulose synthesis in actions of gibberellin and kinetin on
    cell expansion. Gibberellin-coumarin and kinetin-coumarin
    interactions on stem elongation. Plant Cell Physiology, 15,
    265-272.
Hogetsu, T., Shibaoka, H. & Shimkoriyama, M. (1974b). Involvement of
    cellulose synthesis in actions of gibberellin and kinetin on
    cell expansion. 2,4-Dichlorobenzonitrile as a new
    cellulose-synthesis inhibitor. Plant Cell Physiology, 15,
    389-393.
Inoue, H., Yamamoto, R. & Masuda, Y. (1984). Auxin-induced changes in
    the molecular weight distribution of cell wall xyloglucans
    in Avena coleoptiles. Plant Cell Physiology (in press).
Jaccard, M. & Pilet, P.E. (1975). Extensibility and rheology of
    collenchyma. I. Creep relaxation and viscoelasticity of
    young and senescent cells. Plant Cell Physiology, 16, 113-120.
Jaccard, M. & Pilet, P.M. (1977). Extensibility and rheology of
    collenchyma. II. Low-pH effect on the extension of
    collocytes isolated from high- and low-growing material.
    Plant Cell Physiology, 18, 883-891.
Jaccard, M. & Pilet, P.M. (1979). Growth and rheological changes of
    collenchyma cells: the fusicoccin effect. Plant Cell
    Physiology, 20, 1-7.

Jacobs, M. & Taiz, L. (1980). Vanadate inhibition of auxin-induced H[+] secretion and elongation in pea epicotyls and oat coleoptiles. Proceedings of the National Academy of Sciences of the United Stations of America, 77, 7242-7246.

Kamisaka, S. & Masuda, Y. (1971). Auxin-induced growth of tuber tissue of Jerusalem artichoke. IV. Biochemical changes in chromatin during aging and cell expansion. Plant Cell Physiology, 12, 201-209.

Kato, Y. & Matsuda, K. (1976). Presence of a xyloglucan in the cell wall of Phaseolus aureus hypocotyls. Plant Cell Physiology, 17, 1185-1198.

Kato, Y. & Katsuda, K. (1981). Occurrence of a soluble and low molecular weight xyloglucan and its origin in etiolated mung bean hypocotyls. Agricutural and Biological Chemistry, 45, 1-8.

Kato, Y., Iki, K. & Matsuda, K. (1981). Cell-wall polysaccharides of immature barley plants. II. Characterization of a xyloglucan. Agricultural and Biological Chemistry, 45, 2745-2753.

Kato, Y., Ito, S., Iki, K. & Matsuda, K. (1982). Xyloglucan and β-D-glucan in cell walls of rice seedlings. Plant Cell Physiology, 23, 351-364.

Katsumi, M. & Kazama, H. (1974). Auxin-gibberellin relationships in their effects on hypocotyl elongation of light-grown cucumber seedlings. III. Gibberellin specificity in its enhancing effect on IAA-induced elongation. Plant Cell Physiology, 15, 315-319.

Katsumi, M. & Kazama, H. (1978). Gibberellin control of cell elongation in cucumber hypocotyl sections. Botanical Magazine, Tokyo, Special Issue 1, 141-158.

Kawamura, H., Kamisaka, S. & Masuda, Y. (1976). Regulation of lettuce hypocotyl elongation by gibberellic acid. Correlation between cell elongation, stress-relaxation properties of the cell wall and wall polysaccharide content. Plant Cell Physiology, 17, 23-34.

Kazama, H. & Katsumi, M. (1976). Biphasic response of cucumber hypocotyl sections to auxin. Plant Cell Physiology, 17, 467-473.

Kazama, H. & Katsumi, M. (1978). Effect of light on auxin-induced elongation of light-grown cucumber hypocotyl sections. Plant Cell Physiology, 19, 1137-1144.

Katellapper, H.J. (1953). The mechanism of the action of indole-3-acetic acid on the water absorption by Avena coleoptile sections. Acta Botanica Neerlandica, 2, 387-444.

Key, J.L. & Ingle, J. (1964). Requirement for the synthesis of DNA-like RNA for growth of excised tissue. Proceedings of the National Academy of Sciences of the United States of America, 52, 1382-1388.

Kohji, J., Nishitani, K. & Masuda, Y. (1981). A study on the mechanism of nodding initiation of the flower stalk in a poppy, Papaver rhoeas L. Plant Cell Physiology, 22, 413-422.

Kooiman, P. (1960). A method for the determination of amyloid in plant seeds. Rev. Trav. Chim. Pays-Bas, 79, 675-678.

Koyama, T., Hayashi, T., Kato, Y. & Matsuda, K. (1981). Degradation of xyloglucan by wall-bound enzymes from soybean tissue. I. Occurrence of xyloglucan-degrading enzymes in soybean cell wall. Plant Cell Physiology, 22, 1191-1198.

Labavitch, J.M. & Ray, P.M. (1974). Turnover of cell wall polysaccharides
    in elongating pea stem segments. Plant Physiology, 53, 669-
    673.
Labavitch, J.M. & Ray, P.M. (1978). Structure of hemicellulosic poly-
    saccharides of Avena sativa coleoptile cell walls.
    Phytochemistry, 17, 933-937.
Lockhart, J.A. (1965). Cell wall extension. In Plant Biochemistry, eds.
    J. Bonner & J.E. Varner, pp. 827-849. New York: Academic
    Press.
Loescher, D.J. & Nevins, D.J. (1972). Auxin-induced changes in Avena
    coleoptile cell wall composition. Plant Physiology, 50, 556-
    563.
Loescher, D.J. & Nevins, D.J. (1973). Tugor-dependent changes in Avena
    coleoptile cell wall composition. Plant Physiology, 52, 248-
    251.
Marrè, E., Lado, P., Caldogno, F.R. & Colombo, R. (1973). Correlation
    between cell enlargement in pea internode segments and
    decrease in the pH of the medium of incubation. I. Effects
    of fusicoccin, natural and synthetic auxins and mannitol.
    Plant Science Letters, 1, 179-184.
Masuda, Y. (1961). Effect of auxin and oxalic acid on the cell wall
    property of Avena coleoptile. Plant Cell Physiology, 2, 129-
    138.
Masuda, Y. (1965). RNA in relation to the effect of auxin, kinetin and
    gibberellic acid on the tuber tissue of Jerusalem artichoke.
    Physiologia Plantarum, 18, 15-23.
Masuda, Y. (1966). Auxin-induced growth of tuber tissue of Jerusalem
    artichoke. II. The relation to protein and nucleic acid
    metabolism. Plant Cell Physiology, 7, 75-91.
Masuda, Y. & Wada, S. (1966). Requirement of RNA for the auxin-induced
    elongation of oat coleoptiles. Physiologia Plantarum, 19,
    1055-1063.
Masuda, Y., Tanimoto, E. & Wada, S. (1967). Auxin-stimulated RNA
    synthesis in oat coleoptile cells. Physiologia Plantarum, 20,
    713-719.
Masuda, Y. (1968). Role of cell-wall-degrading enzymes in cell wall
    loosening in oat coleoptiles. Planta, 83, 171-184.
Masuda, Y. (1969). Auxin-induced cell expansion in relation to cell wall
    extensibility. Plant Cell Physiology, 10, 1-9.
Masuda, Y. & Kamisaka, S. (1969). Rapid stimulation of RNA bio-
    synthesis by auxin. Plant Cell Physiology, 10, 79-86.
Masuda, Y. & Yamamoto, R. (1972). Control of auxin-induced stem
    elongation by the epidermis. Physiologia Plantarum, 27, 109-
    115.
Masuda, Y., Yamamoto, R. & Tanimoto, E. (1972). Auxin-induced changes in
    cell wall properties and growth of Avena coleoptiles and
    green pea epicotyls. In Plant Growth Substances 1970, ed.
    D.J. Carr, pp. 17-22. Berlin, Heidelberg, New York:
    Springer-Verlag.
Masuda, Y., Yamamoto, R., Maki, K. & Yamagata, Y. (1974a). Effect of
    auxin and hydrogen ions on cell extension and cell wall
    properties of light-grown pea epicotyls. In Plant Growth
    Substances 1973, pp. 806-813. Tokyo, Hirokawa.

Masuda, Y., Yamamoto, R., Kawamura, H. & Yamagata, Y. (1974b). Stress
relaxation properties of the cell wall of tissue segments
under different growth conditions. Plant Cell Physiology,
15, 1083-1092.
Masuda, Y. (1977). Wall extensibility in relation to auxin effects. In
Plant Growth Regulation, ed. P.E. Pilet, pp. 21-26. Berlin,
Heidelberg, New York: Springer-Verlag.
Masuda, Y. (1980). Auxin-induced changes in noncellulosic polysaccharides
of cell walls of monocot coleoptiles and dicot stems. In
Plant Growth Substances 1979, ed. F. Skoog, pp. 79-89.
Berlin, Heidelberg, New York: Springer-Verlag.
Masuda, Y., Kamisaka, S., Yanagisawa, H. & Suzuki, Y. (1981). Effect of
light on growth and metabolic activities in pea seedlings.
I. Changes in cell wall polysaccharides during growth in the
dark and in the light. Biochem. Physiol. Pflanzen, 176, 23-
34.
Masuda, Y. & Pilet, P.E. (1983). Mechanical properties and polysaccharide nature of the cell walls of maize root.
Physiologia Plantarum, 59, 512-517.
Nakamura, T., Sekine, S., Arai, K. & Takahashi, N. (1975). Effects of
gibberellic acid and indole-3-acetic acid on stress-
relaxation properties of pea hook cell wall. Plant Cell
Physiology, 16, 127-138.
Nevins, D.J. (1975). The effect of nojirimycin on plant growth and its
implications concerning a role for exo-β-glucanases in
auxin-induced cell expansion. Plant Cell Physiology, 16,
347-356.
Nevins, D.J., Huber, D.J., Yamamoto, R. & Loescher, W.H. (1977). β-D-
glucan of Avena coleoptile cell walls. Plant Physiology, 60,
617-621.
Nevins, D.J., Yamamoto, R. & Huber, D.J. (1978). Cell wall of β-glucans
of five grass species. Phytochemistry, 17, 1503-1505.
Nishitani, K. & Masuda, Y. (1979). Growth and cell wall changes in azuki
bean epicotyls. I. Changes in wall polysaccharides during
intact growth. Plant Cell Physiology, 20, 63-74.
Nishitani, K. & Masuda, Y. (1979). Growth and cell wall changes in azuki
bean epicotyls. II. Changes in wall polysaccharides during
auxin-induced growth of excised segments. Plant Cell
Physiology, 20, 463-472.
Nishitani, K. & Masuda, Y. (1980). Modifications of cell wall polysaccharides during auxin-induced growth in asuki bean
epicotyl segments. Plant Cell Physiology, 21, 169-181.
Nishitani, K. & Masuda, Y. (1981). Auxin-induced changes in the cell wall
structure: changes in the sugar compositions, intrinsic
viscosity and molecular weight distributions of matrix polysaccharides of the epicotyl cell wall of Vigna angularis.
Physiologia Plantarum, 52, 482-494.
Nishitani, K. & Masuda, Y. (1982). Roles of auxin and gibberellic acid
in growth and maturation of epicotyls of Vigna angularis:
cell wall changes. Physiologia Plantarum, 56, 38-45.
Nishitani, K. & Masuda, Y. (1983). Acid pH-induced structural changes in
cell wall xyloglucans in Vigna angularis epicotyl segments.
Plant Science Letters, 28, 87-94.
Noodén, L.D. & Thimann, K.V. (1966). Action of inhibitors of RNA and
protein synthesis on cell enlargement. Plant Physiology, 41,
157-164.

Ockerse, R. & Galston, A.W. (1967). Gibberellin-auxin interaction in
    pea stem elongation. Plant Physiology, 42, 47-54.
Ockerse, R. (1970). The dependence of auxin-induced pea stem growth on
    gibberellin. Botanical Gazette, 131, 95-97.
Olson, A.C., Bonner, J. & Morré, D.J. (1965). Force extension analysis
    of Avena coleoptile cell wall. Planta, 66, 126-134.
Ordin, L. & Bonner, J. (1957). Effect of galactose on growth and
    metabolism of Avena coleoptile sections. Plant Physiology,
    32, 212-215.
Penny, D. Miller, K.F. & Penny, P. (1972). Studies on the mechanism of
    cell elongation of lupin hypocotyls. New Zealand Journal of
    Botany, 10, 97-111.
Preston, R.D. & Hepton, J. (1960). The effect of indoleacetic acid on
    cell wall extensibility in Avena coleoptiles. Journal of
    Experimental Botany, 11, 13-27.
Purves, W.K. & Hillman, W.S. (1958). Response of pea stem sections to
    indoleacetic acid, gibberellic acid, and sucrose as
    affected by length and distance from apex. Physiologia
    Plantarum, 11, 29-35.
Ray, P.M. (1962). Cell wall synthesis and cell elongation in oat
    coleoptile tissue. American Journal of Botany, 49, 928-939.
Ray, P.M. (1973a). Regulation of β-glucan synthetase activity by auxin
    in pea stem tissue. I. Kinetic aspects. Plant Physiology,
    51, 601-608.
Ray, P.M. (1973b). Regulation of β-glucan synthetase activity by auxin in
    pea stem tissue. II. Metabolic requirements. Plant
    Physiology, 51, 609-614.
Rayle, D.L. & Cleland, R. (1970). Enhancement of wall loosening and
    elongation by acid solutions. Plant Physiology, 46, 250-253.
Rayle, D.L. Haughton, P.M. & Cleland, R. (1970). An in vitro system that
    stimulates plant cell extension growth. Proceedings of the
    National Academy of Sciences of the United States of America,
    67, 1814-1817.
Rayle, D.L. & Cleland, R. (1972). The in vitro acid-growth response:
    relation to in vivo growth responses and auxin action.
    Planta, 104, 282-296.
Sakurai, N., Nevins, D.J. & Masuda, Y. (1977). Auxin- and hydrogen
    ion-induced cell wall loosening and cell extension in Avena
    coleoptile segments. Plant Cell Physiology, 18, 371-380.
Sakurai, N. & Masuda, Y. (1977). Effect of indole-3-acetic acid on cell
    wall loosening: changes in mechanical properties and non-
    cellulosic glucose content of Avena coleoptile cell wall.
    Plant Cell Physiology, 18, 587-594.
Sakurai, N. & Masuda, Y. (1978a). Auxin-induced changes in barley
    coleoptile cell wall composition. Plant Cell Physiology, 19,
    1217-1223.
Sakurai, N. & Masuda, Y. (1978a). Auxin-induced extension, cell wall
    loosening and changes in the wall polysaccharide content of
    barley coleoptile segments. Plant Cell Physiology, 19, 1225-
    1233.
Sakurai, N., Nishitani, K. & Masuda, Y. (1979). Auxin-induced changes in
    the molecular weight of hemicellulosic polysaccharides of the
    Avena coleoptile cell wall. Plant Cell Physiology, 20, 1349-
    1357.

Sakurai, N., Fujihara, S., Yamamoto, R. & Masuda, Y. (1982). A stress-
    relaxation parameter b of the oat coleoptile cell wall and
    its implication in cell wall loosening. Journal of Plant
    Growth Regulation, 1, 75-83.
Setterfield, G. (1963). Growth regulation in excised slices of
    Jerusalem artichoke tuber tissue. Symposium of the Society
    for Experimental Biology, 17, 98-126.
Shen-Miller, J. & Masuda, Y. (1973). Kinetics of stress-relaxation
    properties of oat coleoptile cell wall after geotropic
    stimulation. Plant Physiology, 51, 464-467.
Shibaoka, H. (1972). Gibberellin-colchicine interaction in elongation of
    asuki bean epicotyl sections. Plant Cell Physiology, 13, 461-
    469.
Shibaoka, H. (1974). Involvement of wall microtubules in gibberellin
    promotion and kinetin inhibition of stem expansion. Plant
    Cell Physiology, 15, 255-263.
Tagawa, T. & Bonner, J. (1957). Mechanical properties of the Avena
    coleoptile as related to auxin and to ionic interactions.
    Plant Physiology, 32, 207-212.
Takeda, K. & Shibaoka, H. (1981). Effects of gibberellin and colchicine
    on microfibril arrangement in epidermal cell walls of Vigna
    angularis Ohwi et Ohashi epicotyls. Planta, 151, 393-398.
Tanimoto, E., Yanagishima, N. & Masuda, Y. (1967). Effect of gibberellic
    acid on dwarf and normal pea plants. Physiologia Plantarum,
    20, 291-298.
Tanimoto, E. & Masuda, Y. (1971). Role of the epidermis in auxin-induced
    elongation of light-grown pea stem segments. Plant Cell
    Physiology, 12, 663-673.
Terry, M.E. & Bonner, B.A. (1980). An examination of centrifugation as a
    method of extracting and extracellular solution from peas,
    and its use for the study of indoleacetic acid-induced
    growth. Plant Physiology, 66, 321-325.
Terry, M.E., Jones, R.L. & Bonner, B.A. (1981). Soluble cell wall
    polysaccharides released from pea stems by centrifugation.
    I. Effect of auxin. Plant Physiology, 68, 531-537.
Thimann, K.V. & Schneider, C.L. (1938). Differential growth in plant
    tissues. American Journal of Botany, 25, 627-641.
Tobolsky, A. & Eyring, H. (1943). Mechanical properties of polymeric
    materials.
Vanderhoef, L.N., Stahl, C.A., Williams, C.A. & Brinkmann, K.A. (1976).
    Additional evidence for separable responses to auxin in
    soybean hypocotyl. Plant Physiology, 57, 817-819.
Wada, S. & Ray, P.M. (1978). Matrix polysaccharides of oat coleoptile
    cell walls. Phytochemistry, 17, 923-931.
Yamagata, Y. & Masuda, Y. (1975). Comparative studies on auxin and
    fusicoccin actions on plant growth. Plant Cell Physiology,
    16, 41-52.
Yamagata, Y. & Masuda, Y. (1976). Auxin-induced extension of the isolated
    epidermis of light-grown pea epicotyls. Plant Cell Physiology,
    17, 1235-1242.
Yamamoto, R., Shinozaki, K. & Masuda, Y. (1970). Stress-relaxation
    properties of plant cell walls with special reference to
    auxin action. Plant Cell Physiology, 11, 947-956.
Yamamoto, R. & Masuda, Y. (1971). Stress-relaxation properties of the
    Avena coleoptile cell wall. Physiologia Plantarum, 25, 330-
    335.

Yamamoto, R., Kawamura, H. & Masuda, Y. (1974a). Stress-relaxation
    properties of the cell wall of growing intact plants. Plant
    Cell Physiology, 15, 1073-1082.
Yamamoto, R., Maki, K., Yamagata, Y. & Masuda, Y. (1974b). Auxin and
    hydrogen ion actions on light-grown pea epicotyl segments.
    I. Tissue specificity of auxin and hydrogen ion actions.
    Plant Cell Physiology, 15, 823-831.
Yamamoto, R., Maki, K. & Masuda, Y. (1974c). Auxin and hydrogen ion
    actions on light-grown pea epicotyl segments. III. Effect
    of auxin and hydrogen ions on stress-relaxation properties.
    Plant Cell Physiology, 15, 1027-1038.
Yamamoto, R., Fujihara, S. & Masuda, Y. (1974d). Measurement of stress-
    relaxation properties of plant cell walls. In Plant Growth
    Substances 1973, pp. 798-805. Tokyo, Hirokawa.
Yamamoto, R. & Nevins, D.J. (1978). Structural studies on the β-D-glucan
    of the Avena coleoptile cell-wall. Carbohydrate Research, 67,
    275-280.
Yamamoto, R., Fujihara, S. & Masuda, Y. (1981a). Compression method for
    measurement of stress-relaxation properties of plant cell
    walls with reference to plant hormone actions. Biorheology,
    18, 643-652.
Yamamoto, R., Sakurai, N. & Masuda, Y. (1981b). Inhibition of auxin-
    induced cell elongation by galactose. Physiologia
    Plantarum, 53, 543-547.
Yamamoto, R. & Masuda, Y. (1984a). Galactose inhibition of auxin-
    induced cell elongation in oat coleoptile segments.
    Physiologia Plantarum (in press).
Yamamoto, R. & Masuda, Y. (1984b). Auxin-induced modifications of cell
    wall polysaccharides in oat coleoptile segments. Effect of
    galactose. Symposium of the University of California,
    Riverside (in press).
Yoda, S. & Ashida, J. (1960). Effect of gibberellin and auxin on the
    extensibility of the pea stem. Plant Cell Physiology, 1, 99-
    108.
Zarra, I. & Masuda, Y. (1979a). Growth and cell wall changes in rice
    coleoptiles growing under different conditions. I. Changes
    in tugor pressure and cell wall polysaccharides during
    intact growth. Plant Cell Physiology, 20, 1117-1124.
Zarra, I. & Masuda, Y. (1979b). Growth and cell wall changes in rice
    coleoptiles growing under different conditions. II. Auxin-
    induced growth in coleoptile segments. Plant Cell Physiology,
    20, 1125-1133.

SOCIETY FOR EXPERIMENTAL BIOLOGY SEMINAR SERIES

A series of multi-author volumes developed from seminars held by the Society
for Experimental Biology. Each volume serves not only as an introductory
review of a specific topic, but also introduces the reader to experimental
evidence to support the theories and principles discussed, and points the
way to new research.